Advances in Plastic Forming of Metals

Advances in Plastic Forming of Metals

Special Issue Editors

Myoung-Gyu Lee
Yannis P. Korkolis

MDPI • Basel • Beijing • Wuhan • Barcelona • Belgrade

MDPI

Special Issue Editors

Myoung-Gyu Lee
Seoul National University
Korea

Yannis P. Korkolis
University of New Hampshire
USA

Editorial Office
MDPI
St. Alban-Anlage 66
Basel, Switzerland

This is a reprint of articles from the Special Issue published online in the open access journal *Metals* (ISSN 2075-4701) from 2017 to 2018 (available at: http://www.mdpi.com/journal/metals/special_issues/plastic_forming_metals)

For citation purposes, cite each article independently as indicated on the article page online and as indicated below:

LastName, A.A.; LastName, B.B.; LastName, C.C. Article Title. *Journal Name* **Year**, *Article Number*, Page Range.

ISBN 978-3-03897-260-0 (Pbk)
ISBN 978-3-03897-261-7 (PDF)

Cover image courtesy of Rongting Li.

Contents

About the Special Issue Editors

Myoung-Gyu Lee is a Professor of Materials Science and Engineering at Seoul National University. He received his PhD from Seoul National University. His research interests focus on computational plasticity, the mechanics of materials including the finite element modeling of advanced structure materials, multi-scale modeling and experiments that reveal the deformation mechanisms of materials under complex strain path changes. He has co-authored over 200 journal articles and two proceedings volumes, three book chapters in the areas of plasticity theory, advanced constitutive modeling and finite element simulations. He received the 2014 *International Journal of Plasticity* Award for excellent contributions to the field of plasticity.

Yannis P. Korkolis is an Associate Professor of Mechanical Engineering at the University of New Hampshire. His research is at the interface of constitutive modeling, formability and ductile fracture, and manufacturing processes. He has a PhD in Engineering Mechanics from the University of Texas at Austin. He joined the University of New Hampshire in 2009, where he has taught courses in Solid Mechanics, Manufacturing and Design. He has also held visiting appointments at Kyoto University and Tokyo University of Agriculture and Technology in Japan. He has published over 35 peer-reviewed journal papers and has delivered over 100 conference papers, presentations, posters and invited talks. Currently, he is serving as an Associate Editor for the ASME *Journal of Manufacturing Science and Engineering*, and is the Chair of the 2019 NUMIFORM (Numerical Methods for Industrial Forming Processes) conference. He was recently awarded the Ralph R. Teetor Award from the Society of Automotive Engineers.

metals

MDPI

Editorial

Advances in Plastic Forming of Metals

Myoung-Gyu Lee [1,*] and Yannis P. Korkolis [2,*]

[1]   Department of Materials Science and Engineering, Seoul National University, Seoul 08826, Korea
[2]   Department of Mechanical Engineering, University of New Hampshire, Durham, NH 03824, USA
*   Correspondence: myounglee@snu.ac.kr (M.-G.L.); Yannis.Korkolis@unh.edu (Y.P.K.);
     Tel.: +82-2-880-1711 (M.-G.L.); +1-603-862-2772 (Y.P.K.)

Received: 3 April 2018; Accepted: 6 April 2018; Published: 16 April 2018

1. Introduction

Forming of metals through plastic deformation is a family of methods that produce components through re-shaping of input stock, oftentimes with little waste. Therefore, forming is one of the most efficient and economical manufacturing process families available. A myriad of forming processes exist in this family. In conjunction with their countless existing successful applications and their relatively low energy requirements, these processes are an indispensable part of our future. However, despite the vast accumulated know-how, research challenges remain, be they related to forming of new materials (e.g., for transportation lightweighting applications), pushing the envelope of what is doable, reducing the intermediate steps and/or the scrap, or further enhancing the environmental friendliness. The purpose of this Special Issue is to collect expert views and contributions on the current state-of-the-art on plastic forming, and in this way to highlight contemporary challenges and to offer ideas and solutions were possible.

2. Contributions

Our thought at the onset of this effort was to attract contributions to enhance the understanding of metal deformation processes; discuss improved material models available for simulating forming; improve the traditional and lightweight metal forming processes and modeling capability; and promote research on forming of new materials and/or new forming technologies at various length scales, from microscale to macroscale. The contributions we received can be classified under two major categories: bulk forming and sheet/tube forming.

2.1. Bulk Metal Forming

The papers on bulk forming fall under two themes: processing studies and material characterization and modeling.

Du et al. [1] use finite element analysis to simulate hydrostatic extrusion under pressure- and displacement-control. They use these models to examine the relationships between extrusion pressure, extrusion ratio, and die cone angle. Amigo and Camacho [2] use finite element analysis to study the central-burst defect in extrusion of DP800 steels. They use the modeling tool to design multiple-pass dies as an alternative to single-pass extrusions which would be prone to central-burst. Behrens et al. [3] numerically examine the formation of an oxide scale during hot forging of steel and it effect on material flow and frictional conditions. Alexandrov et al. [4] examine the formation of a severely-deformed layer near the surface due to friction-induced shearing. They propose a new criterion for determining the boundary between the layer of severe plastic deformation and the bulk of the material.

Shifting now to material characterization and modeling, Shun et al. [5] use an artificial neural network to model the hot deformation behavior of an Al-Si-Mg alloy with an Arrhenius-type constitutive model. Zhou et al. [6] identify the optimum hot-working parameters of an as-cast

30Cr2Ni4MoV steel at high temperatures and intermediate strain rates using processing maps. Zhang and Jiang [7] use electron back-scatter diffraction to understand the grain refinement during equal-channel angular extrusion (ECAE) of a NiTiFe shape-memory alloy. Salcedo et al. [8] investigate the production and properties of ultrafine-grained cams from AA5083 by isothermal forging of a billet that first underwent an ECAE process. Zhao et al. [9] examine magnesium metal matrix composites, where the Mg matrix is reinforced by silicon carbide particles (SiCp). In their work, they assess the influence of different sizes and percentages of SiCp particles on microstructural evolution during deformation, as well as on strength, ductility and formability. Li et al. [10] discuss the ultra-low cycle fatigue (e.g., as in an earthquake) of an X65 steel pipeline using experiments and finite element models. A range of material models are used in these simulations, and the material characterization experiments are supplemented by texture-based multiscale simulations, e.g., for calibrating the anisotropic yield locus.

2.2. Sheet/Tube Metal Forming

The papers in this category are mainly discussing process limits and defects. Morales-Palma et al. [11] discuss the extension of the maximum force principle to predict localized necking in stretch-bending. Chalal and Abed-Meraim [12] examine the open question of necking prediction by considering three numerical necking criteria. These are used to predict the forming limit diagrams for sheet metals. Shifting attention to springback, Jung et al. [13] examine the anisotropic springback recovery of advanced high-strength steels using a combined isotropic–kinematic hardening model and applying it to a U-bending process. Seo et al. [14] evaluate the effect of the material models on springback predictions for TRIP 1180 steel. In particular, they use the Hill 1948 and Yld2000-2D yield criteria along with the Yoshida-Uemori kinematic hardening model in finite element simulations of U-bending and T-shape drawing. Trzepiecinski and Lemu [15] examine the effect of anisotropy on the springback predictions for DC04 automotive steel sheets and the impact of the simulation parameters on the accuracy of the predictions. Phanitwong et al. [16] use a combination of finite element analysis and statistical analysis to ascertain the effect of U-bending geometry parameters on springback.

Some of the contributions examine forming limits and defects in the context of actual manufacturing processes. Centeno et al. [17] examine formability and failure in single point incremental forming (SPIF) of AISI304-H111 sheets and compare it to conventional forming conditions, e.g., the Nakajima and stretch-bending tests. Among other things, they determine the conditions upon which necking is suppressed, so that failure in SPIF is by ductile fracture. Abebe et al. [18] examine springback in multi-point dieless forming, especially in the context of reducing computational time. They propose to replace numerical simulations of springback with statistical analyses based on design of experiments.

3. Closing Remarks

In the process of creating this Issue, we were fortunate to have the expert assistance of the Beijing office of *Metals*. To the staff who expertly coordinated the reviews and processing of the papers, we express our sincere thanks. We also express our gratitude to the anonymous reviewers who provided timely and constructive reviews of the submitted manuscripts.

This Special Issue attracted 18 contributions from 12 countries, indicating that advancing research in manufacturing in general, and plastic forming in particular, is a truly global affair. We are looking forward to the research advances in plastic forming in the years to come, and hope that this Special Issue has contributed to a small extent to a greener and more prosperous future for all.

Conflicts of Interest: The authors declare no conflict of interest.

References

1. Du, S.; Zan, X.; Li, P.; Luo, L.; Zhu, X.; Wu, Y. Comparison of Hydrostatic Extrusion between Pressure-Load and Displacement-Load Models. *Metals* **2017**, *7*, 78. [CrossRef]
2. Amigo, F.J.; Camacho, A.M. Reduction of Induced Central Damage in Cold Extrusion of Dual-Phase Steel DP800 Using Double-Pass Dies. *Metals* **2017**, *7*, 335. [CrossRef]
3. Behrens, B.-A.; Chugreev, A.; Awiszus, B.; Graf, M.; Kawalla, R.; Ullmann, M.; Korpala, G.; Wester, H. Sensitivity Analysis of Oxide Scale Influence on General Carbon Steels during Hot Forging. *Metals* **2018**, *8*, 140. [CrossRef]
4. Alexandrov, S.; Šidjanin, L.; Vilotić, D.; Movrin, D.; Lang, L. Generation of a Layer of Severe Plastic Deformation near Friction Surfaces in Upsetting of Steel Specimens. *Metals* **2018**, *8*, 71. [CrossRef]
5. Han, Y.; Yan, S.; Sun, Y.; Chen, H. Modeling the Constitutive Relationship of Al–0.62Mg–0.73Si Alloy Based on Artificial Neural Network. *Metals* **2017**, *7*, 114. [CrossRef]
6. Zhou, P.; Ma, Q.; Luo, J. Hot Deformation Behavior of As-Cast 30Cr2Ni4MoV Steel Using Processing Maps. *Metals* **2017**, *7*, 50. [CrossRef]
7. Zhang, Y.; Jiang, S. The Mechanism of Inhomogeneous Grain Refinement in a NiTiFe Shape Memory Alloy Subjected to Single-Pass Equal-Channel Angular Extrusion. *Metals* **2017**, *7*, 400. [CrossRef]
8. Salcedo, D.; Luis, C.J.; Luri, R.; Puertas, I.; León, J.; Fuertes, J.P. Design and Mechanical Properties Analysis of AA5083 Ultrafine Grained Cams. *Metals* **2017**, *7*, 116. [CrossRef]
9. Zhao, W.; Huang, S.-J.; Wu, Y.-J.; Kang, C.-W. Particle Size and Particle Percentage Effect of AZ61/SiCp Magnesium Matrix Micro- and Nano-Composites on Their Mechanical Properties Due to Extrusion and Subsequent Annealing. *Metals* **2017**, *7*, 293. [CrossRef]
10. Li, R.; Eyckens, P.; E, D.; Gawad, J.; Poucke, M.V.; Cooreman, S.; Bael, A.V. Advanced Plasticity Modeling for Ultra-Low-Cycle-Fatigue Simulation of Steel Pipe. *Metals* **2017**, *7*, 140. [CrossRef]
11. Morales-Palma, D.; Martínez-Donaire, A.J.; Vallellano, C. On the Use of Maximum Force Criteria to Predict Localised Necking in Metal Sheets under Stretch-Bending. *Metals* **2017**, *7*, 469. [CrossRef]
12. Chalal, H.; Abed-Meraim, F. Numerical Predictions of the Occurrence of Necking in Deep Drawing Processes. *Metals* **2017**, *7*, 455. [CrossRef]
13. Jung, J.; Jun, S.; Lee, H.-S.; Kim, B.-M.; Lee, M.-G.; Kim, J.H. Anisotropic Hardening Behaviour and Springback of Advanced High-Strength Steels. *Metals* **2017**, *7*, 480. [CrossRef]
14. Seo, K.-Y.; Kim, J.-H.; Lee, H.-S.; Kim, J.H.; Kim, B.-M. Effect of Constitutive Equations on Springback Prediction Accuracy in the TRIP1180 Cold Stamping. *Metals* **2018**, *8*, 18. [CrossRef]
15. Trzepiecinski, T.; Lemu, H.G. Effect of Computational Parameters on Springback Prediction by Numerical Simulation. *Metals* **2017**, *7*, 380. [CrossRef]
16. Phanitwong, W.; Boochakul, U.; Thipprakmas, S. Design of U-Geometry Parameters Using Statistical Analysis Techniques in the U-Bending Process. *Metals* **2017**, *7*, 235. [CrossRef]
17. Centeno, G.; Martínez-Donaire, A.J.; Bagudanch, I.; Morales-Palma, D.; Garcia-Romeu, M.L.; Vallellano, C. Revisiting Formability and Failure of AISI304 Sheets in SPIF: Experimental Approach and Numerical Validation. *Metals* **2017**, *7*, 531. [CrossRef]
18. Abebe, M.; Yoon, J.-S.; Kang, B.-S. Radial Basis Functional Model of Multi-Point Dieless Forming Process for Springback Reduction and Compensation. *Metals* **2017**, *7*, 528. [CrossRef]

metals [MDPI]

Article

Reduction of Induced Central Damage in Cold Extrusion of Dual-Phase Steel DP800 Using Double-Pass Dies

Francisco Javier Amigo and Ana María Camacho *

Department of Manufacturing Engineering, Universidad Nacional de Educación a Distancia (UNED), 28040 Madrid, Spain; famigo4@alumno.uned.es
* Correspondence: amcamacho@ind.uned.es; Tel.: +34-913-988-660

Received: 25 July 2017; Accepted: 30 August 2017; Published: 31 August 2017

Abstract: Advanced High Strength Steels (AHSS) are a promising family of materials for applications where a high strength-to-weight ratio is required. Central burst is a typical defect commonly found in parts formed by extrusion and it can be a serious problem for the in-service performance of the extrudate. The finite element method is a very useful tool to predict this type of internal defect. In this work, the software DEFORM-F2 has been used to choose the best configurations of multiple-pass dies, proposed as an alternative to single-pass extrusions in order to minimize the central damage that can lead to central burst in extruded parts of AHSS, particularly, the dual-phase steel DP800. It has been demonstrated that some geometrical configurations in double-pass dies lead to a minimum value of the central damage, much lower than the one obtained in single-pass extrusion. As a general rule, the position of the minimum damage leads to choosing higher values of the contacting length between partial reductions (L) for high die semiangles (α) and to lower values of the reduction in the first pass (R_A) for low total reductions (R_T). This methodology could be extended to find the best configurations for other outstanding materials.

Keywords: AHSS; dual-phase steels; cold extrusion; multi-pass dies; damage; central burst; finite element analysis (FEA)

---

1. Introduction

Advanced High Strength Steels (AHSS) are an emerging family of materials for applications where a high strength-to-weight ratio is required, such as aeronautical and automotive ones [1]. The interest of these steels is not only focused on the in-service behavior of the components, but also in the response of machines and tools to support the high forces required to produce the final shapes; this problem has been faced, for example, in previous studies where the finite element simulation of the system press-tool behavior in the stamping processes was used to define criteria for the best design of high-cost dies and punches [2]. The die is a critical part of the system press-tool in forming processes, as it is in direct contact with the workpiece to be formed. Die design has to be optimal in order to increase the tool life and to produce products of the required quality; however, studies about other related topics such as the optimization of multi-axis high-speed milling are also becoming very important when dies of complex shape have to be manufactured [3], as well as the improvement in finishing operations of forming tools, as in the work of López de Lacalle et al. [4], especially focused in the machining of AHSS. Dual-Phase steels (DP) are one interesting group of AHSS, whose microstructure is mainly composed of soft ferrite, with islands of hard martensite dispersed throughout [5]. Thus, the strength level of these steels is related to the amount of martensite in the microstructure. A wide variety of DP grades exhibiting different strength and ductility levels are currently industrially produced; however, it is still a challenge to improve their formability during their processing. As stated by

Moeini et al. [6], a lot of scientific work is being done to improve the knowledge about the effect of microstructure on the mechanical properties of DP steels [7–9]. Due to their different mechanical properties compared to conventional steels, it becomes necessary to know the behavior of these advanced materials under different processing techniques to determine the best operating conditions that ensure a good quality of the final product [10,11]. Some structural components of car bodies in the automotive industry are obtained by extrusion processes, which are commonly classified in direct/forward and indirect/backward ones. In direct extrusion, the directions of work piece and tool movement are identical, and the most relevant parameters are the die semiangle, the extrusion ratio, the friction, and material properties [12]. On the other hand, the most typical defect encountered in extruded parts is "chevron cracking", also called "central burst". Parghazeh and Haghighat [13] have recently developed an upper bound model to predict the appearance of central bursting defects in rod extrusion processes. This defect, that can also be associated with drawing operations [14], can seriously affect the quality of the extrudate and its in-service performance; and it can be especially problematic because central burst is an internal defect and it cannot be detected by visual inspection techniques. As explained in [15], this was a serious problem in the mid-1960s for automotive companies which encountered important problems of axle shaft breakage leading to 100% inspection. Although fractures are important, there is a growing interest of the scientific community to study the appearance and evolution of damage in general, particularly in dual-phase steels [16–18], as it can lead to failure. Damage can affect the mechanical properties of a component under service loads [19]. Reduction of damage that can lead to central burst appearance in DP800 steel obtained by cold forward extrusion is investigated in this paper. Central bursts are internal fractures caused by high hydrostatic tension in combination with internal material weaknesses, mainly porosity [20]. The hydrostatic stress criterion (HSC) has been typically used to predict central burst occurrence [14,21]. This criterion states that "whenever hydrostatic stress at a point on the center line in the deformation zone becomes zero and it is compressive elsewhere, there is fracture initiation leading to central burst" [22]. However, if the level of hydrostatic tension can be kept below a critical level, bursting can likely be avoided. This may be accomplished by a change in lubricant, die profile, temperature, deformation level, or process rate [20]. In multi-pass extrusions, each forming pass plays an important role in decreasing the hydrostatic stress due to the counter pressure effect; previous studies [23,24] have demonstrated that the application of counter pressure decreases the central damage accumulation, which leads to an increase of the material formability, even for brittle materials. When fracture is already presented in the material with the appearance of cracks, the increase of the counter pressure results in a reduction of the crack size and, at a certain level of counter pressure, central burst can even disappear. In Figure 1, a comparison between a single and a double reduction with a double-pass die can be observed. In this last case (double-pass extrusion), the strain rate diagram along the longitudinal axis is divided in to two regions of a lower magnitude than in the case of single pass, resulting in different values of central damage. Partial reductions will determine the increase or reduction of the central damage in the final part compared to a single reduction. In this paper, we investigated which geometrical configurations lead to a decrease in central damage for the material DP800 considering double-pass extrusions; the methodology followed is presented in detail in order to be used for the analysis of other emerging materials.

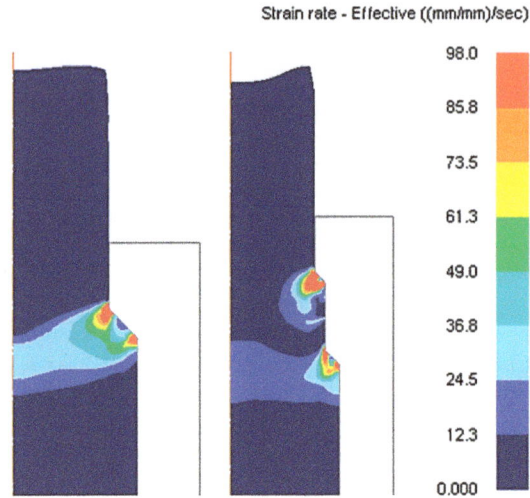

Figure 1. Scheme of single and double-pass extrusions showing strain rate contour diagrams by FEA (finite element analysis).

2. Materials and Methods

2.1. Finite Element Modelling with DEFORM F2™

Unlike general purpose FEM codes, DEFORM is tailored for deformation modeling. This study has been realized using the finite element software DEFORM F2™ (Scientific Forming Technologies Corporation, Columbus, OH, USA); this code is especially designed to simulate axisymmetric metal forming operations such as the ones approached in this study [25].

DEFORM F2™ (Scientific Forming Technologies Corporation, Columbus, OH, USA) preprocessor uses a graphical user interface to integrate the data required to run the simulation. Input data includes:

- Object description: all data associated with an object, including geometry, mesh, temperature, material, etc.
- Material data: data describing the behavior of the material under the conditions which will experience during deformation.
- Inter object conditions: describes how the objects interact with each other, including contact and friction conditions between objects.
- Simulation controls: definition of parameters such as discrete time steps to model the process.

Extrusion dies are modelled as rigid parts and the workpiece is modelled as a deformable body. Regarding the material, the workpiece has been modelled with the dual-phase steel DP800, whose flow curve according to the Swift model is presented in Figure 2. The yield criterion adopted is von Mises, as it is the default setting for an isotropic material model and anisotropy influence has not been considered in this study.

The same geometry of the workpiece has been considered in all the simulations: one billet of initial diameter $d_0 = 20$ mm and length $L_0 = 50$ mm, assuming axisymmetric conditions, so only half of the model has been analyzed. The workpiece has been meshed with first order continuum elements of quadrilateral shape. A key component of this software is a fully automatic, optimized remeshing system tailored for large deformation problems, as in the case of extrusion processes. Contact boundary conditions with robust remeshing allow the simulations to finish without convergence problems [25], even when complex geometries are involved. In Figure 4, it is possible to see a finer mesh close to the initial contact surfaces.

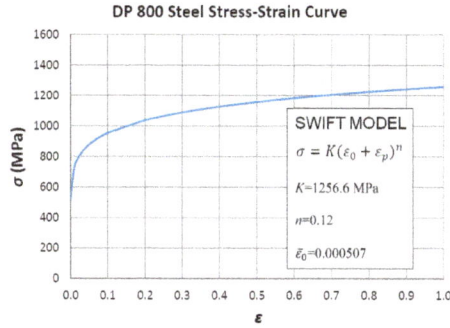

Figure 2. Flow curve of dual-phase steel DP800.

The die geometry is different in each simulation and the different configurations will be defined subsequently. In extrusions with single-pass dies, the chosen die semiangles are defined in the range $0° < \alpha < 90°$ (Table 1), and reductions in the range $0 < R_T < 1$ (Table 2), where the cross-section reduction is defined as $R_T = 1 - A_f/A_0$, being A_0 and A_f the initial and final cross-sections, respectively.

Table 1. Die semi-angles in extrusions with single-pass dies (α).

α (°)					
15	30	45	60	75	90

Table 2. Cross-section reductions in extrusions with single-pass dies (R_T).

R_T			
0.2	0.4	0.6	0.8

In extrusions with double-pass dies, the total cross-section reduction, R_T, is divided into two consecutive partial reductions, R_A and R_B, connected by one cylindrical surface of length, L. The position of this connecting surface is given by the value of R_A, (Figure 3). The calculation of R_A and R_B is as follows: $R_A = 1 - A_1/A_0$, where A_1 is the resulting area after the first pass; and $R_B = 1 - A_f/A_1$, so $R_B = (R_T - R_A)/(1 - R_A)$.

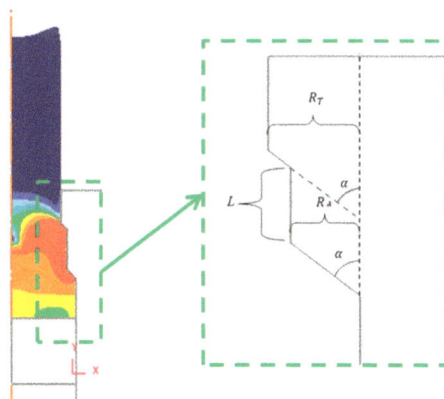

Figure 3. Geometrical definition of double-pass die.

The die semiangles are the same as in extrusions with single pass dies (Table 1); as well as the total reductions, R_T (Table 2). Values of R_A (as a fraction of R_T) and non-dimensional length, L/d_0, are shown in Tables 3 and 4, respectively. With this set of cases, it is possible to find the geometrical configurations where R_A and L induce a lower central damage for each value of R_T and α; however, this is not enough to predict the configuration of minimum damage location, so for each particular case the search is arranged with values of L and R_A conveniently chosen.

Table 3. Cross-section reductions in the first pass relative to the total reduction (R_A/R_T).

R_A/R_T					
0.0	0.2	0.4	0.6	0.8	1.0

Table 4. Non-dimensional length (L/d_0) in extrusions with double-pass dies.

L/d_0						
0.0	0.1	0.2	0.3	0.4	0.5	0.6

The initial geometrical configuration of the model for a particular case ($\alpha = 30°$, $R_T = 0.6$, $R_A/R_T = 60\%$, $L/d_0 = 0.4$) is presented in Figure 4.

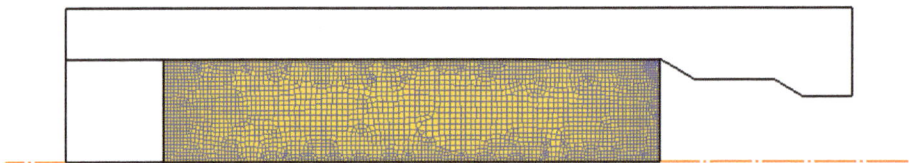

Figure 4. Mesh and geometrical configuration at the initial step before extrusion starts ($\alpha = 30°$, $R_T = 0.6$, $R_A/R_T = 60\%$, $L/d_0 = 0.4$).

In this paper cold forming conditions are considered, so the flow stress does not depend on the strain rate as the temperature is considered constant and equal to 20 °C.

Typical values for ram velocity in extrusion processes can reach up to 500 mm/s [26], so the punch has been modelled to move at a ram velocity of 200 mm/s in all the simulations, considering that it is a cold forming process.

The shear friction model has been assumed, which considers a constant friction factor, m, and its analytical expression is (Equation (1))

$$\tau = m \cdot k, \tag{1}$$

This model assumes that friction stress (τ) is constant and it only depends on the shear flow stress, k; it has been demonstrated to be more realistic than Coulomb's friction model in modelling forming operations, so it is specially recommended in metal forming analysis.

Regarding simulation controls, DEFORM F2™ is a numerical code of implicit methodology that uses the Newton-Raphson method for solving the equations. The model includes 200 steps and the step increment is defined as 10. The number of steps is given by Equation (2)

$$n = \frac{x}{V \cdot \Delta t}, \tag{2}$$

where,

n: number of steps;

x: total movement of the primary die;

V: ram velocity;

Δ*t*: time increment per step.

DEFORM F2™ allows choosing different levels of shape complexity and accuracy, offering a different range for the number of elements of the mesh. Higher defined values means more accurate final results from the simulation; however, the computation time will increase accordingly.

To determine the best combination of these parameters, and previously to the set of simulations planned, a brief study has been developed using a cross section reduction of $R_T = 0.5$ and a die semiangle of $\alpha = 45°$. The reason of choosing an intermediate situation is that DEFORM is a software specially designed to simulate metal forming operations; the study of the mesh is a key issue in other simulation programs of general purpose, being mandatory in these cases to realize a mesh sensitivity analysis. As the software used in this work includes a fully automatic, optimized remeshing algorithm, and direct extrusion of cylindrical billets is not a complex problem from a numerical point of view compared to other geometries (extrusion of complex profiles), the selection of accuracy and shape complexity parameters can be extrapolated to the other configurations and no important differences are expected to occur. Moreover, as explained before and indicated in the DEFORM user's manual, the program implements a contact boundary condition with robust remeshing, so the mesh at the contact zone will be remeshed automatically in every case.

A moderate accuracy level and 3000 mesh elements have been determined as the best options because the central damage factors are similar to those obtained for higher levels (Figure 5a,b), and the computation times are adequate (Figure 5c).

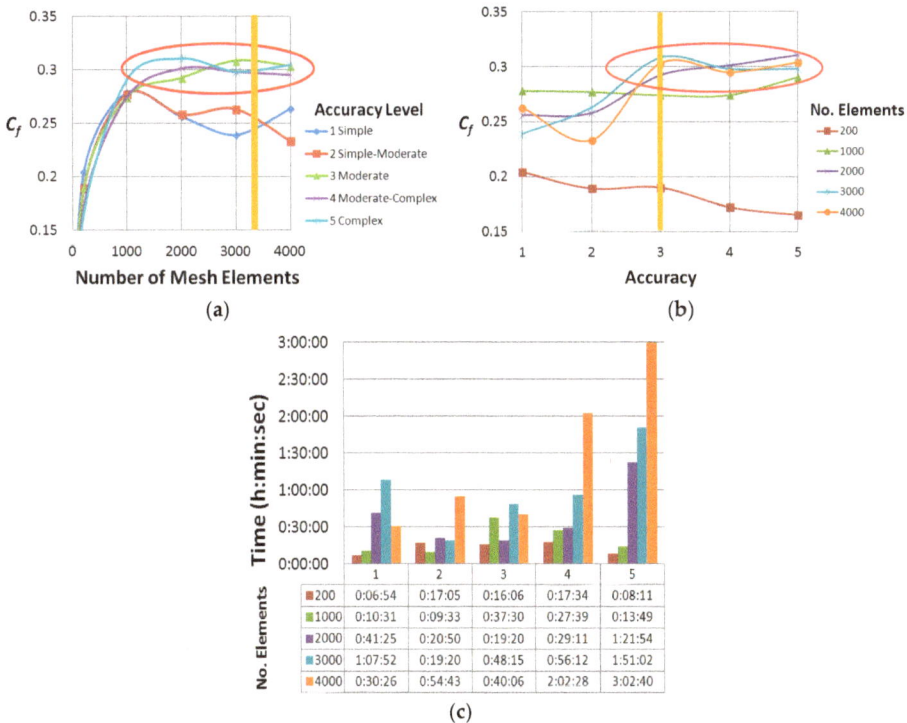

(a)

(b)

(c)

Figure 5. (**a**) Selection of number of elements of the mesh; (**b**) Selection of the accuracy level; (**c**) Computation times.

The damage factor used by DEFORM F2™ is based on the Cockcroft-Latham criterion [27] and it establishes that fracture occurs in a ductile material when the integral in Equation (3) reaches a constant value, C, for a given temperature and strain rate

$$C = \int_0^{\overline{\varepsilon}_f} \sigma * d\overline{\varepsilon},$$ (3)

where $\sigma*$ is the maximum principal stress, $\overline{\varepsilon}$ is the equivalent strain, and $\overline{\varepsilon}_f$ is the equivalent strain to fracture. Damage in graphs specifies the damage factor at each element, C_f, and it is defined as

$$C_f = \int_0^{\overline{\varepsilon}_f} \frac{\sigma*}{\overline{\sigma}} d\overline{\varepsilon},$$ (4)

where $\overline{\sigma}$ is the effective stress. The damage factor is a non-dimensional parameter and can be used to predict fracture in cold forming operations [25].

2.2. Finite Element Model Validation

In order to validate the finite element model developed in DEFORM F2™, some results are going to be compared to the ones obtained by Soyarslan [23]. To this aim, the extrusion forces to extrude a billet of Cf53 steel (UNS G10550) are calculated for a double-pass die. This steel is an unalloyed high carbon steel with high stability and hardness, low deformation, and good wear resistance; the geometrical dimensions of the billet and the double-pass die are presented in Figure 6.

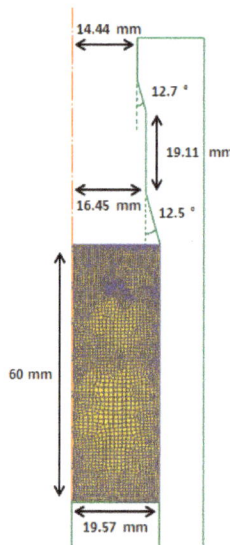

Figure 6. Mesh and geometrical details defined in [23].

The extrusion force versus the ram stroke is represented in Figure 7 and compared to the work of Soyarslan [23]. The extrusion force at the first pass reaches the value of 600 kN; followed by a second pass where the maximum force reached is around 1100 kN.

Effective strain distributions and deformation pattern (Figure 8) in different stages of the simulation have also been compared to the ones presented in [23], having a perfect concordance; maximum residual strains are reached at the surface, and the deformation pattern at the die exit is the same.

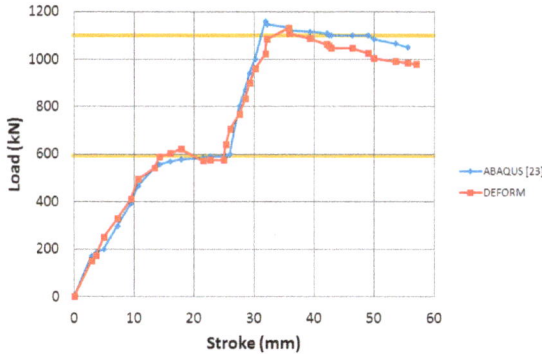

Figure 7. Extrusion force in a double-pass extrusion for Cf53 steel obtained by the finite element software ABAQUS [23] and DEFORM-F2. Yellow horizontal lines show approximate values of the maximum forces required to extrude the workpiece in each pass.

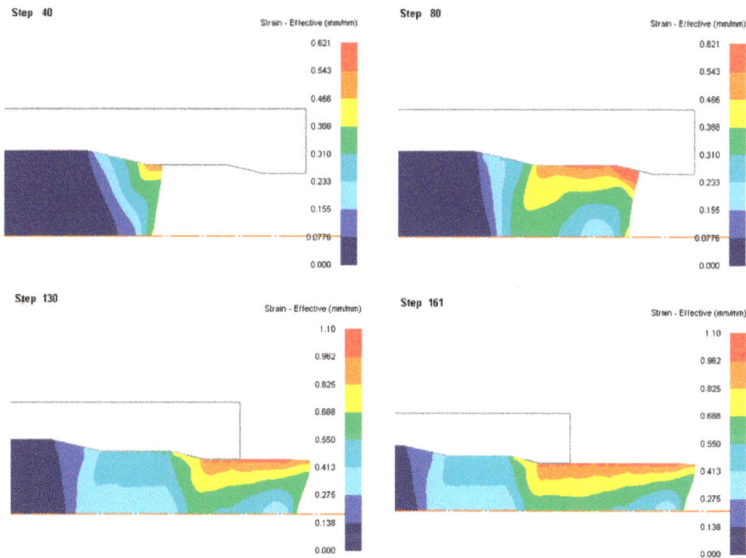

Figure 8. Effective strain distributions in different stages of the simulation used in the validation of the finite element model.

Results are in good agreement with those obtained in [23], so the finite element model is considered validated.

3. Results and Discussion

3.1. Single-Pass Dies

Results of damage factors for single-pass dies are shown in Figure 9 as a function of R_T and α for a friction factor of $m = 0.08$. This is the value suggested by DEFORM for a general cold forming operation. To confirm that this value is in accordance to the industrial practice of extrusion of steels, we have checked that this value is also in the range of values found for the shear factor, m, obtained from double cup backward extrusion tests conducted in steels at room temperature and presented in [28]. Concretely, the values are in the range: $0.035 < m < 0.075$ for different steels and lubrication systems, so the value $m = 0.08$ can be considered acceptable as a reference value where no lubrication system is defined, as in the case of this paper.

In indentation the non-homogeneity in forming causes secondary stresses and they depend on the ratio h/b, h being the height of the workpiece and b the width of the punch in contact with the workpiece. According to previous work about extrusion [21], the hydrostatic stress becomes positive and so leads to fracture, when h/b reaches the value 1.8. The theoretical limit of formability described by the hydrostatic stress criterion can be approached by the equation indicated in Figure 9 (blue curve), that represents the combination of cross-section reduction and die semiangle where $h/b = 1.8$. Considering this, central burst is not expected to appear to the left of the curve, whereas it could take place to the right depending of the microstructural characteristics.

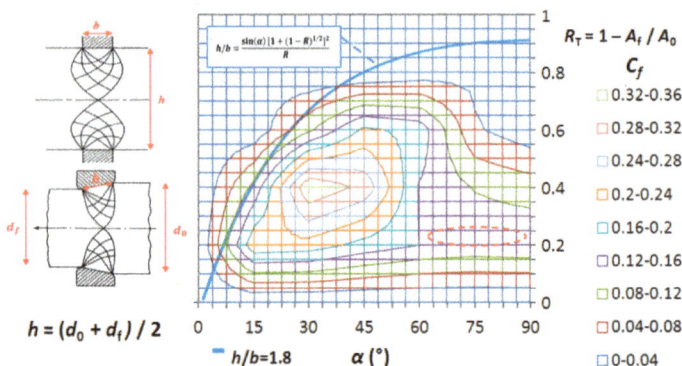

Figure 9. Central damage factors (C_f) map in single-pass extrusion of dual-phase steel DP800 for $m = 0.08$ and theoretical limit of central burst appearance (blue curve) according to [21].

The most damaged zones are located in the range of reductions $0.2 < R_T < 0.6$ and die semiangles $15° < \alpha < 55°$. For the highest semiangles ($60° < \alpha < 90°$) damage becomes constant due to the dead zone appearance. Avitzur explained this effect in his work from 1968 [29], concluding that at high semiangles the dead zone formation is energetically more favorable than the central burst appearance and then the extrusion force experiences an asymptotic behavior (Figure 10a).

Additionally, as an example of friction influence on damage appearance, results have been analyzed for $\alpha = 30°$. As it can be seen in Figure 10b, the maximum damage factor is expected when there is not friction at the contact surfaces, and the damage diminishes when the friction factor increases.

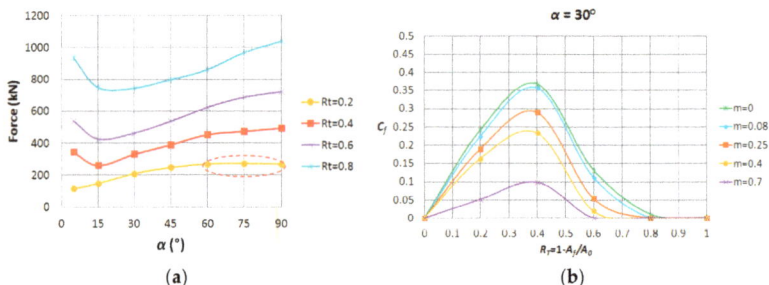

Figure 10. (a) Extrusion force at different cross-section reductions for $m = 0.08$ and dead zone effect; (b) Central damage (C_f) versus cross-section reduction for $\alpha = 30°$ and different friction factors.

3.2. Double-Pass Dies

In Figure 11, damage factor curves and extrusion forces are represented in the range $0\% < R_A/R_T < 100\%$ for $L/d_0 = 0.3$. The extrusion force is divided into two levels: the green dashed line is the force required to extrude the billet through the first pass, whereas the continuous line shows the total extrusion force. As expected, both lines are coincident when $R_A/R_T = 100\%$, as it is the same case than a single-pass die. The red line shows the central damage behavior with R_A. Again, this curve has the same values at $R_A/R_T = 0\%$ and $R_A/R_T = 100\%$ because it is a single-pass situation.

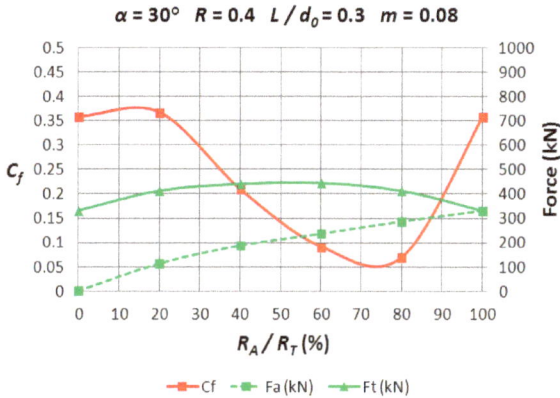

Figure 11. Damage factors (C_f) and forces versus R_A/R_T.

As an example of the results obtained, the behavior of damage factor with R_A and L is presented in Figure 12 for a particular case ($\alpha = 15°$, $R_T = 0.4$, $m = 0.08$).

According to Figure 12, all the cases where $L/d_0 = 0$, $R_A/R_T = 0\%$ and $R_A/R_T = 100\%$ show the same values of damage factors as they represent the single-pass extrusion. Generally speaking, as the length L grows, the central damage factor also increases; however, there is an intermediate zone where the central damage is minimum and this is the most interesting area of the graph because it shows the optimal geometrical configurations in order to reduce damage in the final part. Figure 13 sums up the results in two-dimensional graphs, using the same scale than in Figure 12, and showing the central damage factors contours versus R_A/R_T and L/d_0 for each total reduction and semiangle. Simulations of double-pass extrusions have not been realized for a total reduction of $R_T = 0.8$ because damage is low for all the semiangles in single-pass extrusion.

In each graph of Figure 13, a blue point indicating the absolute minimum has been included; the position of the minimum damage factor moves to higher values of L as the semiangle increases and to lower values of R_A as the total reduction diminishes (show dashed red lines in Figure 13).

Gathering in a graph the minimum points (points in blue) corresponding to each semiangle and total reduction, a new central damage factors map can be depicted (Figure 14), and these values of damage are much lower than the ones represented in Figure 9 for a single-pass die.

The decrease of the central damage in the new map for double-pass extrusions (Figure 14) can contribute to avoiding central burst appearance for the whole range of values.

Based in Figure 13, the best geometrical configurations of double-pass dies (blue points) defined by R_A/R_T and L/d_0, are summarized in Figure 15 for each combination of R_T and α considered in this study.

Effective strain distributions have been also obtained for the best configurations, and they are presented with the same scale in Figure 16.

$$\alpha = 15° \quad R_T = 0.4 \quad m = 0.08$$

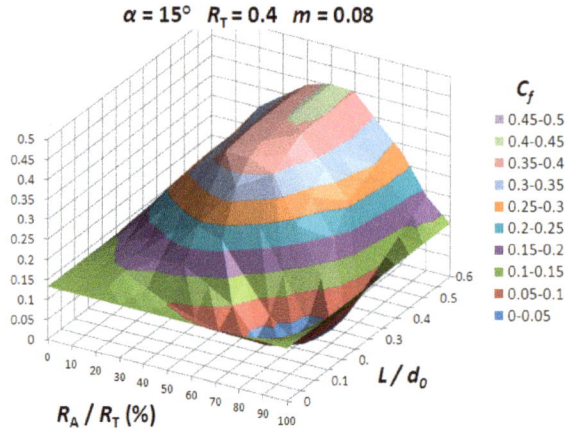

Figure 12. Surface of damage (C_f) as a function of L and R_A.

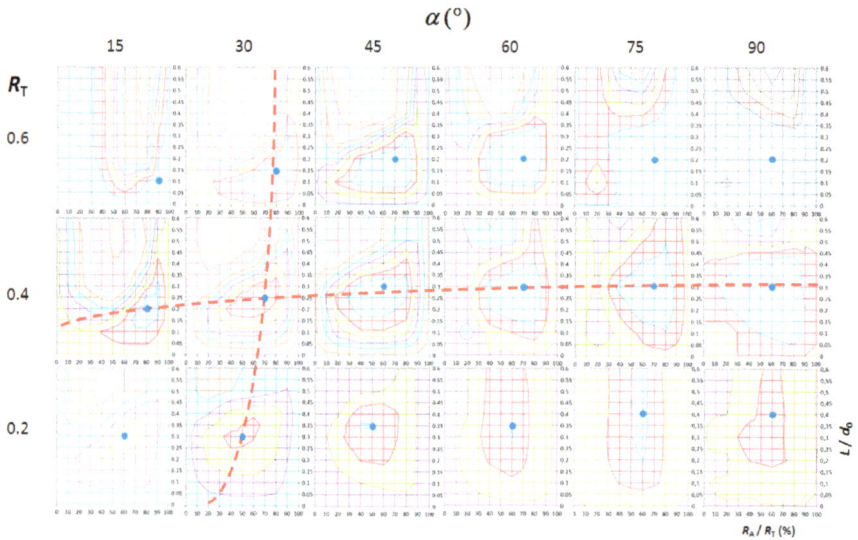

Figure 13. Contours of central damage factors (C_f) in double-pass extrusions and location of the point of minimum damage.

As expected, the highest values of strains are obtained for the highest total reduction ($R_T = 0.6$). For the same value of total reduction, the higher die semiangle, the higher the strains at the surface of the final part. The deformation pattern at the die exit changes, showing a dependency with the geometrical conditions.

In order to show the degree of agreement of the deformation pattern, a comparison with the case considered in the validation subsection has been developed (Figure 17), considering the steel DP800.

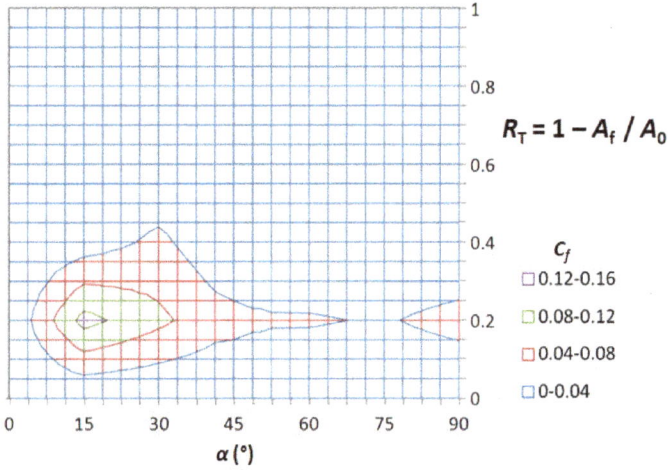

Figure 14. Central damage factors (C_f) map in double-pass extrusion of dual-phase steel DP800 for $m = 0.08$, showing the minimum values.

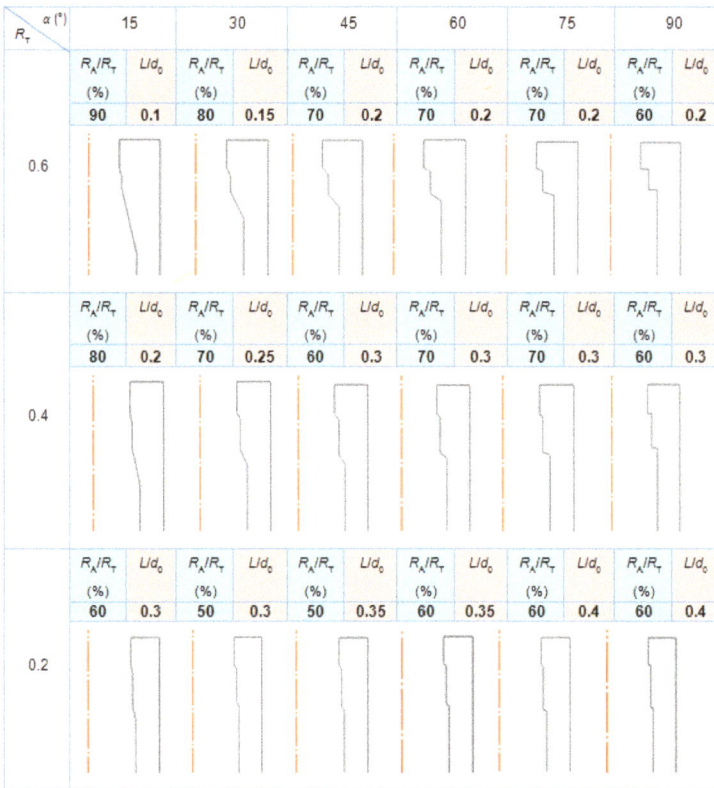

$$R_T = 1 - A_f / A_0$$

C_f
- ☐ 0.12-0.16
- ☐ 0.08-0.12
- ☐ 0.04-0.08
- ☐ 0-0.04

R_T \ α (°)	15		30		45		60		75		90	
	R_A/R_T (%)	L/d_o	R_A/R_T (%)	L/d_o	R_A/R_T (%)	L/d_o	R_A/R_T (%)	L/d_o	R_A/R_T (%)	L/d_o	R_A/R_T (%)	L/d_o
0.6	90	0.1	80	0.15	70	0.2	70	0.2	70	0.2	60	0.2
0.4	80	0.2	70	0.25	60	0.3	70	0.3	70	0.3	60	0.3
0.2	60	0.3	50	0.3	50	0.35	60	0.35	60	0.4	60	0.4

Figure 15. Summary of the best configurations of double-pass dies in cold extrusion of dual-phase steel DP800.

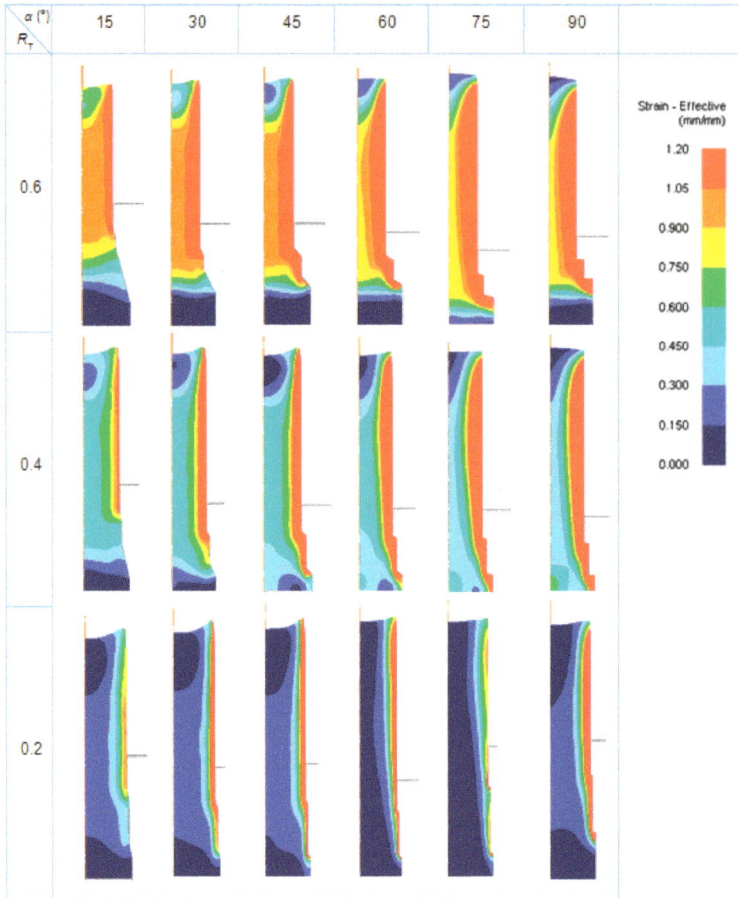

Figure 16. Effective strain diagrams for the best configurations of double-pass dies in cold extrusion of dual-phase steel DP800.

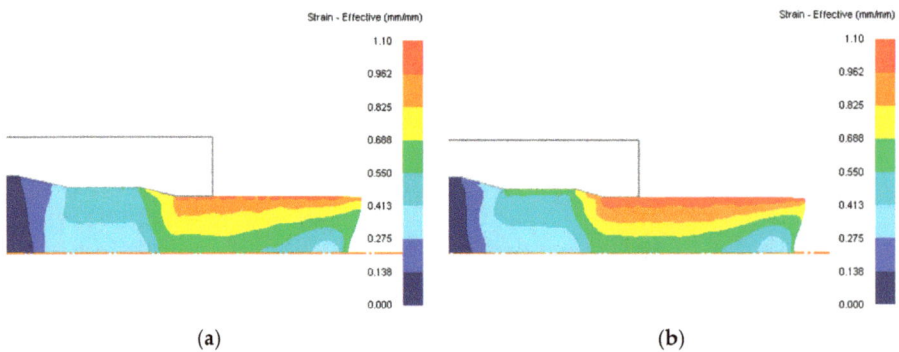

(a)   (b)

Figure 17. Comparison of strain distributions (R_T = 45.55%, $\alpha \cong 12.6°$ [23]); (a) Steel UNS G10550; (b) Steel DP800.

According to Figure 17, the deformation pattern presents a good degree of agreement as the profile of the workpiece at the die exit is coincident and the strain distributions show the maximum values of strain at the surface, and a minimum at the centerline.

3.3. Single-Pass Dies vs. Double-Pass Dies

In order to clearly show the reduction of central damage reached by the use of double-pass dies, a comparison of the results for single-pass and double-pass dies is presented in this section.

In Figure 18a,b, central damage factors maps for both cases are presented in the same figure, together with some graphs (Figure 18c) that specify in detail the values of minimum damage factor reached for all the combinations of die semiangle and total reduction in both cases (single and double-pass dies).

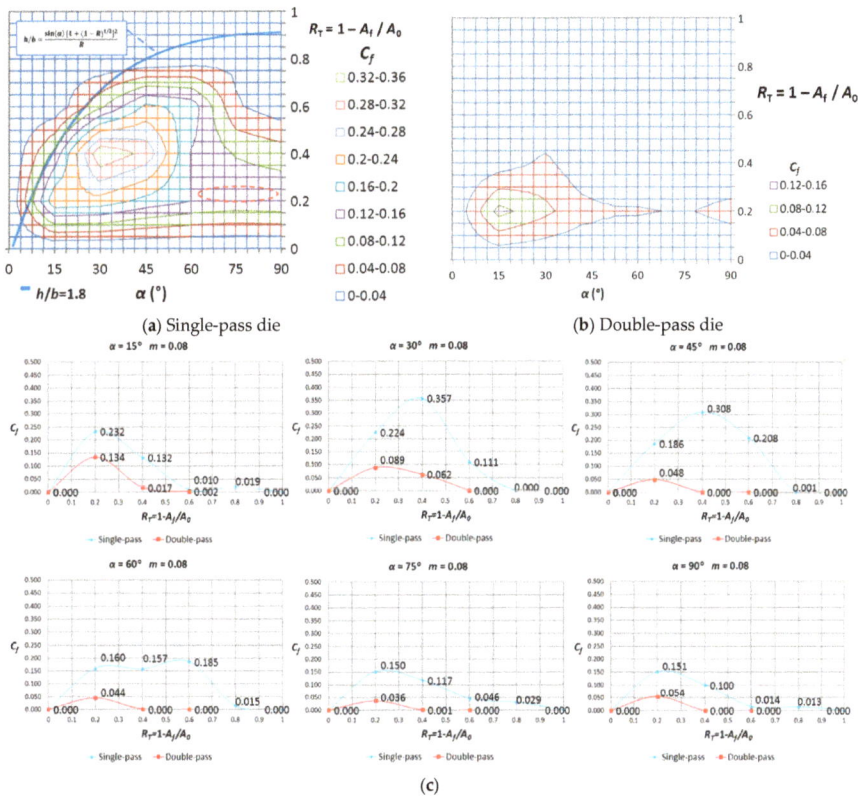

Figure 18. Comparison of minimum damage (C_f) induced with (**a**) single-pass dies vs. (**b**) double pass dies in cold extrusion of dual-phase steel DP800. (**c**) Values of minimum damage factor reached for all the combinations of die semiangle and total reduction for single and double-pass dies.

According to this figure, we can appreciate that the minimum damage is reached with double-pass dies in all the cases, when compared with single-pass dies. In single-pass extrusion, a clear tendency of damage factor with die semiangle cannot be defined; whereas in double-pass extrusion, the higher the die semiangle, the lower the induced damage.

In Figures 19–21, a comparison of induced damage distributions (considering the parameter damage factor) with single-pass dies vs. the best configurations of double-pass dies is presented.

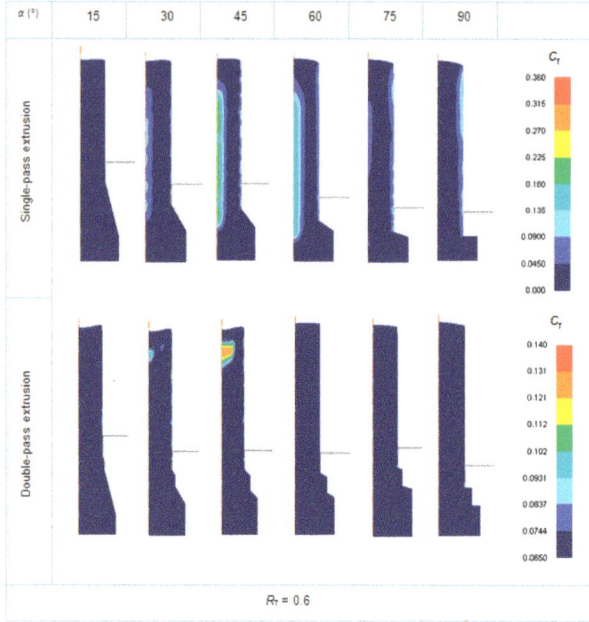

Figure 19. Comparison of damage (C_f) induced with single-pass dies vs. the best configurations of double-pass dies with $R_T = 0.6$.

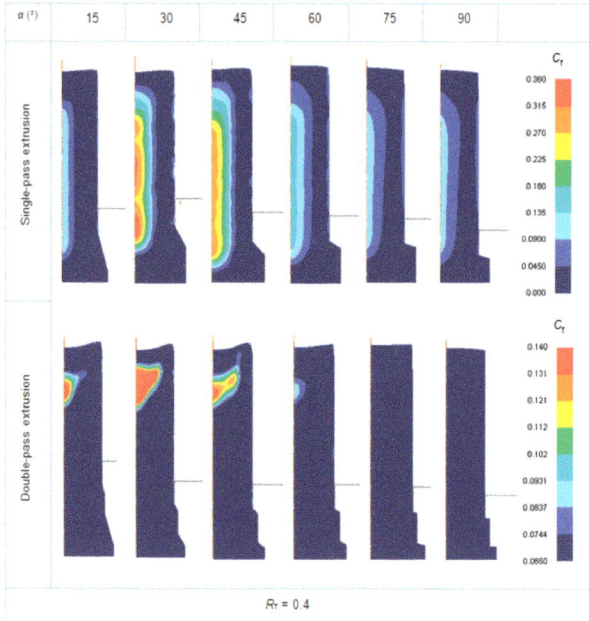

Figure 20. Comparison of damage (C_f) induced with single-pass dies vs. the best configurations of double-pass dies with $R_T = 0.4$.

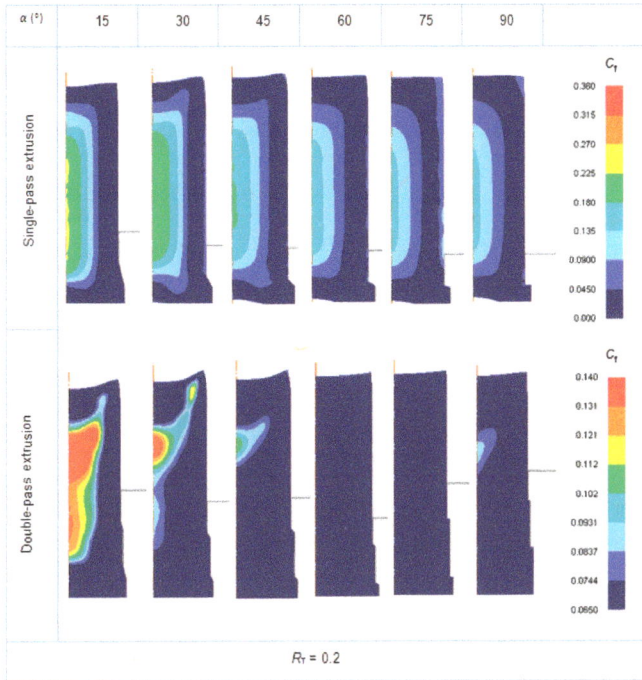

Figure 21. Comparison of damage (C_f) induced with single-pass dies vs. the best configurations of double-pass dies with $R_T = 0.2$.

A clear reduction of induced central damage is achieved with a change in the geometrical design of the die. This is a really important observation from an industrial point of view, because the final product of the extrusion process can significantly improve its quality and in-service behavior thanks to a change in the die design.

4. Conclusions and Future Work

In this work, the selection of the best geometrical configurations of double-pass dies is proposed as an alternative to single-pass extrusions in order to minimize the central damage that can lead to central burst in extruded parts. A methodology has been proposed using finite element simulation for the dual-phase steel DP800; around 500 simulations have been realized to take into account the combination of parameters in single and double-pass extrusions. Accordingly, some interesting conclusions have been extracted. First of all, simulation of single-pass extrusions was able obtain a map of central damage factors depending on the die semiangle and the total cross-section reduction, and it is consistent with the hydrostatic stress criterion. Simulations demonstrate that friction phenomenon reduces central damage. In double-pass extrusions, for each pair (R_T, α), there are combinations of R_A and L that cause a minimum value of damage, even lower than the one obtained in single-pass extrusions. Choosing these two parameters (R_A and L), which means selecting the best die design, it is possible to perform an extrusion where central damage is significantly reduced compared to single extrusion, as it has been shown through all the induced damage diagrams and central damage factors maps presented when comparing single and double-pass dies extrusions. These kinds of maps are quite useful to avoid defective products in industrial practice, as for example, forming limit diagrams typically used in sheet metal forming.

Therefore, in this paper, the best designs of die geometry to reduce central damage have been determined. This is a really important contribution from an industrial point of view, because the final product of the extrusion process can significantly improve its quality and in-service behavior thanks to a change in the die design.

As a general rule, the position of the minimum damage leads to choose higher values of L for high semiangles and to lower values of R_A for low total reductions.

In this paper, as a preliminary work in this field, we have focused on considering the most general conditions in a cold extrusion process of a particular dual-phase steel in order to determine if it is possible to establish a general trend in the reduction of damage when modifying the most relevant parameters; once it has been demonstrated by the results presented in the paper, and given that there are other parameters affecting the results, future work should be done to optimize the process, by searching for the most appropriate combination of parameters that leads to the best results (considering not only geometrical parameters, but also operating conditions and microstructural and tribological aspects); for example, taking into account different die semiangles in each pass and/or friction conditions, or analyzing the effect of this complex microstructure. In this regard, micromechanical modeling of damage is one promising research field; the use of 3D simulation software will be highly recommended in view of the results of Ayatollahi et al. [18], in order to accurately reproduce the different phases and their distribution in the microstructure of the workpiece. 3D finite element modelling will be also required in the case of analysis of extrusions of asymmetrical parts, where 2D modelling is not suitable.

Additionally, we think this methodology can be used to determine the optimal configuration of multiple-pass dies for other AHSS, whose formability currently under study.

Acknowledgments: This work has been financially supported by funds provided through the Annual Grant Call of the E.T.S.I.I. of UNED of reference 2017-ICF04 and the Department of Construction and Manufacturing Engineering of UNED. The authors would like to take this opportunity to thank the Research Group of the UNED "Industrial Production and Manufacturing Engineering (IPME)" for the support provided during the development of this work.

Author Contributions: Ana María Camacho conceived the problem; Francisco Javier Amigo designed the cases to be analyzed and performed the finite element simulations; Ana María Camacho and Francisco Javier Amigo analyzed the results; Francisco Javier Amigo wrote the paper.

Conflicts of Interest: The authors declare no conflict of interest.

References

1. Bhargava, M.; Tewari, A.; Mishra, S.K. Forming limit diagram of advanced high strength steels (AHSS) based on strain-path diagram. *Mater. Des.* **2015**, *85*, 149–155. [CrossRef]
2. Del Pozo, D.; López de Lacalle, L.N.; López, J.M.; Hernández, A. Prediction of press/die deformation for an accurate manufacturing of drawing dies. *Int. J. Adv. Manuf. Technol.* **2008**, *37*, 649–656. [CrossRef]
3. López de Lacalle, L.N.; Lamikiz, A.; Muñoa, J.; Sánchez, J.A. The cam as the centre of gravity of the five-axis high speed milling of complex parts. *Int. J. Prod. Res.* **2005**, *43*, 1983–1999. [CrossRef]
4. López de Lacalle, L.N.; Lamikiz, A.; Sánchez, J.A. Improving the high-speed finishing of forming tools for advanced high-strength steels (ahss). *Int. J. Adv. Manuf. Technol.* **2005**, *29*, 49–63. [CrossRef]
5. Nasser, A.; Yadav, A.; Pathak, P.; Altan, T. Determination of the flow stress of five ahss sheet materials (DP 600, DP 780, DP 780-CR, DP 780-HY and TRIP 780) using the uniaxial tensile and the biaxial viscous pressure bulge (VPB) tests. *J. Mater. Process. Technol.* **2010**, *210*, 429–436. [CrossRef]
6. Moeini, G.; Ramazani, A.; Myslicki, S.; Sundararaghavan, V.; Könke, C. Low cycle fatigue behaviour of DP steels: Micromechanical modelling vs. validation. *Metals* **2017**, *7*, 265. [CrossRef]
7. Nam, W.J.; Bae, C.M. Microstructural evolution and its relation to mechanical properties in a drawn dual-phase steel. *J. Mater. Sci.* **1999**, *34*, 5661–5668. [CrossRef]
8. Ramazani, A.; Kazemiabnavi, S.; Larson, R. Quantification of ferrite-martensite interface in dual phase steels: A first-principles study. *Acta Mater.* **2016**, *116*, 231–237. [CrossRef]

9. Ramazani, A.; Mukherjee, K.; Schwedt, A.; Goravanchi, P.; Prahl, U.; Bleck, W. Quantification of the effect of transformation-induced geometrically necessary dislocations on the flow-curve modelling of dual-phase steels. *Int. J. Plast.* **2013**, *43*, 128–152. [CrossRef]
10. Evin, E.; Tomáš, M. The influence of laser welding on the mechanical properties of dual phase and trip steels. *Metals* **2017**, *7*, 239. [CrossRef]
11. Gutiérrez-Regueras, J.M.; Camacho, A.M. Investigations on the influence of blank thickness (*t*) and length/wide punch ratio (*LD*) in rectangular deep drawing of dual-phase steels. *Comput. Mater. Sci.* **2014**, *91*, 134–145. [CrossRef]
12. García-Domínguez, A.; Claver, J.; Camacho, A.M.; Sebastián, M.A. Comparative analysis of extrusion processes by finite element analysis. *Procedia Eng.* **2015**, *100*, 74–83. [CrossRef]
13. Parghazeh, A.; Haghighat, H. Prediction of central bursting defects in rod extrusión process with upper bound analysis method. *Trans. Nonferrous Met. Soc. China* **2016**, *26*, 2892–2899. [CrossRef]
14. Camacho, A.M.; Gonzalez, C.; Rubio, E.M.; Sebastian, M.A. Influence of geometrical conditions on central burst appearance in axisymmetrical drawing processes. *J. Mater. Process. Technol.* **2006**, *177*, 304–306. [CrossRef]
15. Oh, S.-I.; Walters, J.; Wu, W.-T. Finite element method applications in bulk forming. In *ASM Handbook. Metals Process Simulation*; Furrer, D.U., Semiatin, S.L., Eds.; ASM International: Materials Park, OH, USA, 2010; Volume 22B, pp. 267–289. ISBN 978-1-61503-005-7.
16. Pathak, N.; Butcher, C.; Worswick, M.J.; Bellhouse, E.; Gao, J. Damage evolution in complex-phase and dual-phase steels during edge stretching. *Materials* **2017**, *10*, 346. [CrossRef] [PubMed]
17. Tasan, C.C.; Hoefnagels, J.P.M.; Diehl, M.; Yan, D.R.F.; Raabe, D. Strain localization and damage in dual phase steels investigated by coupled in-situ deformation experiments and crystal plasticity simulations. *Int. J. Plast.* **2014**, *63*, 198–210. [CrossRef]
18. Ayatollahi, M.R.; Darabi, A.C.; Chamani, H.R.; Kadkhodapour, J. 3D micromechanical modeling of failure and damage evolution in dual phase steel based on a real 2D microstructure. *Acta Mech. Solida Sin.* **2016**, *29*, 95–110. [CrossRef]
19. Tekkaya, A.E.; Khalifa, N.B.; Hering, O.; Meya, R.; Myslicki, S.; Walther, F. Forming-induced damage and its effects on product properties. *CIRP Ann. Manuf. Technol.* **2017**, *66*, 281–284. [CrossRef]
20. Dieter, G.E.; Kuhn, H.A.; Semiatin, S.L. *Handbook of Workability and Process Design*; ASM International: Materials Park, OH, USA, 2003; ISBN 978-0-87170-778-9.
21. Reddy, N.V.; Dixit, P.M.; Lal, G.K. Ductile fracture criteria and its prediction in axisymmetric drawing. *Int. J. Mach. Tool Manuf.* **2000**, *40*, 95–111. [CrossRef]
22. Reddy, N.V.; Dixit, P.M.; Lal, G.K. Central bursting and optimal die profile for axisymmetric extrusion. *J. Manuf. Sci. Eng.-Trans. ASME* **1996**, *118*, 579–584. [CrossRef]
23. Soyarslan, C. Modelling Damage for Elastoplasticity. Ph.D. Thesis, Middle East Technical University, Çankaya Ankara, Turkey, 2008.
24. Wagener, H.W.; Haats, J. Crack prevention and increase of workability of brittle materials by cold extrusion. *Stud. Appl. Mech.* **1995**, *43*, 373–386.
25. Scientific Forming Technologies Corporation. *DEFORM-F2 v11.0 User's Manual*; Scientific Forming Technologies Corporation: Columbus, OH, USA, 2014.
26. Kalpakjian, S.; Schmid, S.R. *Manufacturing Engineering and Technology*, 7th ed.; Pearson: Mexico, 2014; ISBN 978-0133128741.
27. Cockroft, M.; Latham, D. Ductility and the workability of metals. *J. Inst. Met.* **1968**, *96*, 33–39.
28. Gariety, M.; Ngaile, G. *Cold and Hot Forging. Fundamentals and Applications*; ASM International: Materials Park, OH, USA, 2005.
29. Avitzur, B. Analysis of central bursting defects in extrusion and wire drawing. *J. Eng. Ind.* **1968**, *90*, 79–91. [CrossRef]

![metals logo] *metals*

MDPI

Article

Effect of Computational Parameters on Springback Prediction by Numerical Simulation

Tomasz Trzepiecinski [1] and Hirpa G. Lemu [2,*]

[1]   Depertment of Materials Forming and Processing, Rzeszow University of Technology, Al. Powst. Warszawy 8, 35-959 Rzeszow, Poland; tomtrz@prz.edu.pl
[2]   Department of Mechanical and Structural Engineering, University of Stavanger, N-4036 Stavanger, Norway
*   Correspondence: hirpa.g.lemu@uis.no; Tel.: +47-51-83-21-73

Received: 27 August 2017; Accepted: 15 September 2017; Published: 19 September 2017

Abstract: Elastic recovery of the material, called springback, is one of the problems in sheet metal forming of drawpieces, especially with a complex shape. The springback can be influenced by various technological, geometrical, and material parameters. In this paper the results of experimental testing and numerical study are presented. The experiments are conducted on DC04 steel sheets, commonly used in the automotive industry. The numerical analysis of V-die air bending tests is carried out with the finite element method (FEM)-based ABAQUS/Standard 2016 program. A quadratic Hill anisotropic yield criterion is compared with an isotropic material described by the von Mises yield criterion. The effect of a number of integration points and integration rules on the springback amount and computation time is also considered. Two integration rules available in ABAQUS: the Gauss' integration rule and Simpson's integration rule are considered. The effect of sample orientation according to the sheet rolling direction and friction contact behaviour on the prediction of springback is also analysed. It is observed that the width of the sample bend in the V-bending test influences the stress-state in the cross-section of the sample. Different stress-states in the sample bend of the V-shaped die cause that the sheet undergoes springback in different planes. Friction contact phenomena slightly influences the springback behaviour.

Keywords: anisotropy; bending; numerical simulation; sheet metal forming; springback

---

1. Introduction

Bending is one of the sheet forming methods and is a plastic deformation of the material subjected to bending moment. Plastic forming of the sheets requires, at the design stage of manufacturing, taking into account specific properties of the sheet material, i.e., Young's modulus, yield stress, ratio of yield stress to ultimate tensile stress, and microstructure of the material [1]. The non-uniform strain state at the section of bent material leads to existence of residual stress after load releasing. This stress produces springback which is manifested by unintended changes in the shape of the element after forming. The measure of the springback value is a springback coefficient or angle of springback. The value of springback coefficient depends on, among others, the value of angle and radius of bending, thickness and width of the sheet strip, the mechanical properties of the sheet material, the temperature of bending process, and strain rate [1]. The investigations of Caden et al. [2] proved the effect of coefficient of friction on the springback amount.

Elastic recovery of material is one of the main sources of shape and dimensional accuracy of drawpieces. Springback cannot be eliminated, but there arc a few methods to minimize elastic return of the stamped part due to elastic recovery of sheet metal after forming. One of the methods is a suitable design of the die which takes into consideration the amount of springback. Furthermore, the change in selected bending process parameters can minimize the springback. The idea of correction of the die shape consists in additional overbending of the material [3]. Among the many advanced methods of predicting the final shape of the drawpiece, the finite element method (FEM) is the most often used [4]. FEM is the main technique used to simulate sheet metal forming processes in order to determine the distribution of stresses and deformations in the material, forming forces and potential locations of the defects.

For simple problems analytical methods for bending process analysis may be used. Due to the assumed simplifications, however, the analytical methods are not sufficiently general to accommodate the material and the geometrical influences [5]. Although, the experiments are time-consuming, they are still needed to better understand the elastic deformations of materials. To study the springback of sheet metals, several forms of experimental tests are used, including U-bending [1], V-bending [6], cylindrical bending, three-point bending, rotary bending, and flanging. Karağaç [7] estimated springback by using fuzzy logic based on the results of the V-bending test conducted at different holding times and bending angles. Leu and Hsieh [8] explored the influence of the coining force on the spring-back reduction in the V-die bending process. The effects of various process parameters, including the material anisotropy and coefficient of friction, on the spring-back reduction were confirmed. Bakhshi-Jooybari et al. [9] investigated the effects of significant parameters on spring-back in U-die and V-die bending of anisotropic steel sheet. Based on the comparison of experimental results with the numerical ones it was found that the bending angle to the rolling direction will influence the spring-back, where the greater the angle to the rolling direction, the greater the springback.

Results of study on the effect of the speed of deformation on the spring value of the sheet springback have been discussed by Firat et al. [10]. Hang and Leu [11] conducted experimental studies for steel sheets and presented the impact of variable process parameters such as the radius of the punch, die radius, punch speed, friction coefficient, and normal anisotropy on the sheet springback amount during the V-bending process. Garcia-Romeu et al. [12] conducted bending experiments on aluminium and stainless steel sheets for analysis of effects of bending angle on springback. Ragai et al. [13] presented experimental and numerical results of the anisotropy of the mechanical properties of stainless steel 410 sheet metal on springback. Vin et al. [14] investigated the effect of Young's modulus on the springback in the air V-bending process. Thipprakmas and Rojananan [15] examined the springback and spring-go phenomena on the V-bending process using the finite element method (FEM). Tekiner [16] examined the effect of bending angle on springback of six types of materials with different thicknesses in V-die bending.

The springback phenomenon of the sheets is also affected by the accuracy of manufacturing the stamping dies. Lingbeek et al. [17] presented a method for springback compensation in the tools for sheet metal products, concluding that for industrial deep drawing products the accuracy of the results has not yet reached an acceptable level. Del Pozo et al. [18] presented a method for the reduction of both the try-out and lead-time of complex dies. Furthermore, López de Lacalle [19] concluded that the two main problems have to be overcome in high-speed finishing of forming tools. The first problem is the simultaneous finishing of surfaces with different hardness in the same operation and with the same computer numerical control (CNC) program; and the second one is the unacceptable dimensional error resulting from tool deflection due to cutting forces.

Taking into account the numerical strength of the deformation of the sheet metal, it was possible to improve the convergence of experimental and simulation results, indicating the validity of the Bauschinger effect in simulating the springback problem. In addition to the Bauschinger effect, the elastic stress-strain relation is also important behaviour, especially given that springback is an elastic recovery phenomenon [20]. Experiments by Yoshida et al. [21] were carried out on how to utilize reverse bending that takes place in the forming process, how to improve uneven stress by applying a stress in sheet thickness direction, and how to reduce the plastic strain of a die shoulder without applying blankholder force thus to study the influences of those methods on springback. Besides elastic behaviour of material, the plastic anisotropic properties of material and hardening rule have to be taken into consideration in FEM analysis of springback. The isotropic hardening model and the kinematic hardening model are widely used in FEM analysis of sheet metal forming. While the former model can describe hardening, the latter, on the other hand, can describe the Bauschinger effect qualitatively, but cannot describe hardening [22]. To model the Bauschinger effect, several other models have been proposed [23].

To reflect the nonlinear strain and stress during elastic-plastic deformation of the sheet material, the crucial point in the computational modelling of springback is the proper choice of finite element formulation, the element size and a number of integration points through the sheet thickness. Suitable mesh density, especially in the region of contact of the tools with the sheet is a balance between computational time and springback prediction accuracy. Many publications deal with the determination of the optimal number of integration points through the sheet thickness and the proposed number of integrations points varies from 5 to 51. In the case of non-linear analysis, five integration points are sufficient to provide accurate results [24], while Xu et al. [25] concluded that usually seven integration points are sufficient. On the contrary, Wagoner and Li [26] found that to analyse the springback with 1% computational error, up to 51 points are required for shell type elements. Thus, as noticed by Banabic [27], the choice of a number of integration points is still an open issue in the simulation of springback.

The aim of this paper is investigation of the effect of some numerical approaches on prediction accuracy of the springback phenomenon. Experimental and numerical investigations of springback were carried out in V-bending test. Finite element (FE) elastic-plastic model of V-bending is built in ABAQUS software (Dassault Systèmes Simulia Corp., Providence, RI, USA). The numerical analyses took into account the sample orientation, material anisotropy, and work hardening phenomenon. Furthermore, the number of integration points and sensitivity to the friction coefficient are considered.

2. Materials and Methods

The experiments are conducted on DC04 steel sheets of 2 mm thickness cut along the rolling direction of the sheet and transverse to the rolling direction. To characterize the material properties, specimens for uniaxial tensile test steel sheets were cut at different orientations to the rolling directions ($0°$, $45°$, and $90°$). Three specimens were tested for each direction and average value of basic mechanical parameters (Table 1) were determined using the formula:

$$X_{av} = \frac{X_0 + 2X_{45} + X_{90}}{4} \tag{1}$$

where X is the mechanical parameter, and the subscripts denote the orientation of the sample with respect to the rolling direction of sheet.

Table 1. Mechanical properties of the tested sheets.

Orientation	Yield Stress σ_y (MPa)	Ultimate Tensile Strength σ_m (MPa)	Strain Hardening Coefficient C (MPa)	Strain Hardening Exponent n	Lankford's Coefficient r
0°	182.1	322.5	549.3	0.214	1.751
45°	196	336.2	564.9	0.205	1.124
90°	190	320.9	555.2	0.209	1.846
Average value	191.02	328.9	558.57	0.208	1.461

The representative true stress vs. true strain relations for three analysed sample cut directions are presented in Figure 1. The tested sheets are cold rolled, so the manufacturing process induces a particular anisotropy characterised by the symmetry of the mechanical properties with respect to three orthogonal planes. Furthermore, the method of trimming technology of standard sheet-type tensile test specimens can influence the surface state of the specimen. However, tensile tests of the sample prepared using milling, abrasive water jet, punching, wire electro discharge machining, and milling conducted by Martínez Krahmer et al. [27] show that some changes on the surface state appeared, but the effect on tensile strength was lower than 5%.

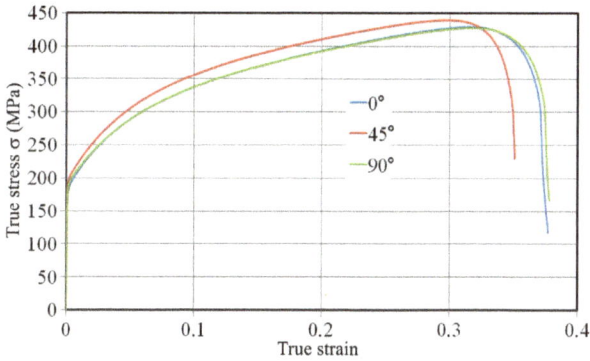

Figure 1. True stress-true strain relation of DC04 sheet.

The anisotropy of plastic behaviour of sheet metals is characterized by the Lankford's coefficient *r*, which is determined using the formula:

$$r = \frac{\ln \frac{w}{w_0}}{\ln \frac{l_0 \cdot w_0}{l \cdot w}} \qquad (2)$$

where w_0 and w are the initial and final widths, while l_0 and l are the initial and final gage lengths, respectively.

If the value of *r*-coefficient is greater than 1, the width strains are dominant, which is a characteristic of isotropic materials. On the other hand, a value of *r* < 1 indicates that the thickness strains will dominate.

The values of the parameters *C* and *n* in Hollomon equation [28] were determined from the logarithmic true stress-true strain plot by linear regression. The mean value of *n* exponent for the whole range of strain is usually assumed in numerical simulations. The strain hardening exponent can be determined using following formula:

$$n = \frac{d \log \sigma}{d \log \varepsilon} = \frac{d\sigma}{d\varepsilon} \frac{\varepsilon}{\sigma} \qquad (3)$$

The average values of Lankford's *r*-coefficient for different directions in the plane of sheet metal represent the average coefficient of normal anisotropy $\bar{r}$ Hovewer, the variation of the normal anisotropy with the angle to the rolling direction is given by the coefficient of planar anisotropy Δr. For the tested sheets $\Delta r = 0.67$, which indicates existence of material flow in 0° and 90° directions. In other words, if the value of Δr is positive then ears are formed in the direction of sheet rolling and in the direction perpendicular to rolling direction.

Air bending experiments were carried out in a designed semi closed 90° V-shaped die (Figure 2). Bending tests were carried out on rectangular samples of dimension 20 × 110 mm². The die assembly consists of a die with $R_d = 10$ mm rounded edge, and a punch with a nose radius of $R_s = 10$ mm. During the tests, punch bend depth under loading f_1 and punch bend depth under unloading f_{ul} (Figure 3) were measured. The values of these parameters were registered by the QuantumX Assistant V.1.1 program (V.1.1, Hottinger Baldwin Messtechnik GmbH, Darmstadt, Germany, 2011) for processing the signals of both force and punch stroke transducers. To measure the bending force, the HBM U9B force transducer (Hottinger Baldwin Messtechnik GmbH, Darmstadt, Germany) with nominal force 5 kN is used. The amount of springback was then evaluated as:

$$K = f_{ul} / f_1 \qquad (4)$$

where f_{ul} is the punch bend depth under unloading and f_1 is the punch bend depth under loading (Figure 3).

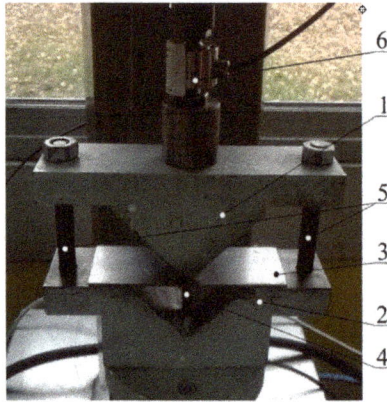

Figure 2. The experimental setup: 1—punch; 2—die; 3—sample; 4—punch stroke controller; 5—guide columns; and 6—force transducer.

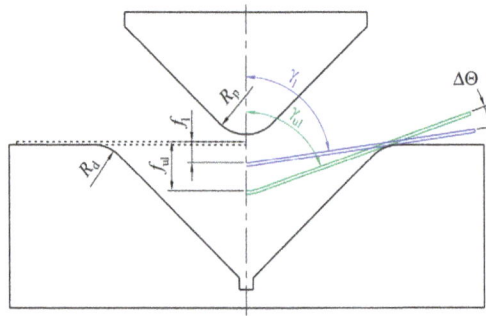

Figure 3. The schematic for the measurement of springback.

The other coefficient which may be used to analyse springback is defined as:

$$K_\gamma = \gamma_1 / \gamma_{ul} \tag{5}$$

where γ_1 is the bend angle under loading and γ_{ul} is the bend angle under unloading (Figure 3).

Three bending tests were conducted for all punch depths under loading and then the average value of springback was evaluated.

3. Numerical Model

The springback computations were conducted using ABAQUS/Standard 3D Experience® 2016 HF2 (2016 HF2, Dassault Systèmes Simulia Corp., Providence, RI, USA, 2016) which is used in springback prediction [29]. The numerical model of the problem (Figure 4) corresponds to the experimental set-up shown in Figure 2. The blank was modelled with an eight-node quadratic, doubly curved shell elements S8R [30]. The analytical discrete rigid tools are meshed using four-node 3D bilinear rigid quadrilateral R3D4 elements. The meshed model of the tools consists of 9586 elements.

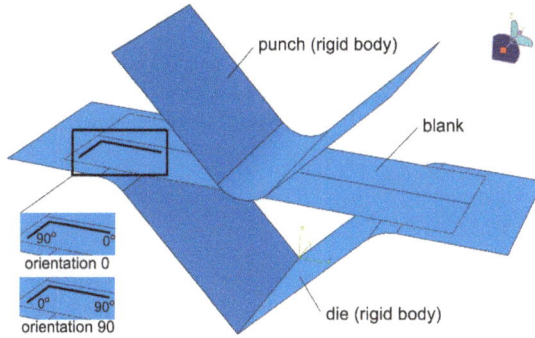

Figure 4. Boundary conditions in V-bending model of the sample cut along the sheet rolling direction.

The elastic behaviour of the sheet metal is specified in the numerical simulations by the value of Young's modulus, E = 210 GPa, and of Poisson's ratio ν = 0.3. The sheet material density is set to 7860 kg/m^3.

In the numerical model, the anisotropy of the material has been established using Hill (1948) [31] yield criterion (Equation (6)) with strain hardening behaviour that uses Hollomon power-type law [29]:

$$\bar{\sigma}^2 = F\left(\sigma_{22} - \sigma_{33}\right)^2 + G\left(\sigma_{33} - \sigma_{11}\right)^2 + H(\sigma_{11} - \sigma_{22})^2 + 2L\sigma_{23}^2 + 2M\sigma_{31}^2 + 2N\sigma_{12}^2 \tag{6}$$

where σ is the equivalent stress, and indices 1, 2, 3 represent the rolling, transverse, and normal direction to the sheet surface. Constants F, G, H, L, M, and N define the anisotropy state of the material and can be computed based on Lankford's coefficients [31]. The major advantage of the Hill (1948) function is that it gives an accurate description of yielding of steel sheets [32]. To investigate the effect of material model on the deformation of the sample material in the width direction, the isotropic material behaviour described by von Mises [33] yield criterion is also considered. For ideal case of isotropic materials, von Mises yield condition is expressed as:

$$\bar{\sigma}^2 = \frac{1}{2}[(\sigma_{11} - \sigma_{12})^2 + (\sigma_{22} - \sigma_{33})^2 + (\sigma_{33} - \sigma_{11})^2] + 3\left(\sigma_{12}^2 + \sigma_{23}^2 + \sigma_{31}^2\right) \tag{7}$$

The shell elements integrated in ABAQUS must be assigned with a method of integration rule and a number of integration points through the sheet thickness. Two integration rules available in ABAQUS—the Gauss' integration rule and Simpson's integration rule—are analysed in this paper. The number of integration points must be odd in order that one point can be in the middle surface of the shell element [24]. In order to study how the number of integration points influence springback prediction and computation time, the integration points 3, 5, 7, 11, 15, 19, and 25 were analysed in the case of Simpson's rule. Due to upper limits of the integration points in Gauss' rule built in ABAQUS, the simulations were carried out for the range of integration points from 3 to 15.

In addition to the number of integration points, the size of the shell elements is a critical parameter that influences the accuracy of computations especially in the bending process where the curvature of the sheet material has to be accurately represented. The optimum elements size may be determined based on the results of mesh sensitivity analysis. Such analysis with identical geometry of the sheet material and tool has been previously reported [6] by the authors of this paper. To determine an optimal mesh size, the numerical analyses were carried out for four selected meshes that resulted in the number of elements: 84, 280, 1120, and 4400. Furthermore, the sensitivity analysis of the mesh size was done for four punch strokes f_1: 3, 6, 12, and 18 mm. It was found that the increase in the number of elements stabilizes the springback measured as the difference between punch bend depth under loading f_l and unloading f_{ul}:

$$K_s = f_l - f_{ul} \text{ mm} \tag{8}$$

The criterion to assess the effect of the number of elements on springback, prediction of accuracy of the absolute mean error E_s^{abs} is assumed:

$$E_s^{abs} = \frac{1}{4} \sum_{i=1}^{i=4} \left| E_s^{(i)} \right| \% \tag{9}$$

where i is the level of punch stroke i = 1, 2, 3, and 4 corresponding to punch strokes 3 mm, 6 mm, 12 mm, and 18 mm, respectively.

The absolute mean error value for all analysed punch strokes f_g is equal to 2.64% (after increasing number of elements from 84 to 280), 2.56% (from 280 to 1120), and 1.14% (from 1120 to 4400). In this study, the accepted E_s^{abs} value is assumed to be 1.5% and, hence, the number of elements 1120 is acceptable in the conducted numerical models.

The contact between the assumed rigid bodies (the die and punch) and the deforming workpiece was defined by the penalty method [30]. To study the effect of the number of integration points, material model and sample orientation, the friction coefficient between the sheet metal and tools was assumed to be 0.01 [34]. However, in the numerical analyses of the effect of the friction coefficient value on the springback behaviour, five friction coefficient values (0.01, 0.03, 0.06, 0.1, 0.2) were considered.

4. Results

4.1. Effect of the Number of Integration Points

The number of integration points is a significant parameter for springback simulation using shell elements. The effect of a number of integration points on the computational time and springback coefficient K is presented in Figures 5 and 6, respectively. The change in computational time is evaluated for a reference time of computation of the numerical model of the sheet with the Gauss integration rule and five integration points through the thickness. For these conditions, which are recommended by many authors in the non-linear analysis of homogeneous shells [24], the computational time takes about 14 min on a standard personal computer (HDD SSD, 32 GB RAM, i7-6700HQ CPU@2.6 GHz, AsusTek Computer Inc., Taipei, Taiwan). As observed, the lowest computation time is for both integration rules for the five integration points (Figure 5).

A further increase in the number of integration points through the sheet thickness results in greater time consumption. In the case of the analysed model, the computation time is not very long, but it can be speculated that if the number of elements increases the computation time increases exponentially.

A higher number of integration points results in a decrease of the predicted springback coefficient (Figure 6). When the number of integration points is over 11, the value of the computation time is stabilised. This relation is observed for both the analysed integration rules. It can be concluded that after increasing 19 integration points, the computation time notably increases (Figure 5), however, no further improvement in springback coefficient is observed (Figure 6). In summary, five integration points are the minimum acceptable, considering the computation time and accurateness of springback prediction. The Gauss' rule with five integration points gives better prediction of the springback coefficient, which is in good agreement with the results of Burgoyne and Crisfield [35] and Wagoner and Lee [26]. In the case of Gauss quadrature, an increase of the number of integration points from five to 15 decreases the springback prediction error at 0.24%. However, the computation time increases to 40%. A similar conclusion can be drawn for Simpson's rule. After exceeding seven integration points through the sheet thickness, both rules give similar results (Figure 6). It is well known that if a sheet undergoes plastic deformation in the bending process, points of discontinuity appear in the stress distribution and the number of necessary integration points increases with an increase of the bending radius (depth of punch).

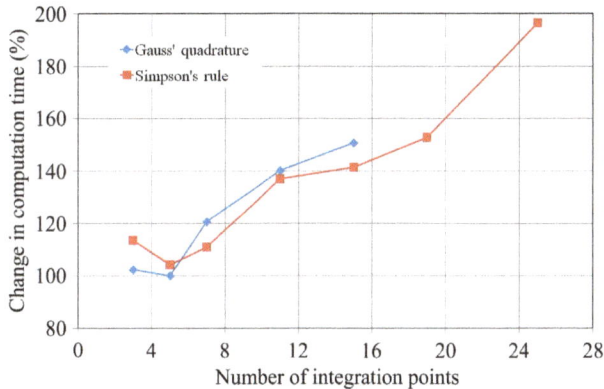

Figure 5. Change of computational time with a number of integration points for Gauss' quadrature and Simpson's rule.

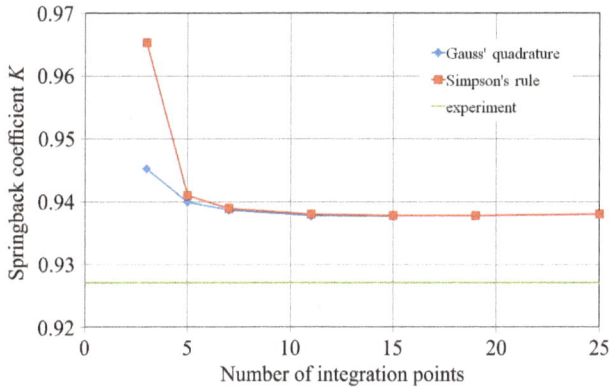

Figure 6. Effect of a number of integration points on the value of the springback coefficient K for Gauss' quadrature and Simpson's rule.

In fact, when material is in an elastic regime, then the stress distribution through the sheet thickness is linear and the number of integration points can be limited. However, when sheet material undergoes the elastic-plastic behaviour, to fit better the nonlinear stress distribution, the required number of integration points must be increased. It can be speculated that a number of integration points depend mainly on the bending radius and the ratio of the inside radius of the bend to material thickness.

4.2. Sample Orientation

The change of the springback coefficient K_γ as a function of the bending angle under loading for samples cut along the rolling direction and transverse to this direction are presented in Figure 7a,b, respectively. According to Equation (4), high springback of the material denotes the lower values of the springback coefficient K_γ. As can be observed in Figure 7b, the samples cut transverse to the rolling direction exhibit lower values of springback coefficient. The relation of springback coefficient for both analysed orientations is almost linear. In all cases, the predicted value of the springback coefficient is higher than the measured ones. The differences in the K_γ value between experimental and numerical results decrease with the increasing bending angle under loading γ_1. The difference in the value of springback for the analysed perpendicular orientations is due to crystallographic structure of the sheet material. The cold rolling of the sheets produces the directional change in the deformation of material microstructure, i.e., the grains are elongated in the direction of the cold rolling process. Thus, the material withstands bending according to rolling direction (orientation 0° in Figure 4). Furthermore, in the case of rolling direction, the grains are only subjected to tensile or bending stresses, but in the case of transverse direction, a significant size of deformation energy is used to change the orientation of grains [36].

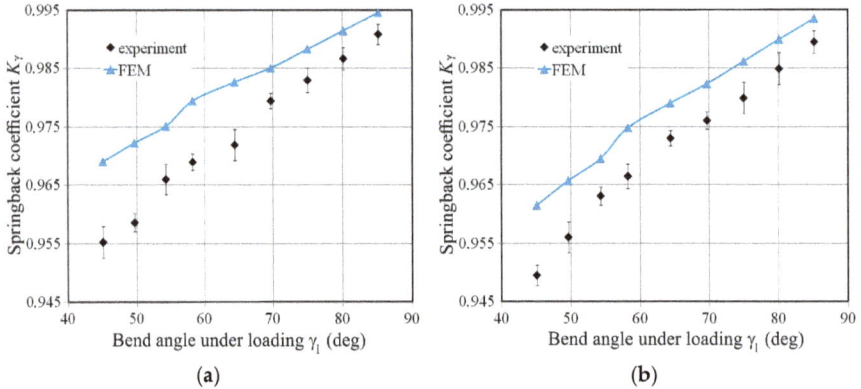

Figure 7. Comparison of the springback coefficient value determined experimentally and by the FEM approach (Gauss' quadrature, five integration points) for samples oriented according to (a) the rolling direction (0°) and (b) transverse to the rolling direction (90°).

During bending of the sheet strip, the outside side of sheet material under the rounded edge of the punch is subjected to elastic stress (Figure 8). If the load reaches the yield point the specimen undergoes plastic deformation and strain hardening phenomenon. Plastic deformation, after unloading, is followed by elastic recovery upon removal of the load. The slope of unloading line is parallel to the elastic characteristics during loading. The inside part of the sheet strip under the rounded edge of the punch is compressed.

4.3. Sheet Deformation during Bending

The large sample width compared to the thickness determines the occurrence of a specific stress state in the sample (Figure 9). It is clear from this figure that in this section of the sample, there is a neutral interlayer on which the sign of the deformation changes. In the middle part of the sample width on the inner side, the longitudinal stresses are compressive and the deformation is negative, and on the external side, the longitudinal stress is negative and the deformation is positive (Figure 9). This phenomenon is, however, disturbed at the edge of the sheet.

In general, if the sheet width is much larger than the sheet thickness, the width of the sheet is not changed during bending. If the ratio of the sheet width to the sheet thickness is lower than 4, the rectangular section of the sheet is changed to a trapezoidal section. At the inner side of the bending curvature the width of the sample increases. The situation at the outer side is the contrary, i.e., the sheet width is reduced.

The numerically-determined deviation of the sample profile at sample width direction is presented in Figure 10. This deviation is evaluated at the middle layer of the shell elements. The profile deviation does not depend on the material description and sample orientation. Thus, the character of the profile deviation in all cases is identical. The presented phenomenon of dependence of the sample width on the stress-state in the sample material is rarely investigated by researchers who studied the springback phenomenon. The different stress-state in the sample bend in V-shaped die causes the sheet to undergo springback in different planes.

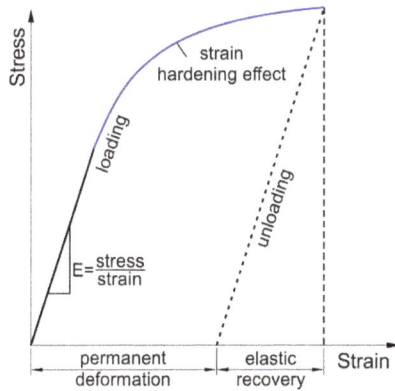

Figure 8. The bending characteristics of material.

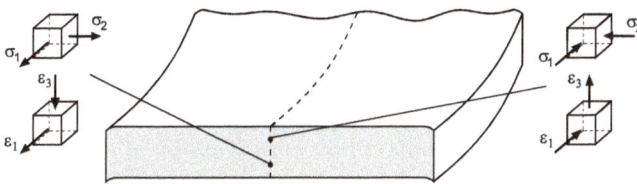

Figure 9. Stress and strain state during V-bending of a sheet strip.

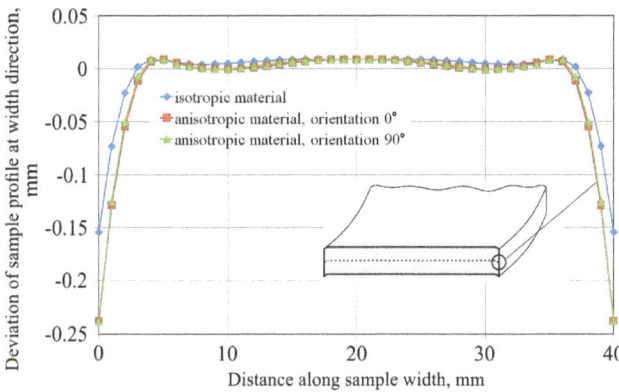

Figure 10. Deviation of the sample profile in width direction.

Springback intensity is influenced by the orientation of the sample in terms of symmetric plane of the punch. It is conceivable that sample orientation in the rolling direction produces variations of the elastic-plastic properties of the sheet metal and residual stress after sample unloading (Figure 11). The distribution of HMH (Huber-Mises-Hencky) stress (Figure 11) along the sheet width direction in the lower part of sample that gets contact with the punch is related to the distribution of sample profile deviation (Figure 10). The lowest values of the HMH stress are observed in the vicinity of the sample edge (Figure 11, right).

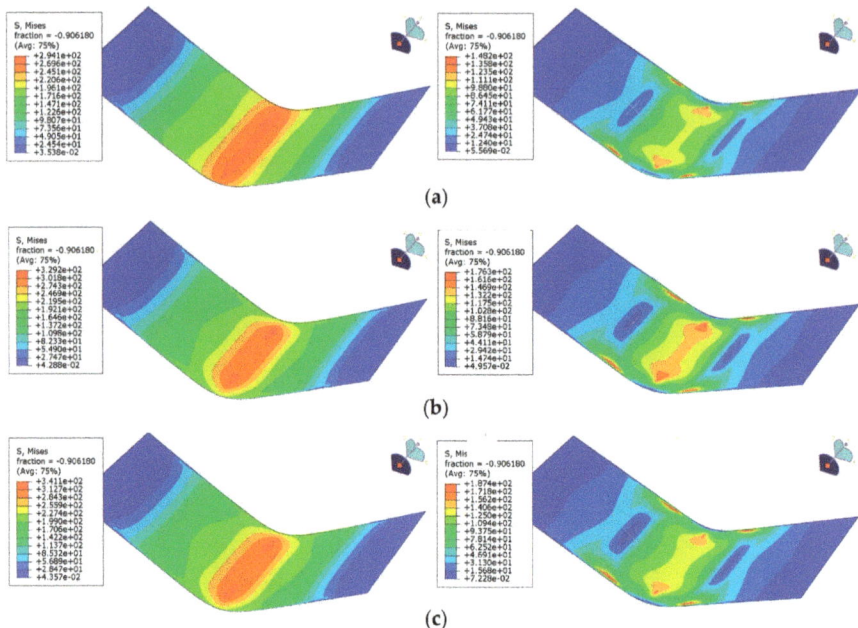

(a)

(b)

(c)

Figure 11. The HMH stress distribution (left) under loading and (right) after unloading, for (**a**) samples with isotropic material and (**b**) anisotropic material oriented at 0° and (**c**) 90°, for a punch depth under loading of 18 mm.

4.4. Effect of Friction Coefficient

The effect of friction coefficient on springback phenomenon is studied for both sample orientations and three bend angles under loading that corresponds to punch bend depths under loading f_1: 6 mm, 12 mm, 18 mm, and 24 mm. The results indicate that the value of coefficient of friction in small scale influences the springback coefficient K_γ (Figure 12). The difference in angle γ_l for all used friction coefficients does not exceed 0.4%. However, increasing of coefficient of friction value causes slight decrease in springback angle under unloading γ_{ul}. Similar results were found for the 90° sample orientation.

Based on conducted numerical investigations, it can be concluded that the friction coefficient slightly influences the springback phenomenon in V-die air bending test. In this test, the contact between the punch and sample is linear in a large distance during punch displacement. Only at final punch depths does the area of contact increase quickly. The geometrical analysis shown in Figure 13 shows that the relation between the punch contact angle and punch depth is linear. Furthermore, linear contact exists between the sheet and samples at the rounded edge of the die, so the friction coefficient slightly influences the springback amount. The friction coefficient is different on the curved and flat parts of both die and punch, and it is very difficult to measure this coefficient experimentally. Thus, in modelling bending tests, the constant friction coefficient value is usually assumed [34,37]. The coefficient of friction is an important factor in processes of sheet metal forming, such as deep drawing, where there exists great mutual displacements in the sheet-tool interface and high changes in surface topography due to plastic deformation of the workpiece. The small effect of changes in friction coefficient on springback amount may be a result of the used elements. The shell element represents the sheet mid-plane. However, bending causes the highest amount of plastic deformation in the extreme fibres of the sheet. The accuracy of contact prediction depends on the number of integration points. Due to the small area of contact between rounded tools and the sheet, the sensitivity of the friction coefficient on the contacting surface on the numerical results of shell elements is smaller than in the case of brick elements. In the case of both elements, the dimensions of the element should be as proportional as possible. However, the usage of many brick elements through the thickness may guarantee the lower element size in contacting surfaces, and a better area of contact prediction.

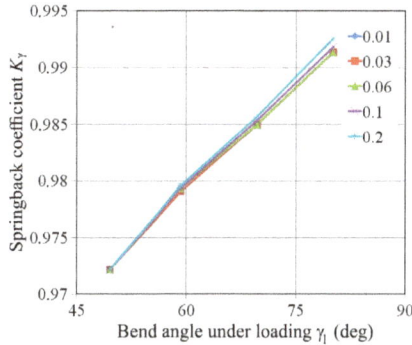

Figure 12. Effect of bend angle under loading on the springback coefficient K_γ determined for sample orientation 0.

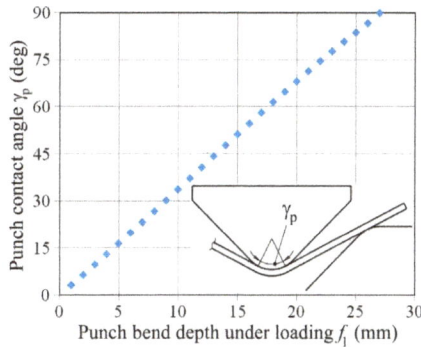

Figure 13. Effect of the bend depth under loading f_l on the punch contact angle γ_p.

5. Conclusions

The study of the influence of computational parameters on the springback phenomenon of sheet metals is presented in this article. Samples cut along the rolling direction and transverse to the rolling directions are employed. Experimental tests with identical geometry and parameters are conducted to compare with the numerical simulation results.

The main conclusions drawn from experimental and numerical analyses are as follows:

- The samples cut along the transverse direction exhibit greater springback than the blank cut in the rolling direction.
- The variation of the springback coefficient value almost shows linear dependence between the springback coefficient value and bend angle under loading.

- The results of the numerical model indicate that five integration points are the minimum acceptable considering the computation time and accuracy of springback prediction.
- Analyses of the sheet material with seven or more integration points through the thickness indicated that both Simpson's and Gauss' rules provide similar accuracy of prediction.
- The Gauss' rule with five integration points is optimal to obtain the accurate results of springback prediction.
- The friction coefficient value slightly influences the springback amounts.

Author Contributions: Tomasz Trzepieciński conceived and designed the experiments; Hirpa G. Lemu performed the review of literature and established the assumptions for numerical modelling; and both authors contributed equally to writing the paper and numerical modelling.

Conflicts of Interest: The authors declare no conflict of interest.

References

1. Dai, H.L.; Jiang, H.J.; Dai, T.; Xu, W.L.; Luo, A.H. Investigation on the influence of damage to springback of U-shape HSLA steel plates. *J. Alloys Compd.* **2017**, *708*, 575–586. [CrossRef]
2. Carden, W.D.; Geng, L.M.; Matlock, D.K.; Wagoner, R.H. Measurement of springback. *Int. J. Mech. Sci.* **2002**, *44*, 79–101. [CrossRef]
3. Yang, X.A.; Ruan, F. A die design method for springback compensation based on displacement adjustment. *Int. J. Mech. Sci.* **2011**, *53*, 399–406. [CrossRef]
4. Eggertsen, P.A.; Mattiasson, K. On constitutive modeling for springback analysis. *Int. J. Mech. Sci.* **2010**, *52*, 804–818. [CrossRef]
5. Nilsson, A.; Melin, L.; Magnusson, C. Finite-element simulation of V-die bending: A comparison with experimental results. *J. Mater. Process. Technol.* **1997**, *65*, 52–58. [CrossRef]
6. Trzepiecinski, T.; Lemu, H.G. Prediction of springback in V-die air bending process by using finite element method. *MATEC Web Conf.* **2017**, *121*, 03023. [CrossRef]
7. Karağaç, I. The experimental investigation of springback in V-bending using the flexforming process. *Arab. J. Sci. Eng.* **2017**, *42*, 1853–1864. [CrossRef]
8. Leu, D.K.; Hsieh, C.M. The influence of coining force on spring-back reduction in V-die bending process. *J. Mater. Process. Technol.* **2008**, *196*, 230–235. [CrossRef]
9. Bakhshi-Jooybari, M.; Rahmani, B.; Daeezadeh, V.; Gorji, A. The study of spring-back of CK67 steel sheet in V-die and U-die bending processes. *Mater. Des.* **2009**, *30*, 2410–2419. [CrossRef]
10. Firat, M.; Osman, M.; Kocabicak, U.; Ozsoy, M. Stamping process design using FEA in conjunction with orthogonal regression. *Finite Elem. Anal. Des.* **2010**, *46*, 992–1000. [CrossRef]
11. Huang, Y.M.; Leu, D.K. Effects of process variables on V-die bending process of steel sheet. *Int. J. Mech. Sci.* **1998**, *40*, 631–650. [CrossRef]
12. Garcia-Romeu, M.L.; Ciurana, J.; Ferrer, I. Springback determination of sheet metals in air bending process based on an experimental work. *J. Mater. Process. Technol.* **2007**, *191*, 174–177. [CrossRef]
13. Ragai, I.; Lazim, D.; Nemes, J.A. Anisotropy and springback in draw bending of stainless steel 410: Experimental and numerical study. *J. Mater. Process. Technol.* **2005**, *166*, 116–127. [CrossRef]
14. Vin, L.J.; Streppel, A.H.; Singh, U.P.; Kals, H.J.J. A process model for air bending. *J. Mater. Process. Technol.* **1996**, *57*, 48–54. [CrossRef]
15. Thipprakmas, S.; Rojananan, S. Investigation of spring-go phenomenon using finite element method. *Mater. Des.* **2008**, *29*, 1526–1532. [CrossRef]
16. Tekıner, Z. An experimental study of the examination of spring-back of sheet metals with several thicknesses and properties in bending dies. *J. Mater. Process. Technol.* **2004**, *145*, 109–117. [CrossRef]
17. Lingbeek, R.; Huétink, J.; Ohnimus, S.; Petzoldt, M.; Weiher, J. The development of a finite elements based springback compensation tool for sheet metal products. *J. Mater. Process. Technol.* **2005**, *169*, 115–125. [CrossRef]
18. Del Pozo, D.; López de Lacalle, L.N.; López, J.M.; Hernández, A. Prediction of press/die deformation for an accurate manufacturing of drawing dies. *Int. J. Adv. Manuf. Technol.* **2008**, *37*, 649–656. [CrossRef]

19. López de Lacalle, L.N.; Lamikiz, A.; Muñoa, J.; Salgado, M.A.; Sánchez, J.A. Improving the high-speed fnishing of forming tools for advanced high-strength steels (AHSS). *Int. J. Adv. Manuf. Technol.* **2006**, *29*, 49–63. [CrossRef]
20. Sumikawa, S.; Ishiwatari, A.; Hiramoto, J.; Urabe, T. Improvement of springback prediction accuracy using material model considering elastoplastic anisotropy and Bauschinger effect. *J. Mater. Process. Technol.* **2016**, *230*, 1–7. [CrossRef]
21. Yoshida, T.; Katayama, K.; Hashimoto, Y.; Kuriyama, Y. Shape control techniques for high strength steel in sheet metal forming. *Nippon Steel Tech. Rep.* **2003**, *88*, 27–32.
22. Hattori, Y.; Furakawa, K.; Hamasaki, H.; Yoshida, F. Experimental and simulated springback after stamping of copper-based spring materials. In Proceedings of the 59th Holm Conference on Electrical Contacts, Newport, RI, USA, 2013; pp. 1–6. [CrossRef]
23. Yoshida, F.; Uemori, T. A model of large-strain cyclic plasticity and its application to springback simulation. *Int. J. Mech. Sci.* **2003**, *45*, 1687–1702. [CrossRef]
24. Rashid, M.; Liebert, C. Finite Element Analysis of a Lifting Portable Offshore Unit. Master's Thesis, Chalmers University of Technology, Goteborg, Sweden, 2015.
25. Xu, W.L.; Ma, C.H.; Li, C.H.; Feng, W.J. Sensitive factors in springback simulation for sheet metal forming. *J. Mater. Proc. Technol.* **2004**, *151*, 217–222. [CrossRef]
26. Wagoner, R.H.; Li, M. Simulation of springback: Through-thickness integration. *Int. J. Plast.* **2007**, *23*, 345–360. [CrossRef]
27. Martínez Krahmer, D.; Polvorosa, R.; López de Lacalle, L.N.; Alonso-Pinillos, U.; Abate, G.; Riu, F. Alternatives for specimen manufacturing in tensile testing of steel plates. *Exp. Tech.* **2016**, *40*, 1555–1565. [CrossRef]
28. Hollomon, J.H. Tensile deformation. *Trans. Metall. Soc. AIME* **1945**, *162*, 268–290.
29. *ABAQUS 2016. Theory Guide*; Dassault Systèmes: Vélizy-Villacoublay Codex, France, 2015.
30. *ABAQUS 2016. Analysis User's Guide. Volume IV: Elements*; Dassault Systèmes: Vélizy-Villacoublay Codex, France, 2015.
31. Hill, R. A theory of the yielding and plastic flow of anisotropic metals. *Proc. R. Soc. A* **1948**, *193*, 281–297. [CrossRef]
32. Trzepieciński, T.; Gelgele, H.L. Investigation of anisotropy problems in sheet metal forming using finite element method. *Int. J. Mater. Form.* **2011**, *4*, 357–369. [CrossRef]
33. Von Mises, R. Mechanik der festen Körper im plastisch-deformablen Zustand. *Nachrichten von der Gesellschaft der Wissenschaften zu Göttingen, Mathematisch-Physikalische Klasse* **1913**, *1913*, 582–592.
34. Frącz, W.; Stachowicz, F. Springback phenomenon in sheet metal V-die air bending—Experimental and numerical study. *Manuf. Ind. Eng.* **2008**, *2*, 34–37.
35. Burgoyne, C.J.; Crisfield, M.A. Numerical integration strategy for plates and shells. *Int. J. Numer. Methods Eng.* **1990**, *29*, 105–121. [CrossRef]
36. Albrut, A.; Brabie, G. The influence of the rolling direction of the joined steel sheets on the springback intensity in the case of Ω-shape parts made from tailor welded strips. *Arch. Civ. Mech. Eng.* **2006**, *6*, 5–12. [CrossRef]
37. Papeleux, L.; Ponthot, J.P. Finite element simulation of springback in sheet metal forming. *J. Mater. Proc. Technol.* **2002**, *125–126*, 785–791. [CrossRef]

![metals logo] *metals*

MDPI

Article

The Mechanism of Inhomogeneous Grain Refinement in a NiTiFe Shape Memory Alloy Subjected to Single-Pass Equal-Channel Angular Extrusion

Yanqiu Zhang and Shuyong Jiang *

College of Mechanical and Electrical Engineering, Harbin Engineering University, Harbin 150001, China; zhangyq@hrbeu.edu.cn
* Correspondence: jiangshuyong@hrbeu.edu.cn; Tel.: +86-0451-8251-9710

Received: 6 September 2017; Accepted: 26 September 2017; Published: 29 September 2017

Abstract: Based on electron backscattered diffraction analysis and transmission electron microscopy observation, the mechanism of inhomogeneous grain refinement in a NiTiFe shape memory alloy (SMA) subjected to single-pass equal-channel angular extrusion (ECAE) was investigated. The results show that refined grains are mainly nucleated near grain boundaries and a small fraction of them emerges in the grain interior. The size of refined grains increases as deformation temperature increases, which indicates that a higher deformation temperature is adverse to grain refinement in the ECAE of NiTiFe SMAs. It is the accumulation and rearrangement of geometrically necessary dislocations as plastic strain increases that leads to the transition of lower angle subgrain boundaries, and finally higher angle subgrain boundaries are induced and finer grains are formed. Due to the limitation of slip systems, the mechanism of grain refinement in a NiTiFe SMA subjected to ECAE is different from that in face-centered cubic and body-centered cubic crystals. Dislocation cells and shear bands are two transition microstructures of grain refinement in the ECAE of NiTiFe SMAs. The nucleation of fine grains mainly occurs along shear bands or grain boundaries, which leads to the inhomogeneity of grain refinement.

Keywords: shape memory alloy (SMA); NiTiFe alloy; grain refinement; equal-channel angular extrusion (ECAE); microstructure

---

1. Introduction

As a functional material, NiTi-based shape memory alloys (SMAs) have been extensively applied in medical and engineering fields because they possess excellent superelasticity and perfect shape memory. It has been accepted that the functional performances of SMAs are closely related to their microstructures, where grain refinement is one of key factors. As a consequence, researchers have investigated the grain refinement of NiTi-based SMAs via various severe plastic deformation (SPD) methods, including cold rolling, high pressure torsion (HPT), surface mechanical attrition treatment (SMAT), local canning compression, cold drawing, and equal-channel angular extrusion (ECAE) [1–6]. Because the process of ECAE is relatively simple and can be used for manufacturing ultrafine-grained metals with relatively large sizes compared with other SPD methods, it is a potential technique for processing commercial ultrafine-grained metals. As a consequence, ECAE has increasingly gained attention over the last two decades [7–11]. So far, researchers have investigated the deformation behavior of ECAE in many conventional metals, such as magnesium alloys [12], steel [13], aluminum alloys [14], pure titanium [15], and pure copper [16]. These investigations mostly focused on the microstructures or mechanical properties of materials that have experienced multi-pass ECAE; the mechanism for grain refinement has not been thoroughly revealed.

It is widely accepted that the plastic deformation of metals results from dislocation slip or twinning, and different crystals possess different slip characteristics. The above conventional metals belong to face-centered cubic (FCC), body-centered cubic (BCC), or hexagonal close-packed (HCP) crystals. However, NiTi-based SMAs possess a B2 structure at the austenite state or a B19' structure at martensite state. As a consequence, there should exist some differences between the NiTi-based SMAs and the above conventional metals. Even though several investigations on the ECAE of NiTi-based SMAs have been conducted, all of them have focused on the microstructures of ECAE at room temperature and the subsequent heat treatment, the effects of second phases on the mechanical properties and phase transformation, the influence of ECAE on shape memory response, or the damage tolerance of ECAE [17–23]. Up to now, no studies on the mechanism of grain refinement in ECAE of NiTi-based SMAs have been reported.

Therefore, the single-pass ECAE of an NiTiFe SMA at three different temperatures was performed in the present study so as to investigate the mechanism of inhomogeneous grain refinement in NiTiFe SMAs, where electron backscattered diffraction (EBSD) analysis and transmission electron microscopy (TEM) observation were conducted to produce the relevant data.

2. Materials and Methods

A commercially available NiTiFe SMA bar provided by Xi'an Saite Metal Materials Development Co., Ltd. (Xi'an, China) was used as the raw material in the current investigation [24]. The composition of the bar was $Ti_{50}Ni_{47}Fe_3$ (atom %), and the diameter of it was 9 mm. Three billets with a cross section of 5 mm × 5 mm and the length of 25 mm were taken from the as-received NiTiFe bar along the axis direction via a DK7725 type electro-discharge machine (EDM) (Jiangsu Dongqing CNC Machine Tool Co. Ltd., Taizhou, China). Subsequently, the NiTiFe billets experienced a single-pass ECAE via a die (Harbin Engineering University, Harbin, China) shown in Figure 1a at 400 °C, 450 °C, and 500 °C, respectively. The ECAE die possesses the channel cross section of 5 mm × 5 mm, and the radii of the upper and lower corners at the intersection between the inlet and outlet channels were set to 1 mm and 2 mm, respectively. The channel angle ϕ of the ECAE die was set to 120°, and the arc angle of the lower corner was set to 24°. The ECAE experiments were carried out via an INSTRON-5500R universal material testing machine (Instron Corporation, Norwood, MA, USA), where the velocity of the punch was set to 1.5 mm/min. Figure 1b illustrates the process of ECAE in the current investigation. In the present study, the equivalent strain ε induced by the single-pass ECAE can be obtained based on the equation proposed by Iwahashi et al. [25]:

$$\varepsilon = \frac{2\cot\left(\frac{\phi}{2} + \frac{\psi}{2}\right) + \psi\cos ec\left(\frac{\phi}{2} + \frac{\psi}{2}\right)}{\sqrt{3}},\tag{1}$$

where ϕ is channel angle and ψ is the arc angle of the lower corner, as shown in Figure 1b. As a consequence, the equivalent strain in the present study was calculated as 0.63.

After the extrusion, slice specimens with a size of 5 mm × 5 mm were taken from the as-received sample and the three ECAE samples, respectively, where the sampling planes are parallel to the longitudinal sections of the samples, as shown in Figure 1b,c. Subsequently, they were processed for EBSD analysis and TEM observation. Before the EBSD analysis, the slice specimens experienced mechanical polishing and subsequent electro-polishing so that strain-free surfaces were acquired. The EBSD analysis was conducted via a field-emission scanning electron microscope (SUPRA 55 SAPPHIRE, University of South Carolina, Columbia, SC, USA) equipped with an EBSD detector. During EBSD analysis, the relationship between the crystal coordinate and the sample coordinate of the as-received sample was set to (100)//TD, (010)//RD, and (001)//ND, where TD, RD, and ND represent the transverse direction, the rolling direction, and the normal direction of the as-received sample, respectively. However, the relationship between the crystal coordinate and the sample coordinate of the ECAE samples was set to (100)//ED, (010)//TD, and (001)//ND, where ED, TD, and ND represent the extrusion direction, the transverse direction, and the normal direction of the ECAE sample, respectively. The sample systems for the as-received and ECAE samples are shown in Figure 1b,c, respectively. Before TEM observation, the foils were firstly ground to 70 μm by the mechanical method and then thinned by the twin-jet method in the electrolyte, with a composition of 34% $CH_3(CH_2)_3OH$, 6% $HClO_4$, and 60% CH_3OH (in volume). Afterward, TEM observation was performed via an FEI TECNAI G2 F30 microscope (Field Electron and Ion Company, Hillsboro, OR, USA).

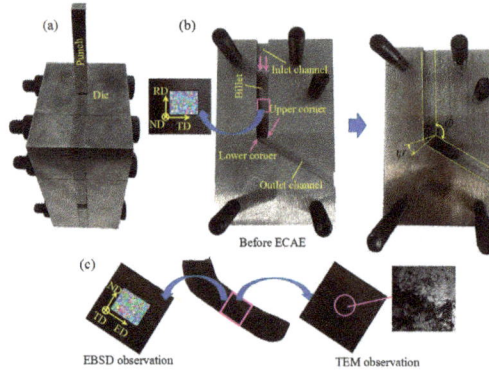

Figure 1. Experimental illustration of equal-channel angular extrusion (ECAE): (**a**) ECAE die; (**b**) process of ECAE; (**c**) sampling location for analysis (In the case of as-received sample, TD, RD, and ND represent the transverse direction, the rolling direction, and the normal direction. In the case of ECAE sample, ED, TD, and ND represent the extrusion direction, the transverse direction, and the normal direction, respectively. TEM: Transmission electron microscopy; EBSD: Electron backscattered diffraction).

3. Results

Figure 2 illustrates the EBSD results of the as-received NiTiFe SMA by means of orientation imaging microscopy (OIM; Technische Universität Bergakademie Freiberg, Freiberg, Germany), where a boundary misorientation criterion of 15° was used in the present study. As seen in Figure 2a, all the grains present an approximately equiaxed shape. In addition, it can be found from Figure 2b that the majority of the grains possess an equivalent grain diameter ranging from 3 μm to 15 μm, and the size of the largest grain is lower than 30 μm.

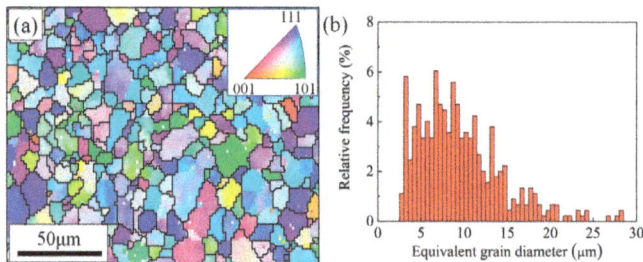

Figure 2. Electron backscattered diffraction (EBSD) results of the as-received NiTiFe shape memory alloy (SMA) by means of orientation imaging microscopy (OIM): (**a**) Microstructure; (**b**) histogram of equivalent grain diameter.

Figure 3 illustrates the OIM microstructures of the NiTiFe SMA subjected to ECAE at various temperatures. It can be seen that the microstructures of the NiTiFe SMA present inhomogeneity after the single-pass ECAE, where a number of elongated or equiaxed fine grains arise in the grain interior and at the grain boundaries of the remained coarse grains. Meanwhile, the remained coarse grains of all three NiTiFe samples are elongated slightly along the shear direction of ECAE, which is inclined by about 60° with respect to the loading direction. It can be also seen that the numbers of the fine grains in the samples subjected to ECAE at 400 °C and 450 °C are much greater than the one in the sample subjected to ECAE at 500 °C. The phenomenon indicates that the deformation temperature has a significant effect on the microstructure of NiTiFe SMA subjected to single-pass ECAE. In order to investigate the effect of deformation temperature on the grain size of the sample subjected to ECAE, histograms of equivalent grain diameter for the three samples are illustrated in Figure 4. It can be found that the number of grains lower than 5 μm decreases as deformation temperature increases, which indicates that lower ECAE temperature is beneficial to the refinement of grains. In addition, it can be observed that the sizes of the largest grains in the three ECAE samples are all larger than the one in the as-received sample. The aforementioned phenomenon indicates that the single-pass ECAE performed at temperatures ranging from 400 °C to 500 °C is not only able to lead to the refinement of some grains, but can also coarsen the remained coarse grains. The driving force for the growth of remained coarse grains properly comes from deformation-stored energy in them, and the mechanism for this phenomenon is unclear and will be investigated in a future study.

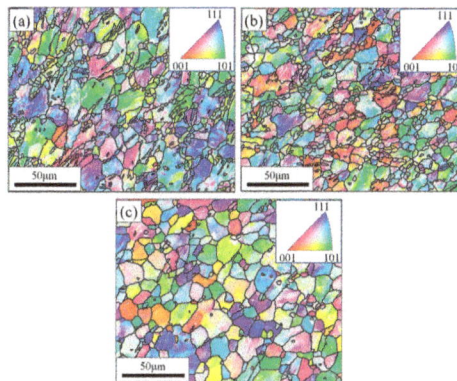

Figure 3. OIM microstructures of the NiTiFe SMA subjected to ECAE at various temperatures: (**a**) 400 °C; (**b**) 450 °C; (**c**) 500 °C.

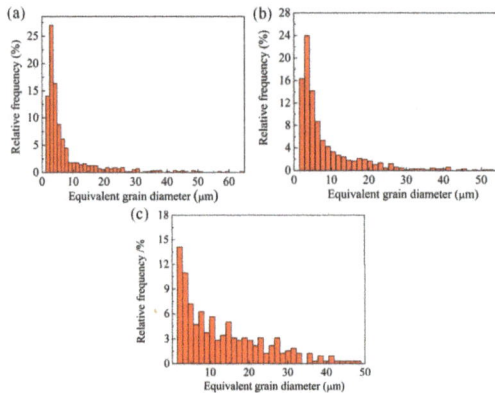

Figure 4. Histograms of equivalent grain diameter for the NiTiFe SMA subjected to ECAE at various temperatures: (**a**) 400 °C; (**b**) 450 °C; (**c**) 500 °C.

In order to reveal the substructure evolution in the grains of NiTiFe SMA subjected to ECAE, distributions of kernel average misorientation (KAM) in the three samples are calculated, as shown in Figure 5. It can be seen that the subgrain boundaries induced in the NiTiFe samples subjected to ECAE at 400 °C and 450 °C are much greater than the ones in the NiTiFe sample subjected to ECAE at 500 °C. In addition, there are many subgrain boundaries with higher misorientation angles in the grain interior and near the grain boundaries of the former two samples. The above phenomena indicate that the grain refinement of ECAE results from the transition of lower angle grain boundaries to higher ones, and lower deformation temperature is able to induce more substructures during the ECAE of NiTiFe SMA. As a consequence, the NiTiFe SMA is not suitable for being processed by ECAE at temperatures above 500 °C due to its weak ability of grain refinement.

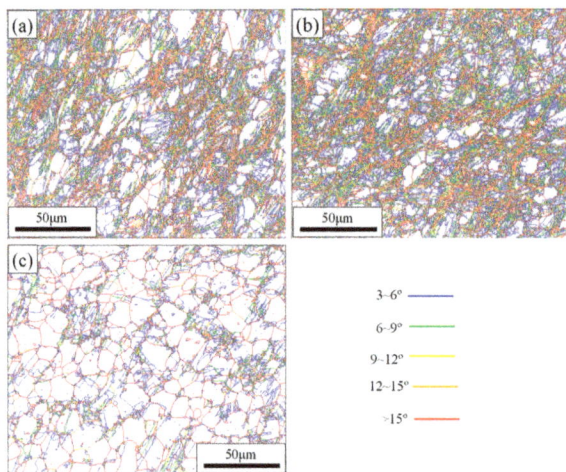

Figure 5. Distributions of kernel average misorientation (KAM) in the NiTiFe SMA subjected to ECAE at various temperatures: (**a**) 400 °C; (**b**) 450 °C; (**c**) 500 °C.

It has been accepted that there is deformation inhomogeneity both among the grains and in the grain interior. Consequently, geometrically necessary dislocations (GNDs) will be induced to accommodate the deformation inhomogeneity so that the material is not fractured. For the purpose of investigating the influence of temperature on the deformation in homogeneity in the NiTiFe SMA subjected to ECAE, geometrically necessary dislocation (GND) distributions in the three samples are calculated based on EBSD data, as shown in Figure 6. According to the concept of GND, the region with a higher density of GNDs indicates a higher inhomogeneity of deformation. It can be seen in Figure 6 that, in the samples subjected to ECAE at 400 °C and 450 °C, the area with a higher density of GNDs is much larger than that in the sample subjected to 500 °C. Furthermore, the regions with a higher density of GNDs are in accordance with those with higher misorientations, as shown in Figure 5. This phenomenon occurs because subgrain boundaries disturb the coordination between the two parts beside the boundary, and more GNDs are required to adjust the deformation coordination. It is the accumulation and rearrangement of these GNDs as plastic strain increases that leads to the transition of lower angle subgrain boundaries. Finally, higher angle subgrain boundaries are induced, and finer grains are formed.

For the purpose of investigating the mechanism of grain refinement at a more microscopic scale, TEM micrographs of the three samples are obtained, as shown in Figure 7. It can be seen that in the case of ECAE at 400 °C, several dislocation cells and some shear bands can be observed in a zone of about $3 \times 3 \ \mu m^2$, and some subgrains and fine grains can be found in a region of about $1.6 \times 1.6 \ \mu m^2$. In the case of ECAE at 450 °C, only two fragmentary dislocation cells and one fine grain can be observed in a region of about $4 \times 4 \ \mu m^2$. However, as for the ECAE at 500 °C, only a fragmentary dislocation cell can be seen in a region of about $3 \times 3 \ \mu m^2$, and no subgrains or fine grains are observed. The aforementioned phenomena indicate that the size of dislocation cells increases as ECAE temperature increases, and the number of subgrains decreases as ECAE temperature increases. The evidence demonstrates that the fine grains of the NiTiFe SMA subjected to single-pass ECAE might be related to the dislocation cells, and a lower deformation temperature leads to a higher density of dislocation cells. Furthermore, the grain refinement during the ECAE of a NiTiFe SMA might also result from the shear bands.

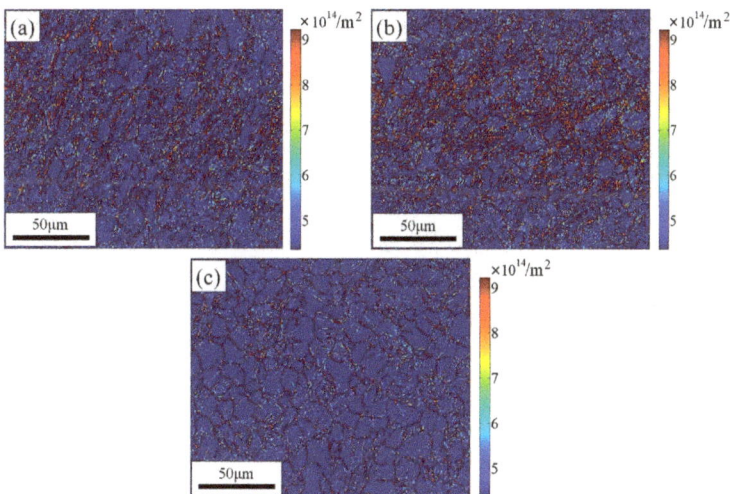

Figure 6. Geometrically necessary dislocation (GND) distributions in the NiTiFe SMA subjected to ECAE at various temperatures: (**a**) 400 °C; (**b**) 450 °C; (**c**) 500 °C.

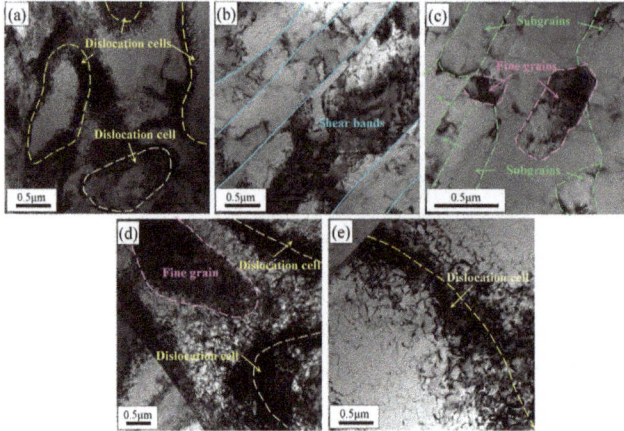

Figure 7. Transmission electron microscopy (TEM) micrographs of the NiTiFe SMA subjected to ECAE at various temperatures: (**a**) Dislocation cells at 400 °C; (**b**) shear bands at 400 °C; subgrains and fine grains at (**c**) 400 °C; (**d**) 450 °C; (**e**) 500 °C.

4. Discussion

Based on the aforementioned results, it can be concluded that single-pass ECAE performed at 400–500 °C is able to refine the grains of NiTiFe SMA. However, the obtained microstructures present a high inhomogeneity, where the refined grains are mainly nucleated near the grain boundaries and a small fraction of them emerges in the grain interior. In addition, the grain size of NiTiFe SMA subjected to ECAE increases as deformation temperature increases. Based on these evidences and the information on substructure distributions, GND density distributions, and TEM micrographs, the mechanism of grain refinement in the NiTiFe SMA subjected to ECAE can be proposed, as shown in Figure 8.

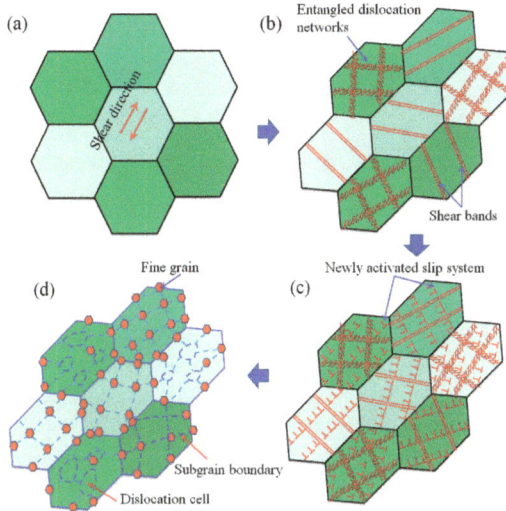

Figure 8. The proposed mechanism for the grain refinement of NiTiFe SMA subjected to single-pass ECAE: (**a**) Equiaxed coarse grains before ECAE; (**b**) microstructure at the early stage of ECAE, showing original coarse grains being elongated along the shear plane and the formation of shear bands and entangled dislocation networks; (**c**) the activation of other slip systems and the block of later induced dislocations; (**d**) rearrangement of dislocations, formation of dislocation cells, subgrains and newly fine grains induced by continuous recovery (In the figure, different background colors represent different orientations).

It is well known that the structure of NiTiFe SMA is B2 austenite above room temperature. It has been accepted that the B2 cubic austenite of NiTi-based SMA is composed of two simple cubic lattices, where each simple cubic lattice consists of eight Ni atoms or eight Ti atoms. According to previous studies, the slip systems found in B2 austenite of NiTi-based SMA are the <100>{001}, <100>{011}, and <111>{110} families, where the <100>{011} family is the most likely to be activated one because it requires the least stacking fault energy and the least critical stress for plastic deformation [26–28]. As a consequence, the <100>{011} family will be activated at an early stage of ECAE. According to the crystallographic characteristics of cubic crystals, there are six {011} planes in B2 austenite of NiTi-based SMAs, and only one direction is allowed to slip in each {011} plane. Therefore, there are only six possible slip systems in the <100>{011} family of the NiTiFe SMA. This number is only a half of those in FCC and BCC crystals, which both possess 12 slip systems. Therefore, the mechanism of ECAE in NiTiFe SMAs is different from that in FCC and BCC crystals. In the two crystals, the SPD leads to the simultaneous activation of two or more slip systems in a grain, which induces numerous complicated networks that are composed of dislocation entanglements. With increasing strain, GNDs are accumulated in these dislocation networks, which increase misorientations among them. As a result, subgrains and subsequent grain refinement can be achieved when the misorientations reach a certain critical value. It has been proved that the <100>{011} family only permits the occurrence of slip in three independent slip systems, including [100](011), [010](101), and [001](110) [28,29]. In addition, according to Schmid's law, the initially activated slip systems are those possessing the largest value of the Schmid factor because minimum force is required [24].The value of the Schmid factor is related to the loading direction and the orientation of an individual grain. In addition, because there are angles among the three independent slip systems, their Schmid factors are unable to reach the maximum value at the same time. As a consequence, in a B2 austenite grain, only one or two of the three systems may be activated at the same time due to the limitation of slip system number. In the case of one activated slip system alone, shear bands are dominant at the early stage of ECAE. In the case of two activated slip systems, entangled dislocation networks are dominant at the initial stage of ECAE. Figure 8a illustrates the equiaxed coarse grains before ECAE, and Figure 8b shows the microstructure at the early stage of ECAE, where original coarse grains are elongated along the shear plane by shear deformation, and shear bands and entangled dislocation networks are induced in the grain interior. It has been widely accepted that both slip and orientation rotation are induced during plastic deformation. As a result, with the proceeding of ECAE, other slip systems will be activated when the orientation rotation causes the inactivated systems of the <100>{011} family to turn to a preferred orientation or when the stress is high enough for activating the slip systems of the <100>{001} and <111>{110} families. At this stage, the slip of dislocations induced later will be blocked by the shear bands, the grain boundaries, and the dislocation networks. Dislocation pile-ups are formed when they are blocked by shear bands or grain boundaries, while the blocked dislocations are absorbed by the dislocation networks when they meet the dislocation networks. Figure 8c illustrates the activation of other slip systems and the block of dislocations induced later. Shortly after the deformation, continuous recovery occurs, which leads to the rearrangement of dislocations, the formation of dislocation cells, subgrains, and newly fine grains, as shown in Figure 8d. The nucleation of the fine grains mainly occurs along the shear bands or the grain boundaries. This is due to the fact that a high density of dislocations is blocked by the shear bands as well as the grain boundaries. These dislocations are continuously trapped at low angle boundaries in these regions when they attempt to pass through them, which leads to the increase in GNDs and the subsequent formation of newly fine grains. The above mechanism illustrates why the grain refinement in NiTiFe SMAs subjected to single-pass ECAE is inhomogeneous.

Although ECAE is able to refine the grains of metals, the deformation temperature should be considered when the parameters are determined. Higher temperature may lower the dislocation densities during ECAE or will lead to fast growth of the refined grains, which both fail to refine the grains. Lower temperatures can produce higher dislocation densities and maintain the overall fine grains, but it faces the risk of material fracture. In the current investigation, the temperature over 500 °C is not suitable for the ECAE processing of NiTiFe SMAs, which can be confirmed by the weak grain refinement shown in Figure 3c.

5. Conclusions

(1) Single-pass ECAE performed at 400–500 °C is able to refine the grains of a NiTiFe SMA. However, the obtained microstructures present a high inhomogeneity, where the refined grains are mainly nucleated near the grain boundaries and a small fraction of them emerges in the grain interior. Furthermore, the size of the refined grains increases as ECAE temperature increases, which indicates that the higher deformation temperature is adverse to the refinement of ECAE.

(2) The grain refinement of ECAE results from the transition of lower angle grain boundaries to higher ones, and a lower deformation temperature is able to induce more substructures during the ECAE of a NiTiFe SMA. The ECAE of NiTiFe SMAs is not suitable for being performed at temperatures above 500 °C due to its weak ability of grain refinement. It is the accumulation and rearrangement of GNDs as plastic strain increases that leads to the transition of lower angle subgrain boundaries. Higher angle subgrain boundaries are finally induced, and finer grains are formed.

(3) Due to the limitation of slip systems, the mechanism of grain refinement in NiTiFe SMA subjected to single-pass ECAE is different from that in FCC and BCC crystals. Dislocation cells and shear bands are two transition microstructures of grain refinement during the ECAE of NiTiFe SMAs. The nucleation of the fine grains mainly occurs along the shear bands or the grain boundaries, which leads to the inhomogeneity of grain refinement.

Acknowledgments: The work was financially supported by the National Natural Science Foundation of China (No. 51475101 and No. 51305091).

Author Contributions: Yanqiu Zhang performed EBSD (electron backscattered diffraction) and TEM (transmission electron microscopy) analysis and wrote the manuscript; Shuyong Jiang supervised the research.

Conflicts of Interest: The authors declare no conflict of interest.

References

1. Tsuchiya, K.; Inuzuka, M.; Tomus, D.; Hosokawa, A.; Nakayama, H.; Morii, K.; Todaka, Y.; Umemoto, M. Martensitic transformation in nanostructured TiNi shape memory alloy formed via severe plastic deformation. *Mater. Sci. Eng. A* **2006**, *438–440*, 643–648. [CrossRef]
2. Khaleghi, F.; Khalil-Allafi, J.; Abbasi-Chianeh, V.; Noori, S. Effect of short-time annealing treatment on the superelastic behavior of cold drawn Ni-rich NiTi shape memory wires. *J. Alloys Compd.* **2013**, *554*, 32–38. [CrossRef]
3. Peterlechner, M.; Waitz, T.; Karnthaler, H.P. Nanoscale amorphization of severely deformed NiTi shape memory alloys. *Scr. Mater.* **2009**, *60*, 1137–1140. [CrossRef]
4. Zhang, Y.; Jiang, S.; Hu, L.; Liang, Y. Deformation mechanism of NiTi shape memory alloy subjected to severe plastic deformation at low temperature. *Mater. Sci. Eng. A* **2013**, *559*, 607–614. [CrossRef]
5. Jiang, S.; Zhao, Y.; Zhang, Y.; Tang, M.; Li, C. Equal channel angular extrusion of NiTi shape memory alloy tube. *Trans. Nonferr. Met. Soc. China* **2013**, *23*, 2021–2028. [CrossRef]
6. Hu, T.; Chen, L.; Wu, S.L.; Chu, C.L.; Wang, L.M.; Yeung, K.W.K.; Chu, P.K. Graded phase structure in the surface layer of NiTi alloy processed by surface severe plastic deformation. *Scr. Mater.* **2011**, *64*, 1011–1014. [CrossRef]
7. Li, S.; Beyerlein, I.J.; Necker, C.T. On the development of microstructure and texture heterogeneity in ECAE via route C. *Acta Mater.* **2006**, *54*, 1397–1408. [CrossRef]
8. Su, C.W.; Lu, L.; Lai, M.O. A model for the grain refinement mechanism in equal channel angular pressing of Mg alloy from microstructural studies. *Mater. Sci. Eng. A* **2006**, *434*, 227–236. [CrossRef]
9. Qarni, M.J.; Sivaswamy, G.; Rosochowski, A.; Boczkal, S. On the evolution of microstructure and texture in commercial purity titanium during multiple passes of incremental equal channel angular pressing (I-ECAP). *Mater. Sci. Eng. A* **2017**, *699*, 31–47. [CrossRef]
10. Frint, S.; Hockauf, M.; Frint, P.; Wagner, M.F.X. Scaling up Segal's principle of equal-channel angular pressing. *Mater. Des.* **2016**, *97*, 502–511. [CrossRef]
11. Li, H.; Li, S.; Zhang, D. On the selection of outlet channel length and billet length in equal channel angular extrusion. *Comput. Mater. Sci.* **2010**, *49*, 293–298. [CrossRef]
12. Cheng, W.; Tian, L.; Wang, H.; Bian, L.; Yu, H. Improved tensile properties of an equal channel angular pressed (ECAPed) Mg-8Sn-6Zn-2Al alloy by prior aging treatment. *Mater. Sci. Eng. A* **2017**, *687*, 148–154. [CrossRef]
13. Hao, T.; Tang, H.; Luo, G.; Wang, X.; Liu, C.; Fang, Q. Enhancement effect of inter-pass annealing during equal channel angular pressing on grain refinement and ductility of 9Cr1Mo steel. *Mater. Sci. Eng. A* **2016**, *667*, 454–458. [CrossRef]
14. Sitdikov, O.; Avtokratova, E.; Sakai, T. Microstructural and texture changes during equal channel angular pressing of an Al-Mg-Sc alloy. *J. Alloys Compd.* **2015**, *648*, 195–204. [CrossRef]
15. Qarni, M.J.; Sivaswamy, G.; Rosochowski, A.; Boczkal, S. Effect of incremental equal channel angular pressing (I-ECAP) on the microstructural characteristics and mechanical behaviour of commercially pure titanium. *Mater. Des.* **2017**, *122*, 385–402. [CrossRef]
16. Gholami, D.; Imantalab, O.; Naseri, M.; Vafaeian, S.; Fattah-alhosseini, A. Assessment of microstructural and electrochemical behavior of severely deformed pure copper through equal channel angular pressing. *J. Alloys Compd.* **2017**, *723*, 856–865. [CrossRef]
17. Shahmir, H.; Nili-Ahmadabadi, M.; Mansouri-Arani, M.; Langdon, T.G. The processing of NiTi shape memory alloys by equal-channel angular pressing at room temperature. *Mater. Sci. Eng. A* **2013**, *576*, 178–184. [CrossRef]
18. Shahmir, H.; Nili-Ahmadabadi, M.; Wang, C.T.; Jung, J.M.; Kim, H.S.; Langdon, T.G. Annealing behavior and shape memory effect in NiTi alloy processed by equal-channel angular pressing at room temperature. *Mater. Sci. Eng. A* **2015**, *629*, 16–22. [CrossRef]

19. Zhang, D.; Guo, B.; Tong, Y.; Tian, B.; Li, L.; Zheng, Y.; Gunderov, D.V.; Valiev, R.Z. Effect of annealing temperature on martensitic transformation of $Ti_{49.2}Ni_{50.8}$ alloy processed by equal channel angular pressing. *Trans. Nonferr. Met. Soc. China* **2016**, *26*, 448–455. [CrossRef]

20. Kockar, B.; Karaman, I.; Kulkarni, A.; Chumlyakov, Y.; Kireeva, I.V. Effect of severe ausforming via equal channel angular extrusion on the shape memory response of a NiTi alloy. *J. Nucl. Mater.* **2007**, *361*, 298–305. [CrossRef]

21. Song, J.; Wang, L.; Zhang, X.; Sun, X.; Jiang, H.; Fan, Z.; Xie, C.; Wu, M.H. Effects of second phases on mechanical properties and martensitic transformations of ECAPed TiNi and Ti-Mo based shape memory alloys. *Trans. Nonferr. Met. Soc. China* **2012**, *22*, 1839–1848. [CrossRef]

22. Leitner, T.; Sabirov, I.; Pippan, R.; Hohenwarter, A. The effect of severe grain refinement on the damage tolerance of a superelastic NiTi shape memory alloy. *J. Mech. Behav. Biomed. Mater.* **2017**, *71*, 337–348. [CrossRef] [PubMed]

23. Kocich, R.; Szurman, I.; Kursa, M.; Fiala, J. Investigation of influence of preparation and heat treatment on deformation behaviour of the alloy NiTi after ECAE. *Mater. Sci. Eng. A* **2009**, *512*, 100–104. [CrossRef]

24. Zhang, Y.; Jiang, S.; Wang, S.; Sun, D.; Hu, L. Influence of partial staticrecrystallization on microstructures and mechanical properties of NiTiFe shape memory alloysubjected to severe plastic deformation. *Mater. Res. Bull.* **2017**, *88*, 226–233. [CrossRef]

25. Iwahashi, Y.; Wang, J.; Horita, Z.; Nemoto, M. Principle of equal-channel angular pressing for the processing of ultra-fine grained materials. *Scr. Mater.* **1996**, *35*, 143–146. [CrossRef]

26. Gall, K.; Dunn, M.L.; Liu, Y.; Labossiere, P.; Sehitoglu, H.; Chumlyakov, Y.I. Micro and macro deformation of single crystal NiTi. *J. Eng. Mater. Technol.* **2002**, *124*, 238–245. [CrossRef]

27. Benafan, O.; Noebe, R.D.; Padula, S.A.; Garg, A.; Clausen, B.; Vogel, S.; Vaidyanathan, R. Temperature dependent deformation of the B2 austenite phase of a NiTi shape memory alloy. *Int. J. Plast.* **2013**, *51*, 103–121. [CrossRef]

28. Ezaz, T.; Wang, J.; Sehitoglu, H.; Maier, H.J. Plastic deformation of NiTi shape memory alloys. *Acta Mater.* **2013**, *61*, 67–78. [CrossRef]

29. Pelton, A.R.; Huang, G.H.; Moinec, P.; Sinclair, R. Effects of thermal cycling on microstructure and properties in Nitinol. *Mater. Sci. Eng. A* **2012**, *532*, 130–138. [CrossRef]

![metals logo] *metals*

MDPI

Article

Numerical Predictions of the Occurrence of Necking in Deep Drawing Processes

Hocine Chalal and Farid Abed-Meraim *

LEM3, UMR CNRS 7239—Arts et Métiers ParisTech, 4, rue Augustin Fresnel, 57078 Metz CEDEX 03, France; hocine.chalal@ensam.eu
* Correspondence: farid.abed-meraim@ensam.eu; Tel.: +33-3-8737-5479

Received: 8 September 2017; Accepted: 23 October 2017; Published: 27 October 2017

Abstract: In this work, three numerical necking criteria based on finite element (FE) simulations are proposed for the prediction of forming limit diagrams (FLDs) for sheet metals. An elastic–plastic constitutive model coupled with the Lemaitre continuum damage theory has been implemented into the ABAQUS/Explicit software to simulate simple sheet stretching tests as well as Erichsen deep drawing tests with various sheet specimen geometries. Three numerical criteria have been investigated in order to establish an appropriate necking criterion for the prediction of formability limits. The first numerical criterion is based on the analysis of the thickness strain evolution in the central part of the specimens. The second numerical criterion is based on the analysis of the second time derivative of the thickness strain. As to the third numerical criterion, it relies on a damage threshold associated with the occurrence of necking. The FLDs thus predicted by numerical simulation of simple sheet stretching with various specimen geometries and Erichsen deep drawing tests are compared with the experimental results.

Keywords: modeling; simulation; sheet metal; necking; damage; forming limit diagrams; deep drawing test

1. Introduction

The formability of sheet metals is usually characterized by forming limit diagrams (FLDs) obtained by the Nakazima or Marciniak deep drawing tests. The concept of FLD was first introduced by Keeler and Backofen [1] and subsequently improved by Goodwin [2]. The FLD is a limiting curve that depicts the in-plane major and minor strains of the sheet at the onset of localized necking, which precedes the final fracture. The FLD determination was originally based on experimental measurements, which turned out to be difficult and time-consuming. To overcome these drawbacks, a number of alternative theoretical and numerical approaches have been developed in the literature for the prediction of FLDs. These approaches are based on the combination of necking criteria with constitutive models for the prediction of necking in sheet metals. Among the theoretical necking criteria that have been developed in the literature for the prediction of necking, Swift [3] proposed an extension to biaxial stretching to the Considère maximum load criterion [4], which was utilized to predict diffuse necking in the expansion domain of the FLD. For localized necking, Hill [5] proposed an alternative criterion based on the bifurcation theory, which states that localized necking occurs along the direction of zero extension. It is worth noting that Hill's criterion is only applicable to the left-hand side of the FLD and, therefore, it was often combined with the Swift criterion to determine a complete FLD. Marciniak and Kuczynski [6] developed another approach for localized necking prediction, which is known as the M–K criterion. The latter is based on the introduction of an initial imperfection, which ultimately triggers the occurrence of localized necking.

Concurrently with the above theoretical criteria, several numerical criteria for the prediction of necking and ductile fracture in sheet metals have been developed in the last few decades. Thanks to the growing progress in computational resources, simulation of complex sheet metal forming processes, such as the Nakazima and Marciniak deep drawing tests, using the finite element method (FEM) has become an interesting alternative to the theoretical approaches. Indeed, the substantial amount of results provided by FEM allows a realistic prediction of necking and fracture as compared to experiments. Using the FEM approach, the evolution of strain fields during loading is analyzed for each finite element within the sheet to detect the onset of necking. Burn et al. [7] analyzed the thinning of sheet metals, by using the Nakazima deep drawing test, in order to predict the onset of necking. Based on the Marciniak deep drawing test, Petek et al. [8] proposed a numerical approach for the prediction of necking, which consists in analyzing the time evolution of thickness strain and its first and second time derivatives. Later, Situ et al. [9–11] applied the same strategy to the Nakazima deep drawing test in order to predict FLDs for sheet metals involving the whole range of strain paths. They have shown that the analysis based on major strain rate (i.e., first time derivative of major strain) predicts the onset of fracture, while the maximum of major strain acceleration (i.e., second time derivative of major strain) corresponds to the occurrence of localized necking. Furthermore, another class of numerical criteria for the prediction of ductile fracture has emerged (see, e.g., [12–16]). These numerical criteria are based on empirical relationships, which depend on the application, and require several parameter calibrations with respect to experiments. They are labeled "fracture criteria", as the associated FLDs are higher than those predicted using necking criteria.

The above numerical approaches for the prediction of necking and fracture are often combined with undamaged elastic–plastic constitutive models, which is not realistic from an experimental point of view. Indeed, the softening regime exhibited by the material behavior prior to fracture cannot be reproduced by elastic–plastic models alone, which requires the coupling of the constitutive equations with damage for a proper description of the material degradation and, thus, reliable prediction of final fracture. In this context, two well-established theories of ductile damage have been developed over the past few decades. The first theory is based on a micromechanical analysis of void growth, which describes the ductile damage mechanisms in porous materials. It was initiated by Gurson [17], modified by Tvergaard and Needleman [18], and subsequently improved by a number of contributors (see, e.g., [19–22]). The second theory, known as continuum damage mechanics (see, e.g., [23,24]), is based on the introduction of a damage variable, which represents the surface density of defects, and can be modeled as isotropic scalar variable (see, e.g., [23,25]), or tensor variable for anisotropic damage (see, e.g., [26–28]).

In this work, numerical necking criteria, based on finite element (FE) simulations, are proposed for the prediction of forming limit diagram for a steel material. The material response is described by an elastic–plastic model coupled with the Lemaitre isotropic damage approach [23]. The resulting constitutive equations have been implemented into the ABAQUS/Explicit code, within the framework of large strain and a three-dimensional formulation. Several specimen geometries have been simulated in order to reproduce all of the strain paths that are typically encountered in sheet metal forming processes. Two different FE models are considered to predict the FLDs of the studied material. First, the FLDs are predicted using simple sheet stretching tests, applied to different specimen geometries, in which no contact with tools is considered. Then, the FE model based on the Erichsen deep drawing test (see, e.g., [29]) is used to predict the FLDs of the steel material. To determine these forming limit curves for the studied material, three numerical criteria are presented in this work to detect the occurrence of necking in the sheet specimens. In the first numerical criterion, necking is detected when a sudden change in the evolution of the thickness strain at the central area of the specimen is observed. The second numerical criterion is based on the evolution of the thickness strain acceleration, which is obtained by computing the second time derivative of the thickness strain. As to the third numerical criterion, it relies on a critical damage threshold, at which is associated the occurrence of necking. All points of the predicted FLDs, which are obtained using the FE simulations combined with the numerical necking criteria, are compared with the experimental results taken from [30].

2. Constitutive Equations of the Ductile Damage Model

In this section, the elastic–plastic behavior law coupled with a ductile damage model is briefly presented. The latter is based on the continuum damage mechanics and, more specifically, on the Lemaitre isotropic damage model [23]. Using the concept of effective stress $\tilde{\sigma}$, and the strain equivalence principle, the continuum damage is introduced via the scalar variable d by the following expression:

$$\sigma = (1-d)\tilde{\sigma} = (1-d)\mathbf{C} : \varepsilon^{\mathbf{e}}, \tag{1}$$

where σ is the Cauchy stress tensor, $\mathbf{C}$ is the fourth-order elasticity tensor, and $\varepsilon^{\mathbf{e}}$ is the elastic strain tensor. The plastic yield function f is written in the following form:

$$f = \overline{\sigma}(\tilde{\sigma}, \mathbf{X}) - \sigma_Y \leq 0, \tag{2}$$

where $\overline{\sigma}(\tilde{\sigma}, \mathbf{X}) = \sqrt{(\tilde{\sigma}' - \mathbf{X}) : \mathbf{M} : (\tilde{\sigma}' - \mathbf{X})}$ is the equivalent stress, and $\tilde{\sigma}'$ is the deviatoric part of the effective stress. The fourth-order tensor $\mathbf{M}$ contains the six anisotropy coefficients of the Hill quadratic yield criterion [31]. The isotropic hardening of the material is described by the size σ_Y of the yield surface, while kinematic hardening is represented by the back-stress tensor $\mathbf{X}$.

The plastic flow rule is given by the normality law, which defines the plastic strain rate $\mathbf{D}^{\mathbf{P}}$ as

$$\mathbf{D}^{\mathbf{P}} = \dot{\lambda}\frac{\partial f}{\partial \sigma} = \frac{\dot{\lambda}}{1-d}\frac{\mathbf{M} : (\tilde{\sigma}' - \mathbf{X})}{\overline{\sigma}} \tag{3}$$

where $\dot{\lambda}$ is the plastic multiplier, and $\partial f/\partial \sigma$ is the flow direction, normal to the yield surface in the stress space. With a special choice of co-rotational frame, which is associated with the Jaumann objective derivative, the Cauchy stress rate is written in the following form:

$$\dot{\sigma} = (1-d)\mathbf{C} : (\mathbf{D} - \mathbf{D}^{\mathbf{P}}) - \frac{\dot{d}}{1-d}\sigma. \tag{4}$$

The evolution law for the damage variable is expressed by the following equation:

$$\dot{d} = \begin{cases} \frac{1}{(1-d)^\beta}\left(\frac{Y_e - Y_{ei}}{S}\right)^s \dot{\lambda} & \text{if } Y_e \geq Y_{ei} \\ 0 & \text{otherwise} \end{cases}, \tag{5}$$

where Y_e is the strain energy density release rate (see, e.g., [25,32]), and β, s, Y_{ei} and S are four damage parameters that need to be identified. The expression of the strain energy density release rate Y_e is given (for linear isotropic elasticity) as follows:

$$Y_e = \frac{\overline{\tilde{\sigma}}_{vM}^2}{2E}\left[\frac{2}{3}(1+v) + 3(1-2v)\left(\frac{\tilde{\sigma}^s}{\overline{\tilde{\sigma}}_{vM}}\right)^2\right], \tag{6}$$

where $\overline{\tilde{\sigma}}_{vM} = \sqrt{3\tilde{\sigma}' : \tilde{\sigma}'/2}$ is the von Mises equivalent effective stress, $\tilde{\sigma}^s = \tilde{\sigma} : \mathbf{1}/3$ is the hydrostatic effective stress (with $\mathbf{1}$ being the second-order identity tensor), while E and v are, respectively, the Young's modulus and Poisson's ratio.

The above constitutive equations are implemented into the finite element code ABAQUS/Explicit using a co-rotational frame. The fourth-order Runge–Kutta explicit time integration scheme is used to update the stress state and all internal variables.

3. Numerical Integration of the Model and Its Validation

3.1. Time Integration Scheme

A user-defined material (VUMAT) subroutine is used for the implementation of the above elastic–plastic–damage model into the commercial finite element code ABAQUS/Explicit (Dassault Systèmes, France). For each integration point of the FE model, the stress state and all internal variables of the fully coupled elastic–plastic–damage model are known at the beginning of the loading increment. These stress state and internal variables will be updated through the VUMAT subroutine at the end of the loading increment. In this work, the fourth-order Runge–Kutta explicit time integration scheme is adopted to determine the updated stress state and all internal variables at the end of each loading increment. This straightforward integration algorithm represents a reasonable compromise in terms of computational efficiency, accuracy and convergence. Indeed, explicit time integration does not involve matrix inversion or iterative procedures for convergence, unlike implicit time integration. However, for explicit schemes, the time increment must be kept small enough to ensure accuracy and stability (see, e.g., [33,34]).

The evolution equations of the fully coupled model, which were presented in the previous section, can easily be written in the following compact form of general differential equation:

$$\dot{\mathbf{u}} = \mathbf{h_u}(\mathbf{u}), \tag{7}$$

where vector $\mathbf{u}$ encompasses all of the internal variables and stress state, while vector $\mathbf{h_u}(\mathbf{u})$ includes all evolution laws described in the previous section. The above condensed differential equation is then integrated over each loading increment, using the forward fourth-order Runge–Kutta explicit time integration scheme. The resulting algorithm is implemented into the finite element code ABAQUS/Explicit, via a VUMAT user-defined material subroutine, within the framework of large strains and a fully three-dimensional formulation.

3.2. Numerical Validation

In this section, the implementation of the fully coupled elastic–plastic–damage model described in the previous sections is validated through simulations of uniaxial tensile tests, which are then compared with reference solutions taken from the literature.

The first numerical example is a simple tensile test, which allows for the numerical validation of the elastic–plastic implementation of the model, without taking into account the damage contribution. For this purpose, the undamaged elastic–plastic model is recovered by setting to zero all the damage parameters. In this test, the von Mises yield surface is considered along with the Ludwig isotropic hardening law, which is defined by the following expression:

$$\sigma_Y = \sigma_0 + k \left(\bar{\varepsilon}^{\mathrm{pl}} \right)^n, \tag{8}$$

where σ_Y represents the size of the yield surface, and $\bar{\varepsilon}^{\mathrm{pl}}$ is the equivalent plastic strain. In Equation (8), σ_0 is the initial yield stress, while k and n are hardening parameters. Three standard steel materials, with three different values for the hardening exponent (see Equation (8)), are considered for the simulation of the uniaxial tensile test. The associated elastic–plastic material parameters are summarized in Table 1.

Table 1. Elastic–plastic properties for the studied materials.

Material	E (MPa)	ν	σ_0 (MPa)	K (MPa)	n
Steel	$200,000$	0.3	200	10,000	0.3–0.6–1.0

Figure 1 shows the uniaxial stress–strain curves obtained with the elastic–plastic model implemented in the VUMAT subroutine, which are compared with the numerical results given by the built-in elastic–plastic model available in ABAQUS. From these results, one can observe that the uniaxial stress–strain curves given by the VUMAT subroutine coincide with those provided by the built-in ABAQUS model, which demonstrates the successful implementation of the proposed model.

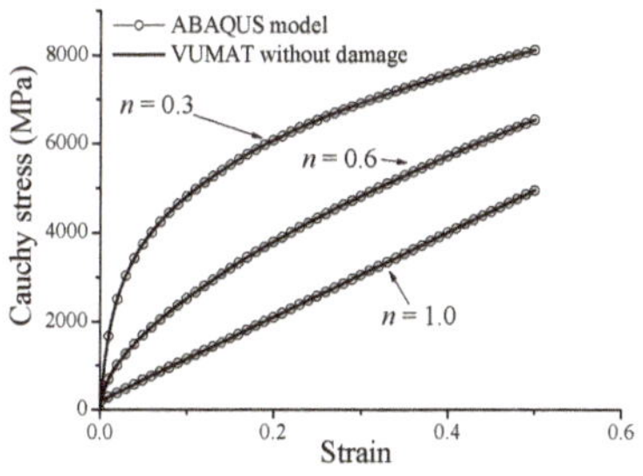

Figure 1. Validation of the numerical implementation of the undamaged elastic–plastic model with respect to the built-in ABAQUS model.

The second numerical test is intended to the validation of the proposed model when the damage behavior is taken into account. To this end, a uniaxial tensile test is simulated, according to the works of Doghri and Billardon [35], where a phenomenological elastic–plastic model with a von Mises plastic yield surface and Ludwig's isotropic hardening is coupled with the Lemaitre damage approach. The studied materials are the same as those used in the previous test (see Table 1 for the elastic and hardening parameters), while the damage parameters used in the simulations are the same for the three materials, and are summarized in Table 2 (see [35]).

Table 2. Lemaitre damage parameters for the studied materials.

Material	β	S (MPa)	s	Y_{ei} (MPa)
Steel	1	0.5	1	0

Figure 2 compares the stress–strain responses and the damage evolution obtained by the implemented fully coupled model with the reference solutions taken from [35]. It can be clearly observed that the simulated stress–strain curves and damage evolution coincide with their counterparts taken from the reference solutions, for the three materials investigated, which validates again the numerical implementation of the present elastic–plastic–damage model.

(a)   (b)

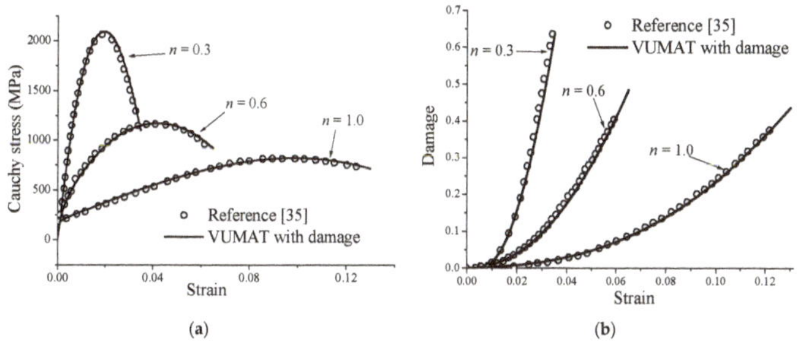

Figure 2. Validation of the numerical implementation of the fully coupled elastic–plastic–damage model with respect to reference solutions taken from [35] for the three studied materials: (**a**) uniaxial tensile stress–strain curves; and (**b**) damage evolution.

4. Identification of the St14 Steel Material Parameters

In this work, we investigate the occurrence of necking in a St14 steel material using the elastic–plastic–damage model described above in conjunction with numerical necking criteria. The mechanical behavior of the St14 steel is based on the Ludwig isotropic hardening law (see Equation (8)) and the von Mises yield surface, which are coupled with the Lemaitre isotropic damage approach. Standard tensile test experiments were performed by Aboutalebi et al. [36] for the investigated St14 steel. The corresponding experimental load–displacement curve is exploited in this work to identify the hardening and damage parameters for the studied St14 steel.

The hardening parameters of the Ludwig isotropic hardening law are first identified in the range of uniform elongation of the uniaxial tensile test. In this range of small to moderate deformations, where the stress and strain fields in the central region of the specimen remain homogeneous, the hardening parameters are accurately identified using a simple regression of the experimental data with the Ludwig power-law. The corresponding elastic and hardening material parameters are summarized in Table 3.

Table 3. Elastic–plastic properties for the St14 steel.

Material	E (MPa)	v	σ_0 (MPa)	K (MPa)	n
St14	180,000	0.3	130	585	0.44

Then, the above-identified elastic–plastic parameters are used for the simulation of the uniaxial tensile test until the final fracture of the specimen. The damage parameters of the Lemaitre model are identified based on the entire experimental load–displacement response of the tensile test using an inverse identification procedure. The latter is based on least-squares minimization of the difference between the experimental and numerical load–displacement response for the uniaxial tensile test. The corresponding identified values for the Lemaitre damage model are summarized in Table 4.

Table 4. Identified damage parameters for the St14 steel.

Material	β	S (MPa)	s	Y_{ei} (MPa)
Steel	4.251	2.648	1.831	0.001

Figure 3 compares the simulated load–displacement response, obtained using the identified material parameters of the Lemaitre damage model, with the experimental counterpart provided by Aboutalebi et al. [36]. This figure clearly shows that the simulated response using the present damage model is in very good agreement with the experimental curve and, in particular, demonstrates the ability of the implemented model to reproduce the sudden load drop that precedes the final fracture.

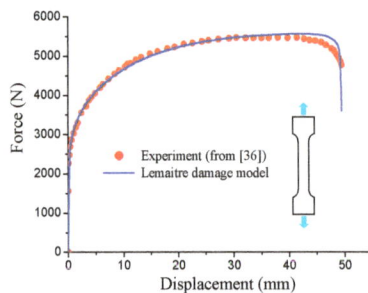

Figure 3. Tensile load–displacement response simulated with the Lemaitre damage model, along with the experimental curve taken from Aboutalebi et al. [36].

5. Description of the Finite Element Models

5.1. Finite Element Simulations

In this work, the predictions of the FLDs for the St14 steel material are carried out using the above elastic–plastic behavior model coupled with ductile damage. The numerical simulations are performed with the ABAQUS/explicit code. Ten specimens of the St14 sheet material with different geometries (a length of 120 mm, and a width varying from 12 mm to 120 mm, with a 12 mm increment in the width direction) are used in the simulations. Each specimen reproduces a particular strain path, which is typically encountered in sheet metal forming processes, and these strain paths range from uniaxial tension to equibiaxial expansion.

Two FE models are used for predicting the FLDs of the studied material. First, simple sheet stretching simulations, based on the different specimens, are performed. Then, the simulation of the deep drawing process, according to the Erichsen test (see, e.g., [29,30]), is conducted with the same specimens described above.

The schematic view of the Erichsen deep drawing test is illustrated in Figure 4.

The geometric parameters used in the simulations are (see [37]):

- Punch diameter D_p = 60 mm;
- Initial sheet thickness t = 0.8 mm;
- Die radius r_d = 3 mm;
- Die opening diameter D_d = 66 mm.

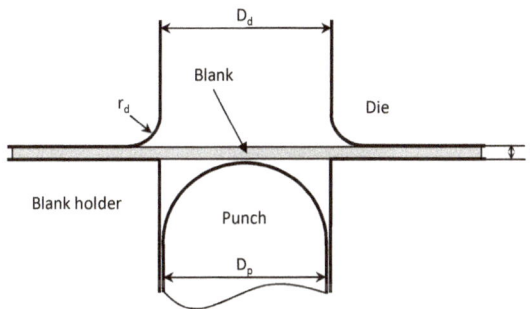

Figure 4. Schematic view of the Erichsen deep drawing test.

Due to the symmetry of the problem, only one quarter of the geometry is discretized for each specimen. Figure 5 provides an illustration of the finite element models for the simple sheet stretching test and the Erichsen deep drawing test. For the particular case of the simple sheet stretching test, two different types of boundary conditions are considered in the simulations in order to reproduce most of the strain paths encountered in the simulations of the Erichsen deep drawing test with the various specimen widths. The first type of boundary conditions corresponds to a simple uniaxial tension, and these boundary conditions are applied to the specimens having a width ranging from 12 mm to 60 mm (see Figure 5a for illustration on the specimen having a width of 12 mm). The second type of boundary conditions corresponds to a proportional biaxial tension, and these boundary conditions are applied to the specimens having a width ranging from 72 mm to 120 mm (see Figure 5b for illustration on the specimen having a width of 84 mm).

The forming tools for the Erichsen deep drawing test are modeled as discrete rigid bodies. The friction coefficient between the tools and the specimen is taken to be equal to 0.15 [30].

(a)

(b)

(c)

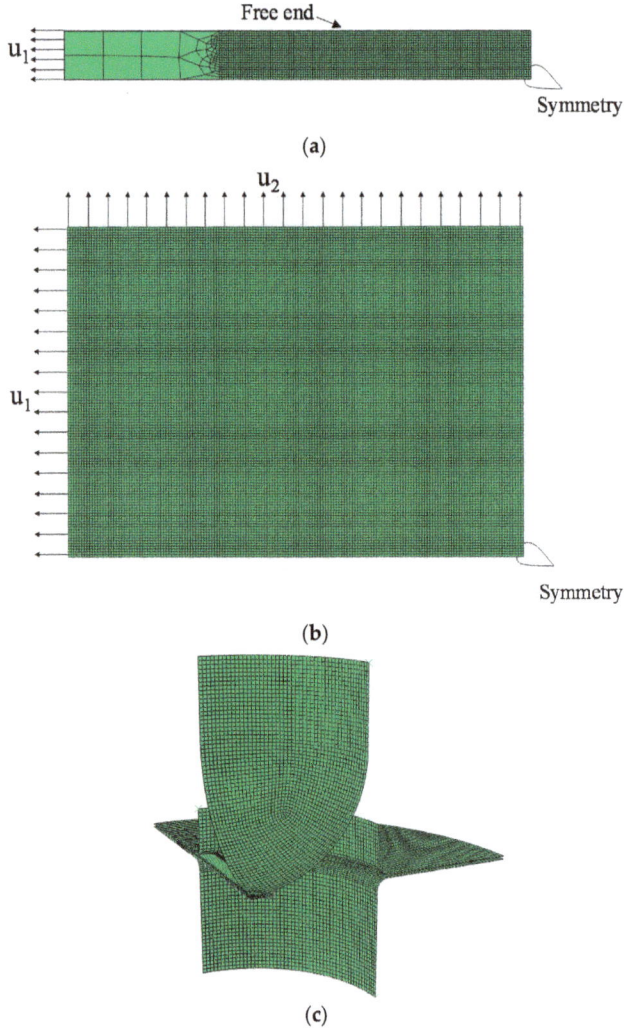

Figure 5. FE models corresponding to (**a**) uniaxial tensile test; (**b**) biaxial tensile test; and (**c**) Erichsen deep drawing test.

5.2. Mesh Sensitivity

For material behavior that exhibits damage-induced softening, it is well known that the numerical solution, and particularly the localization zone, is prone to mesh sensitivity when a local elastic–plastic–damage model is used, which is the case in this work (see, e.g., [18,38,39]). In this section, several finite element models are adopted for the simulation of the uniaxial tensile specimen and the Erichsen deep drawing test, using the specimen with 12 mm width, in order to analyze the mesh-sensitivity effects. In all of the simulations that follow, the specimens are modeled with the eight-node three-dimensional continuum finite element with reduced integration (C3D8R), which is available in the ABAQUS/Explicit software. This element has only one integration point, which means that by considering n layers of elements in the thickness direction, the sheet thickness will be modeled with a total of n integration points.

The effect of the number of elements in the thickness direction is first analyzed by considering, successively, three, then four, and finally five element layers through the thickness. In these three FE models, the same in-plane mesh is used to discretize the useful region of the specimen, with an intermediate in-plane element size of 0.3×0.3 mm^2. Then, the impact of the in-plane FE discretization is analyzed by adopting for the useful region of the specimen four layers of elements through the thickness and three different in-plane element sizes (0.4×0.4 mm^2, 0.3×0.3 mm^2, and 0.2×0.2 mm^2, respectively, as illustrated in Figure 6).

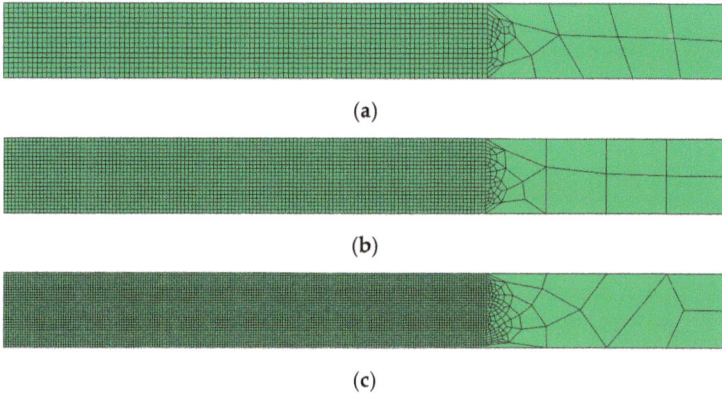

Figure 6. In-plane FE discretization for the central region of the specimen: (**a**) 0.4×0.4 mm^2; (**b**) 0.3×0.3 mm^2; and (**c**) 0.2×0.2 mm^2.

Figures 7 and 8 reveal the influence of the number of elements in the thickness direction on the evolution of the thickness strain and the damage variable in the center of the specimen, as reflected by the simulation of the uniaxial tensile test and the Erichsen deep drawing test, respectively. It is clearly shown that only very small mesh dependence is observed, when varying the mesh refinement in the thickness direction, which suggests that four layers of elements are sufficient to describe the various nonlinear phenomena through the thickness.

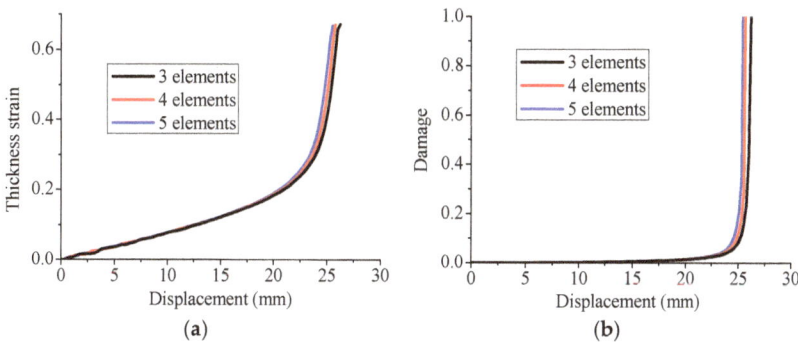

Figure 7. Effect of the number of elements in the thickness direction on the evolution of (**a**) thickness strain and (**b**) damage, during the uniaxial tensile test for the specimen with 12 mm width.

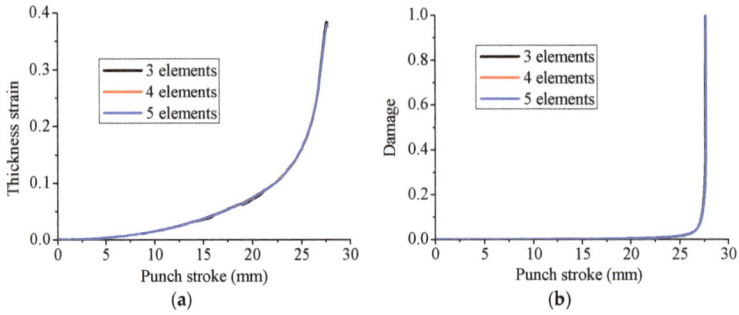

Figure 8. Effect of the number of elements in the thickness direction on the evolution of (**a**) thickness strain and (**b**) damage, during the Erichsen deep drawing test for the specimen with 12 mm width.

Figures 9 and 10 show the evolution of the thickness strain and damage variable in the center of the specimen, as determined by the present constitutive model with three different in-plane mesh sizes, for the uniaxial tensile test and Erichsen deep drawing test, respectively. In contrast to the results obtained with different numbers of element layers, the predicted thickness strain and damage variable reveal more sensitivity to in-plane mesh refinement when the damage variable becomes significant (i.e., $d > 0.3$). More specifically, the numerical results obtained with the intermediate and finer meshes (0.3×0.3 mm^2 and 0.2×0.2 mm^2, respectively) are close to each other, which prompted us to use the in-plane mesh of 0.3×0.3 mm^2 in the subsequent simulations. Indeed, this intermediate mesh involves reasonable computational times for all specimen geometries.

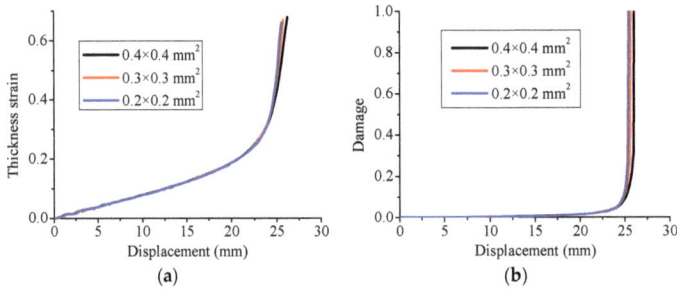

Figure 9. Effect of the in-plane mesh refinement on the evolution of (**a**) thickness strain and (**b**) damage, during the uniaxial tensile test for the specimen with 12 mm width.

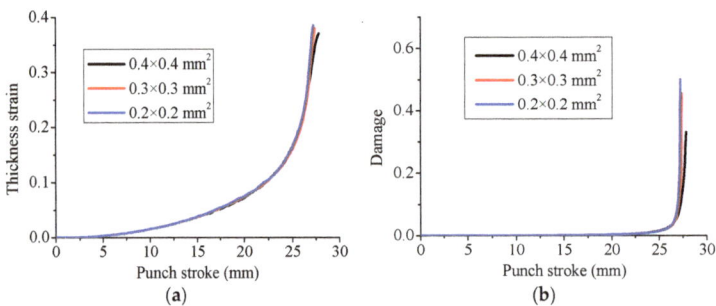

Figure 10. Effect of the in-plane mesh refinement on the evolution of (**a**) thickness strain and (**b**) damage, during the Erichsen deep drawing test for the specimen with 12 mm width.

6. Numerical Criteria for the Prediction of the Occurrence of Necking

In this section, the numerical criteria used for the prediction of FLDs associated with the simple sheet stretching and the Erichsen simulations on various specimen geometries are presented. Three numerical criteria are adopted to predict the critical in-plane strains at the occurrence of necking. The first criterion is based on the analysis of the evolution of the thickness strain for each specimen. The occurrence of necking is detected when a sudden change in the evolution of the thickness strain in the central area is observed (see, e.g., [40]). The minor and major in-plane principal strains, corresponding to the occurrence of necking, are then reported into the FLD. To illustrate this procedure in the case of uniaxial tensile test, Figure 11 shows the evolution of the thickness strain in the central area of the specimen having a width of 12 mm. It is worth noting that, in the case of the Erichsen deep drawing test, the initiation of necking does not necessarily occur in the central area for all specimens. This feature, which depends on the specimen width and the contact between the punch and the specimen, is consistent with the experimental observations (see, e.g., [41]).

The second numerical criterion is based on the analysis of the thickness strain acceleration, which is obtained by computing the second time derivative of thickness strain in the central region of the specimen (see, e.g., [11,40,42]). According to this criterion, the critical minor and major in-plane principal strains at the occurrence of necking are obtained when the second time derivative of the thickness strain, i.e., thickness strain acceleration, reaches a maximum. This is illustrated in Figure 12 for the uniaxial tensile test corresponding to the specimen having a width of 12 mm. Note that the occurrence of localized necking may also be predicted using the first time derivative of thickness strain, which represents the thickness strain rate. However, several works in the literature have shown that the numerical criterion based on the maximum of strain acceleration is more appropriate for the prediction of localized necking than the one based on the maximum of strain rate, as the latter rather indicates the onset of fracture (see, e.g., [8,9]).

A third numerical criterion is analyzed in this work based on a critical damage threshold at which is associated the occurrence of necking. Unlike the numerical criteria described above, the same critical damage value is used here for all simulations using the various specimen geometries. This critical damage value was identified by Aboutalebi et al. [36] using the Vickers micro-hardness test. Using simple sheet stretching tests and Erichsen deep drawing tests, the simulations are performed until the critical damage value of 0.434 is reached at some finite element of the discretized model (see an illustration in Figure 13, in the case of uniaxial tensile test for the specimen with 12 mm width). At this instant, the simulations are stopped and the minor and major in-plane principal strains of the corresponding finite element are plotted into the FLD.

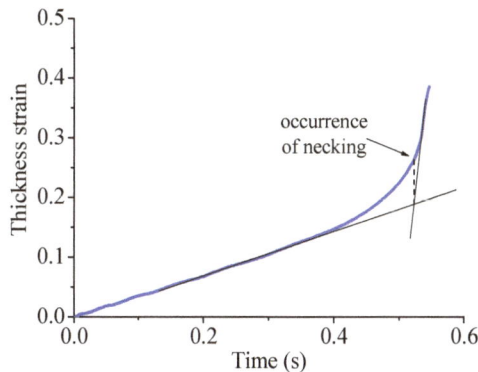

Figure 11. Illustration of the prediction of the occurrence of necking when the thickness strain evolution is taken as indicator.

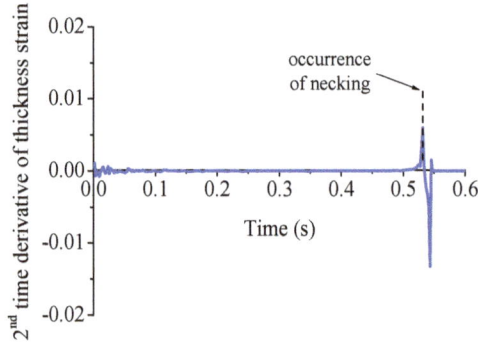

Figure 12. Illustration of the prediction of the occurrence of necking when the second time derivative of thickness strain is taken as indicator.

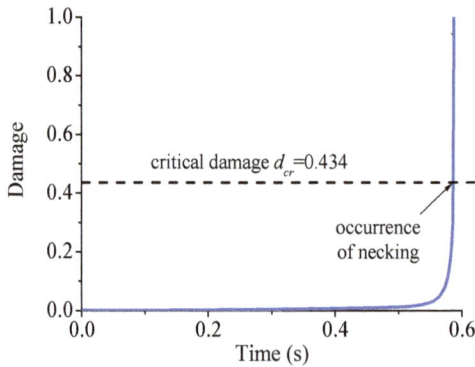

Figure 13. Illustration of the prediction of the occurrence of necking when the critical damage threshold is taken as indicator.

7. Application to the Determination of FLDs

The present numerical methodology, based on the three above-described numerical criteria, is applied in this section to both the simple sheet stretching and the Erichsen simulations with various specimen geometries in order to obtain complete FLDs for the studied material.

Once the simulations of the simple sheet stretching test and the Erichsen test performed for the various specimen geometries, the corresponding numerical load–displacement curves, which are obtained using the previously identified hardening and damage parameters, are plotted in Figure 14. It can be observed that, for both the simple sheet stretching test and the Erichsen test, the fully coupled model reproduces satisfactorily the peak in the load–displacement curves, which is followed by the sudden load drop caused by the damage acceleration at the latest stages of loading.

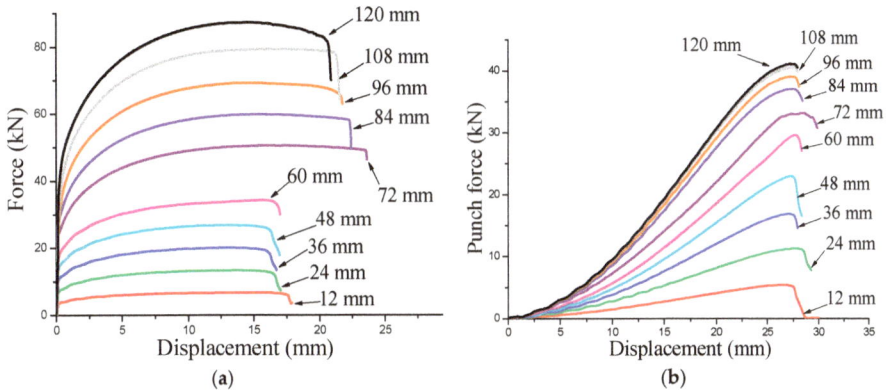

Figure 14. Numerical load–displacement curves for (**a**) the simple sheet stretching test and (**b**) the Erichsen deep drawing test with the various specimen widths.

Moreover, Figures 15 and 16 show the distribution of the damage variable, at different stages of loading, as determined by the numerical simulation of the simple sheet stretching test on the specimen having a width of 12 mm and the Erichsen test on the specimen having a width of 24 mm, respectively. More specifically, for the simple sheet stretching test (i.e., Figure 15), the damage distribution in the central part of the specimen remains uniform until the applied loading reaches its maximum (see Figure 14a). Beyond this limit, the damage distribution becomes heterogeneous, and concentrates gradually in the middle of the specimen in the form of two localization bands (Figure 15e). Finally, the accumulated damage in the narrow bands leads to highly localized necking, which ultimately results in a macrocrack (Figure 15f).

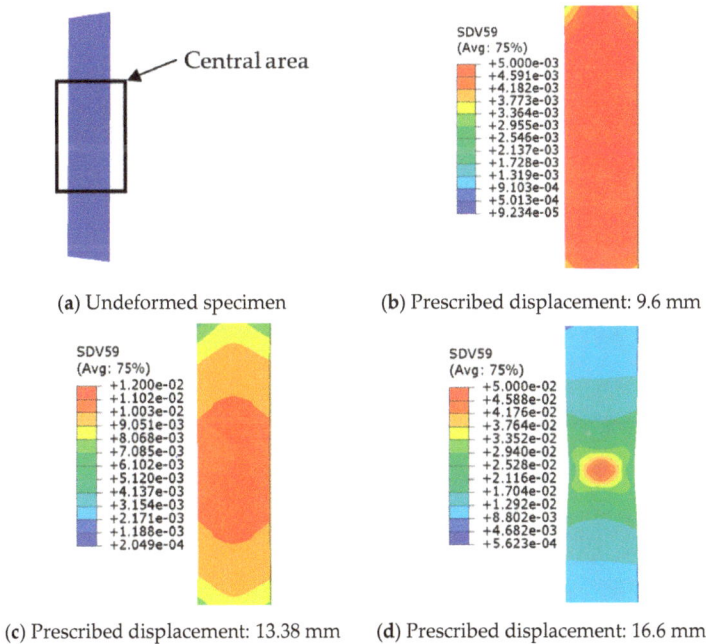

(**a**) Undeformed specimen

(**b**) Prescribed displacement: 9.6 mm

(**c**) Prescribed displacement: 13.38 mm

(**d**) Prescribed displacement: 16.6 mm

Figure 15. *Cont.*

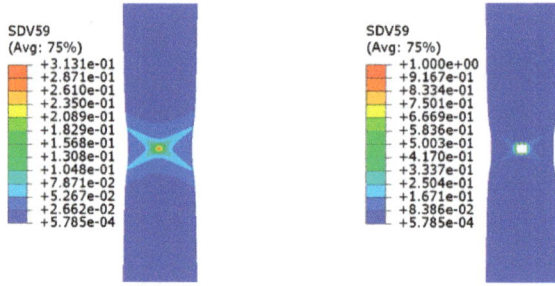

(e) Prescribed displacement: 17.35 mm   (f) Prescribed displacement: 17.53 mm

Figure 15. Distribution of the damage variable, at different stages of loading, as obtained by FE simulation of the simple sheet stretching test with the specimen having a width of 12 mm.

For the Erichsen deep drawing test, the damage distribution is localized around the dome apex of the specimen (i.e., higher specimen point) in the early stages of the deep drawing process, with low damage levels (i.e., $d < 0.001$, see Figure 16a,b). As the punch moves down, a strong localization of damage distribution is observed far from the dome apex, which is due to the frictional contact between the punch and the specimen (see Figure 16d,e). Finally, the damage accumulation leads to localized necking followed by fracture, as shown in Figure 16f.

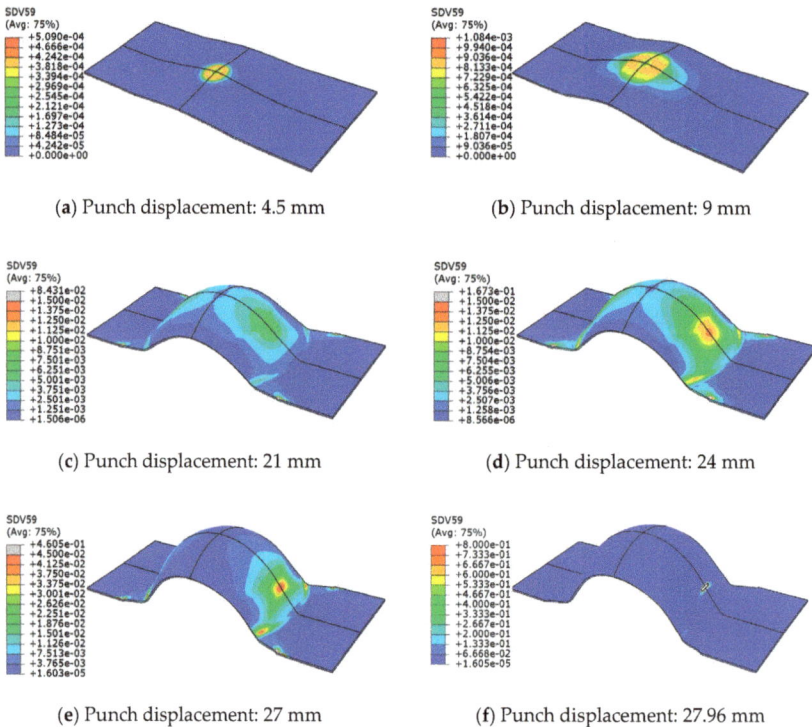

(a) Punch displacement: 4.5 mm

(b) Punch displacement: 9 mm

(c) Punch displacement: 21 mm

(d) Punch displacement: 24 mm

(e) Punch displacement: 27 mm

(f) Punch displacement: 27.96 mm

Figure 16. Distribution of the damage variable, at different stages of loading, as obtained by FE simulation of the Erichsen deep drawing test with the specimen having a width of 24 mm.

The strain paths generated by the simulations of the simple sheet stretching test and the Erichsen test for all specimen widths are reported in Figure 17. Similar trends for the strain-path evolution are observed between both tests. More specifically, the strain paths are almost linear for the specimens having widths ranging from 12 mm to 60 mm (i.e., corresponding to a negative minor strain), while they are clearly nonlinear for the specimens having widths ranging from 72 mm to 120 mm (i.e., corresponding to a positive minor strain). In the latter case, the strain paths remain linear until a sudden transition to some plane-strain state, which indicates the onset of localized necking (see, e.g., [40,43]).

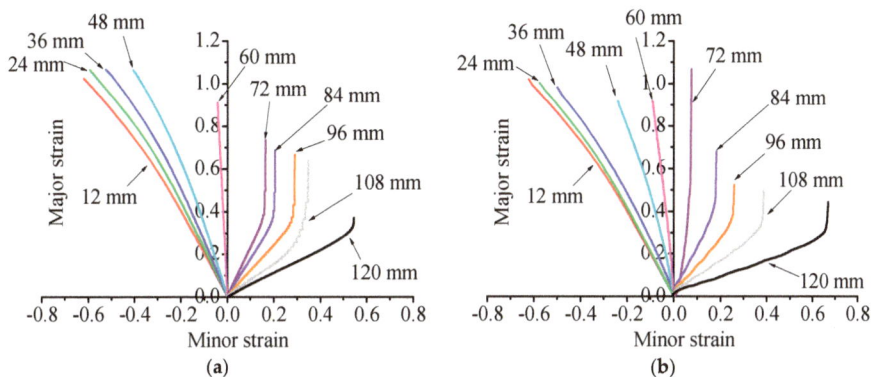

Figure 17. Numerical strain paths predicted by FE simulation of (**a**) the simple sheet stretching test and (**b**) the Erichsen deep drawing test with the various specimen widths.

Figure 18 compares the predicted FLDs for the studied material, based on the simulations of both simple sheet stretching and Erichsen tests, along with the experimental results provided by Aboutalebi et al. [30]. The FLDs based on the analyses of thickness strain, second time derivative of thickness strain, and critical damage threshold are presented in Figure 18a–c, respectively.

Note that the experimental results given by Aboutalebi et al. [30] are obtained using the same Erichsen deep drawing test used in the simulations, which allows consistent comparison between the predicted FLDs and the experimental data. It can be noticed, in general, that the numerical predictions of the FLDs obtained by the Erichsen deep drawing tests are closer to the experimental results than those predicted by simple sheet stretching tests. The fact that the trends obtained by simulation of Erichsen deep drawing tests are more consistent with the experimental results may be explained by the similarity in the mechanical setups used in both cases (i.e., Erichsen deep drawing test), while no forming tools are considered in the simple sheet stretching tests. More specifically, in the left-hand side of the FLDs, the results predicted from Erichsen deep drawing tests using the numerical criteria based on the thickness strain evolution and the second time derivative of thickness strain are in reasonably good agreement with the experimental results, while the predictions overestimate the occurrence of necking when they are based on the critical damage threshold criterion.

In the right-hand side of the FLDs (i.e., positive minor strains), the results predicted from the Erichsen deep drawing tests using the numerical criterion based on the second time derivative of thickness strain show the best agreement with respect to experiments (see Figure 18b). Note that the FLDs predicted from both Erichsen deep drawing tests and simple sheet stretching tests on the basis of the critical damage threshold criterion are overestimated for all strain paths, which proves that this numerical criterion is not suitable for the prediction of the occurrence of necking. Moreover, based on this critical damage threshold criterion, the shapes of the FLDs correspond rather to fracture limit diagrams, which are classically determined in the literature using numerical fracture criteria (see, e.g., [44]). The unsuitability of the latter numerical criterion may be explained by its consideration of a constant critical damage value for all strain paths, which is a strong assumption, since the damage value at the occurrence of necking depends on the stress triaxiality ratio and, therefore, on the loading path (see, e.g., [45]).

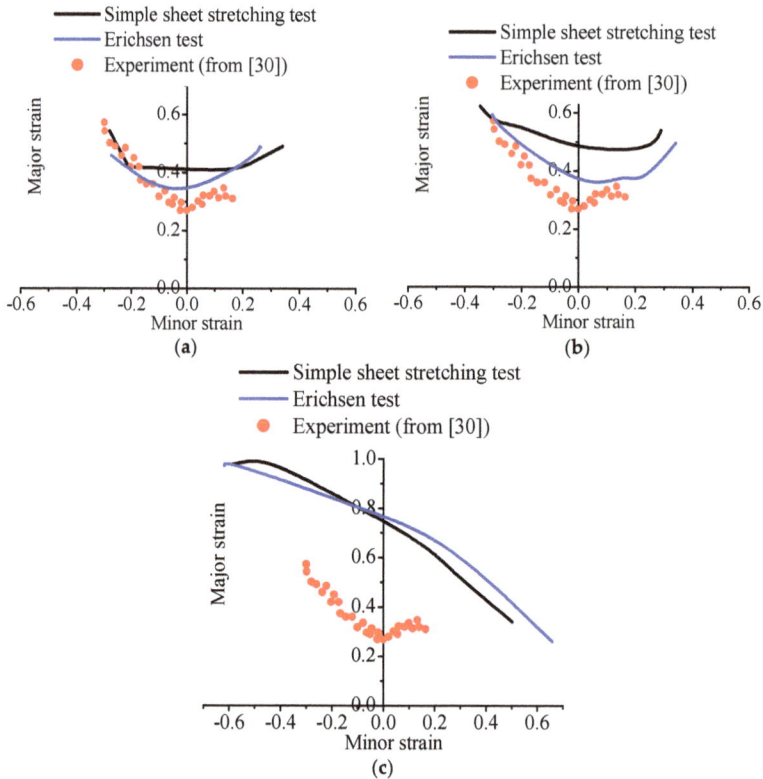

Figure 18. Forming limit diagram predictions based on the analyses of (**a**) thickness strain evolution; (**b**) second time derivative of thickness strain; and (**c**) critical damage threshold.

8. Conclusions

In this paper, an elastic–plastic model has been coupled with the Lemaitre ductile damage approach in order to predict the occurrence of necking in sheet metal forming. The whole set of coupled constitutive equations has been implemented into the finite element code ABAQUS/Explicit in the framework of large strains and a fully three-dimensional formulation. Three numerical necking criteria have been considered for predicting the occurrence of necking in sheet metals. They are based on the analyses of the local thickness strain evolution and its second time derivative during the FE simulations, as well as on a fixed critical damage threshold at which is associated the occurrence of necking. For the FE simulations, simple sheet stretching tests as well as Erichsen deep drawing tests on various specimen geometries, covering all possible strain paths, were used in conjunction with the numerical necking criteria. The numerical results in terms of FLDs were compared to the experimental results taken from [30]. Good agreement between the predicted FLD and the experiments was observed using the Erichsen deep drawing test combined with the numerical criterion based on the second time derivative of thickness strain. Due to the low cost and computational efficiency of the numerical alternative for FLD prediction, as compared to the lengthy and expensive experimental

Metals **2017**, *7*, 455

procedures, the proposed numerical approach can be easily used with different forming setups and a large variety of materials for the prediction of the occurrence of necking in sheet metals.

Author Contributions: Hocine Chalal conceived and performed the simulations. Hocine Chalal and Farid Abed-Meraim analyzed and discussed the results. Both authors contributed to the writing of the manuscript.

Conflicts of Interest: The authors declare no conflict of interest.

References

1. Keeler, S.; Backofen, W.A. Plastic instability and fracture in sheets stretched over rigid punches. *ASM Trans. Q.* **1963**, *56*, 25–48.
2. Goodwin, G. Application of strain analysis to sheet metal forming problems in the press shop. *SAE Tech. Pap.* **1968**. [CrossRef]
3. Swift, H.W. Plastic instability under plane stress. *J. Mech. Phys. Solids* **1952**, *1*, 1–18. [CrossRef]
4. Considère, A. Mémoire sur l'emploi du fer et de l'acier dans les constructions. *Annals Ponts Chaussées* **1885**, *9*, 574.
5. Hill, R. On discontinuous plastic states, with special reference to localized necking in thin sheets. *J. Mech. Phys. Solids* **1952**, *1*, 19–30. [CrossRef]
6. Marciniak, Z.; Kuczyński, K. Limit strains in the processes of stretch-forming sheet metal. *Int. J. Mech. Sci.* **1967**, *9*, 609–620. [CrossRef]
7. Brun, R.; Chambard, A.; Lai, M.; De Luca, P. Actual and virtual testing techniques for a numerical definition of materials. In Proceedings of the 4th International Conference and Workshop NUMISHEET '99, Besançon, France, 13–17 September 1999; pp. 393–398.
8. Petek, A.; Pepelnjak, T.; Kuzman, K. An improved method for determining forming limit diagram in the digital environment. *J. Mech. Eng.* **2005**, *51*, 330–345.
9. Situ, Q.; Jain, M.K.; Bruhis, M. A suitable criterion for precise determination of incipient necking in sheet materials. *Mater. Sci. Forum* **2006**, *519–521*, 111–116. [CrossRef]
10. Situ, Q.; Jain, M.K.; Bruhis, M. Further experimental verification of a proposed localized necking criterion. In Proceedings of the 9th International Conference on Numerical Methods in Industrial Forming Processes, Porto, Portugal, 17–21 June 2007; pp. 907–912.
11. Situ, Q.; Jain, M.K.; Bruhis, M.; Metzger, D.R. Determination of forming limit diagrams of sheet materials with a hybrid experimental-numerical approach. *Int. J. Mech. Sci.* **2011**, *53*, 707–719. [CrossRef]
12. Clift, S.E.; Hartley, P.; Sturgess, C.E.N.; Rowe, G.W. Fracture prediction in plastic deformation processes. *Int. J. Mech. Sci.* **1990**, *32*, 1–17. [CrossRef]
13. Wierzbicki, T.; Bao, Y.; Lee, Y.-W.; Bai, Y. Calibration and evaluation of seven fracture models. *Int. J. Mech. Sci.* **2005**, *47*, 719–743. [CrossRef]
14. Bao, Y.; Wierzbicki, T. On fracture locus in the equivalent strain and stress triaxiality space. *Int. J. Mech. Sci.* **2004**, *46*, 81–98. [CrossRef]
15. Han, H.N.; Kim, K.-H. A ductile fracture criterion in sheet metal forming process. *J. Mater. Process. Technol.* **2003**, *142*, 231–238. [CrossRef]
16. Jain, M.; Allin, J.; Lloyd, D.J. Fracture limit prediction using ductile fracture criteria for forming of an automotive aluminum sheet. *Int. J. Mech. Sci.* **1999**, *41*, 1273–1288. [CrossRef]
17. Gurson, A.L. Continuum theory of ductile rupture by void nucleation and growth: Part I—Yield criteria and flow rules for porous ductile media. *J. Eng. Mater. Technol.* **1977**, *99*, 2–15. [CrossRef]
18. Tvergaard, V.; Needleman, A. Analysis of the cup-cone fracture in a round tensile bar. *Acta Metall. Mater.* **1984**, *32*, 157–169. [CrossRef]
19. Rousselier, G. Ductile fracture models and their potential in local approach of fracture. *Nucl. Eng. Des.* **1987**, *105*, 97–111. [CrossRef]
20. Pardoen, T.; Doghri, I.; Delannay, F. Experimental and numerical comparison of void growth models and void coalescence criteria for the prediction of ductile fracture in copper bars. *Acta Mater.* **1998**, *46*, 541–552. [CrossRef]
21. Benzerga, A.A.; Besson, J. Plastic potentials for anisotropic porous solids. *Eur. J. Mech. A/Solids* **2001**, *20*, 397–434. [CrossRef]

22. Monchiet, V.; Cazacu, O.; Charkaluk, E.; Kondo, D. Macroscopic yield criteria for plastic anisotropic materials containing spheroidal voids. *Int. J. Plast.* **2008**, *24*, 1158–1189. [CrossRef]
23. Lemaitre, J. A continuous damage mechanics model for ductile fracture. *J. Eng. Mater. Technol. ASME* **1985**, *107*, 83–89. [CrossRef]
24. Chaboche, J.L. Thermodynamically founded CDM models for creep and other conditions. In *Creep and Damage in Materials and Structures*; CISM Courses and Lectures No 399, International Centre for Mechanical Sciences; Springer: Vienna, Austria, 1999; Volume 399, pp. 209–283.
25. Lemaitre, J. *A Course on Damage Mechanics*; Springer: Berlin, Germany, 1992.
26. Chow, C.L.; Lu, T.J. On evolution laws of anisotropic damage. *Eng. Fract. Mech.* **1989**, *34*, 679–701. [CrossRef]
27. Chaboche, J.L. Development of continuum damage mechanics for elastic solids sustaining anisotropic and unilateral damage. *Int. J. Damage Mech.* **1993**, *2*, 311–329. [CrossRef]
28. Abu Al-Rub, R.K.; Voyiadjis, G.Z. On the coupling of anisotropic damage and plasticity models for ductile materials. *Int. J. Solids Struct.* **2003**, *40*, 2611–2643. [CrossRef]
29. Takuda, H.; Yoshii, T.; Hatta, N. Finite-element analysis of the formability of a magnesium-based alloy AZ31 sheet. *J. Mater. Process. Technol.* **1999**, *89–90*, 135–140. [CrossRef]
30. Aboutalebi, F.H.; Farzin, M.; Mashayekhi, M. Numerical prediction and experimental validations of ductile damage evolution in sheet metal forming processes. *Acta Mech. Solida Sin.* **2012**, *25*, 638–650. [CrossRef]
31. Hill, R. A theory of the yielding and plastic flow of anisotropic metals. *Proc. R. Soc. A* **1948**, *193*, 281–297. [CrossRef]
32. Lemaitre, J.; Desmorat, R.; Sauzay, M. Anisotropic damage law of evolution. *Eur. J. Mech. A Solids* **2000**, *9*, 187–208. [CrossRef]
33. Li, Y.-F.; Nemat-Nasser, S. An explicit integration scheme for finite deformation plasticity in finite-element methods. *Finite Elem. Anal. Des.* **1993**, *15*, 93–102. [CrossRef]
34. Kojic, M. Stress integration procedures for inelastic material models within finite element method. *ASME Appl. Mech. Rev.* **2002**, *55*, 389–414. [CrossRef]
35. Doghri, I.; Billardon, R. Investigation of localization due to damage in elasto-plastic materials. *Mech. Mater.* **1995**, *19*, 129–149. [CrossRef]
36. Aboutalebi, F.H.; Farzin, M.; Poursina, M. Numerical simulation and experimental validation of a ductile damage model for DIN 1623 St14 steel. *Int. J. Adv. Manuf. Technol.* **2011**, *53*, 157–165. [CrossRef]
37. Aboutalebi, F.H.; Banihashemi, A. Numerical estimation and practical validation of Hooputra's ductile damage parameters. *Int. J. Adv. Manuf. Technol.* **2014**, *75*, 1701–1710. [CrossRef]
38. Besson, J.; Steglich, D.; Brocks, W. Modeling of crack growth in round bars and plane strain specimens. *Int. J. Solids Struct.* **2001**, *38*, 8259–8284. [CrossRef]
39. Peerlings, R.H.G.; Geers, M.G.D.; de Borst, R.; Brekelmans, W.A.M. A critical comparison of nonlocal and gradient-enhanced softening continua. *Int. J. Solids Struct.* **2001**, *38*, 7723–7746. [CrossRef]
40. Zhang, C.; Leotoing, L.; Zhao, G.; Guines, D.; Ragneau, E. A comparative study of different necking criteria for numerical and experimental prediction of FLCs. *J. Mater. Eng. Perform.* **2011**, *20*, 1036–1042. [CrossRef]
41. Kami, A.; Dariani, B.M.; Vanini, A.S.; Comsa, D.S.; Banabic, D. Numerical determination of the forming limit curves of anisotropic sheet metals using GTN damage model. *J. Mater. Process. Technol.* **2015**, *216*, 472–483. [CrossRef]
42. Volk, W.; Hora, P. New algorithm for a robust user-independent evaluation of beginning instability for the experimental FLC determination. *Int. J. Mater. Form.* **2011**, *4*, 339–346. [CrossRef]
43. Liewald, M.; Drotleff, K. Novel punch design for nonlinear strain path generation and evaluation methods. *Key Eng. Mater.* **2015**, *639*, 317–324. [CrossRef]
44. Silva, M.B.; Martínez-Donaireb, A.J.; Centenob, G.; Morales-Palmab, D.; Vallellanob, C.; Martinsa, P.A.F. Recent approaches for the determination of forming limits by necking and fracture in sheet metal forming. *Procedia Eng.* **2015**, *132*, 342–349. [CrossRef]
45. Soyarslan, C.; Richter, H.; Bargmann, S. Variants of Lemaitre's damage model and their use in formability prediction of metallic materials. *Mech. Mater.* **2016**, *92*, 58–79. [CrossRef]

metals

MDPI

Article

On the Use of Maximum Force Criteria to Predict Localised Necking in Metal Sheets under Stretch-Bending

Domingo Morales-Palma *, Andrés J. Martínez-Donaire and Carpóforo Vallellano

Department of Mechanical Engineering and Manufacturing, University of Seville,
Camino de los Descubrimientos s/n, 41092 Seville, Spain; ajmd@us.es (A.J.M.-D.); carpofor@us.es (C.V.)
* Correspondence: dmpalma@us.es; Tel.: +34-954-481355

Received: 10 October 2017; Accepted: 30 October 2017; Published: 2 November 2017

Abstract: The maximum force criteria and their derivatives, the Swift and Hill criteria, have been extensively used in the past to study sheet formability. Many extensions or modifications of these criteria have been proposed to improve necking predictions under only stretching conditions. This work analyses the maximum force principle under stretch-bending conditions and develops two different approaches to predict necking. The first is a generalisation of classical maximum force criteria to stretch-bending processes. The second approach is an extension of a previous work of the authors based on critical distance concepts, suggesting that necking of the sheet is controlled by the damage of a critical material volume located at the inner side of the sheet. An analytical deformation model is proposed to characterise the stretch-bending process under plane-strain conditions. Different parameters are considered, such as the thickness reduction, the gradient of variables through the sheet thickness, the thickness stress and the anisotropy of the material. The proposed necking models have been successfully applied to predict the failure in different materials, such as steel, brass and aluminium.

Keywords: sheet-metal forming; stretch-bending; necking; maximum force criterion; bending effect

1. Introduction

The maximum force principle has been extensively used in the past to study sheet formability. Considère's maximum force criterion (MFC) states that diffuse necking is initiated in a tensile test of a bar when the maximum force is reached. The classical models of Swift [1] and Hill [2] are extensions of the MFC for the determination of necking in metal sheets subjected to different stretching conditions in the sheet plane. The former is applicable to the prediction of diffuse necking in the entire domain of the forming limit diagram (FLD). The latter is coupled with the initiation of strain localisation along a narrow band and is limited to predict localised necking in the left side of the FLD. Later, Hora et al. [3] extended the MFC to predict strain localisation for both sides of the FLD by including the contribution of the minor principal strain.

The simplicity of necking criteria based on the maximum force principle make them very attractive. However, their predictions do not always agree with experimental results. Many extensions and alternatives to the MFC have been proposed to date, to deal with predicting localised necking. For instance, Bressan and Williams [4] suggested that failure occurs when the shear stress reaches a maximum. Hora et al. [5] proposed a phenomenological criterion, which states that once the maximum force is reached, the loading path gradually evolves towards plane strain and then localisation occurs. Brunet and Morestin [6] included the effects of damage to refine the material model for the prediction of the necking curve in the FLD. Recently, Aretz [7] assumed that localised necking does not necessarily occur when the axial force reaches a maximum, but rather when it reaches a critical value. This method aimed to scale the Hill's model so that the major strain predicted by the M–K model [8] or its experimental counterpart at plane-strain conditions (usually referred to as FLD_0) was matched.

The above criteria, which are collectively referred to as maximum force criteria (MFCs), were developed to predict necking in nearly stretching operations, neglecting implicitly the strain and stress gradients through the sheet thickness. However, in sheet metal forming processes such as stretch-bending or stamping with punches of mild or severe radii, it is well known that the occurrence of bending has a "beneficial effect" on the initiation of necking. In practice, the average strain through the sheet thickness (or the strain at the middle of the sheet surface) has been used to characterise failure criteria. However, the predictions of this approach, sometimes called the mid-plane rule (MPR), are in general very conservative.

Research carried out to analyse the sheet failure in simultaneous stretching and bending conditions has followed a different approach from the maximum force principle. Tharrett and Stoughton [9] studied the bending effect on sheet failure and proposed a necking criterion referred to as the concave-side rule (CSR). This criterion establishes that the failure is initiated when the strain at the inner surface (concave side) of the sheet reaches a critical value at the bending zone, which matches the limit strain of in-plane stretching. This criterion notably improved the predictions of sheet failure in some materials compared to the MPR. In a later study, authors recognised that stresses are less sensitive to strain path changes, and consequently they reformulated the CSR in terms of stresses [10].

In a previous work, the present authors proposed a natural improvement of the CSR [11]. The model assumes that necking should be controlled by the development of damage in a material volume located at the inner side of the sheet. The study concludes that the critical size of this volume (represented by a certain critical distance from the sheet surface) should be related to the material microstructure.

This work analyses the maximum force principle under stretch-bending conditions, proposing an analytical deformation model to characterise the stretch-bending process under plane-strain conditions. Different generalisations of the MFC for stretch-bending conditions are presented and discussed. As a result, the study suggests that the effect of the existence of strain and stress gradients on sheet thickness may be assessed by applying the MFCs locally at a given material's critical distance from the inner surface. The proposed necking models are successfully applied to describe the failure by necking in sheets under stretch-bending conditions in different materials, such as 1008 AK steel, 70/30 brass, 6010 aluminium alloy and 7075-O aluminium alloy.

2. Maximum Force Criteria

Considère's analysis of the uniaxial tension of a metal bar states that diffuse necking appears when the maximum force is applied, that is, $dF_1 = 0$, with $F_1 = \sigma_1 l_2 l_3$. Swift extended this model to biaxial loading of a metal sheet of thickness $l_3 = t$, represented schematically in Figure 1, assuming a simultaneous maximum of both components of the force $dF_1 = 0$, $dF_2 = 0$, with $F_1 = \sigma_1 l_2 t$ and $F_2 = \sigma_2 l_1 t$ [1].

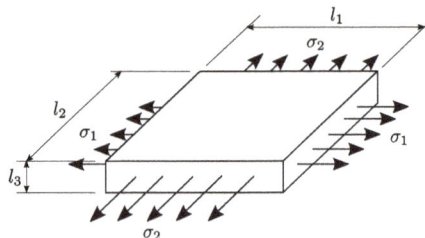

Figure 1. Schema of a stretched element.

Hill also formulated a criterion to predict localised necking in metal sheets [2]. A simplified formulation of this model states that the localisation is initiated when the load per unit of width reaches its maximum: $df_1 = 0$, with $f_1 = \sigma_1 t$. The application of the Hill criterion is restricted to the left-hand side of the FLD, that is, uniaxial-stretching.

To predict localised necking in both sides of the FLD, Hora extended Considère's model to biaxial loading of a metal sheet and considered that the major principal stress (σ_1) is a function of both major (ε_1) and minor (ε_2) principal strains in the plane of the sheet [3].

The above models of plastic instability lead to the formulation of the MFCs in a single and unified expression:

$$\frac{1}{\sigma_Y}\frac{d\sigma_Y}{d\varepsilon_{eq}} = \frac{1}{Z} \tag{1}$$

where the left-hand side is a material property, sometimes referred to as the *non-dimensional strain-hardening* characteristic, whereas Z is the critical value for the subtangent of the stress–strain curve [12]. The former is evaluated using a stress–strain constitutive equation and the latter, by a yield function under plane-stress conditions, which can be characterised by the following parameters:

$$\alpha = \frac{\sigma_2}{\sigma_1}, \quad \beta = \frac{d\varepsilon_2}{d\varepsilon_1}, \quad \varphi = \frac{\sigma_{eq}}{\sigma_1}, \quad \rho = \frac{d\varepsilon_{eq}}{d\varepsilon_1} \tag{2}$$

The expressions for the above parameters are given in Appendix A for both Hosford and Mises yield criteria.

It is not intended in this article to review the formulation of these well-known models. There are several works in the literature that analyse these and provide specific expressions of Equation (1) (e.g., [13–15]). Most of these assume a Hollomon law ($\sigma_Y = K\varepsilon_{eq}^n$) to evaluate the aforementioned material characteristic and a Mises yield function to evaluate $1/Z$. For instance, the Hill necking criterion [2] under proportional load transforms Equation (1) to

$$\frac{1}{\sigma_Y}\frac{d\sigma_Y}{d\varepsilon_{eq}} = \frac{n}{\varepsilon_{eq}} \quad \text{and} \quad \frac{1}{Z} = \frac{1+\beta}{\rho} \quad \rightarrow \quad \frac{n}{\varepsilon_{eq}} = \frac{1+\beta}{\rho} \tag{3}$$

which leads to algebraic expressions to evaluate the principal strains:

$$\varepsilon_1 = \frac{n}{1+\beta}, \quad \varepsilon_2 = \frac{\beta n}{1+\beta} \quad \text{and} \quad \varepsilon_1 + \varepsilon_2 = n \tag{4}$$

Figure 2 presents an illustrative example of the application of the Swift [1], Hill [2] and Hora [3] criteria to predict necking in AA7075-O metals sheets. Both sides of Equation (1) have been evaluated by assuming a Hollomon law and a Mises yield criterion, respectively. Critical values of $1/Z$ are represented in Figure 2a as functions of the strain path β under proportional loading conditions. Figure 2b presents necking predictions in the FLD along with the experimental data of localised necking reported by Martínez-Donaire et al. [16] for AA7075-O sheets of 1.6 mm thickness. The coefficient of the Hollomon law was found to be $n = 0.21$. As can be observed, both Hill and Hora criteria for localised necking reproduce the experimental data in their respective ranges of application reasonably well.

Figure 2 also presents the failure predictions of a modified Hill criterion based on Aretz's approach [7]. Aretz assumed that necking is initiated when the axial force reaches a critical value, rather than its maximum value. The fundamental idea of this approach lies in the calibration of the critical load, which is obtained by scaling the prediction of Hill's model (4) to fit the major strain at necking under plane-strain conditions (FLD_0). Accordingly, Equation (1) is turned into

$$\frac{n}{\varepsilon_{eq}} = \frac{1+\beta}{\rho}\frac{n}{FLD_0} \tag{5}$$

which leads to

$$\varepsilon_1 = \frac{FLD_0}{1+\beta}, \qquad \varepsilon_2 = \frac{\beta\, FLD_0}{1+\beta} \qquad \text{and} \qquad \varepsilon_1 + \varepsilon_2 = FLD_0 \tag{6}$$

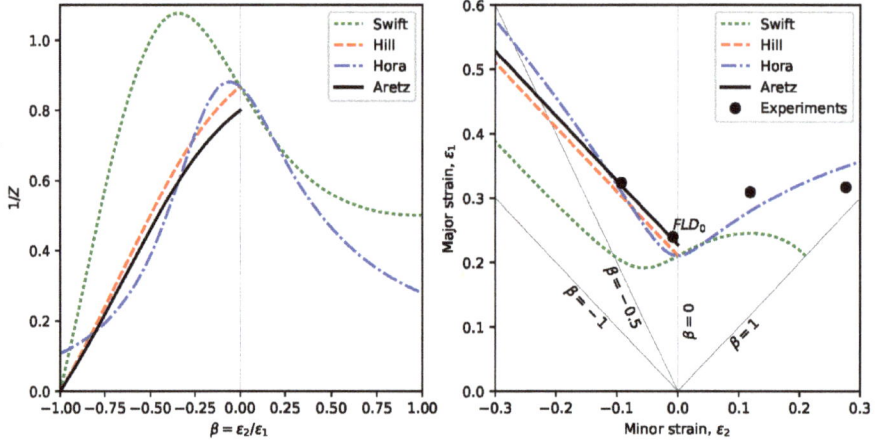

Figure 2. Necking predictions for AA7075-O sheets: (**a**) critical values of $1/Z$, and (**b**) forming limit diagram (FLD) with necking curves and experimental data [16].

3. Stretch-Bending Deformation Model

Under stretch-bending conditions, strain and stress gradients through the sheet thickness are induced in the metal sheet. In this situation, the MFCs must to be reformulated to take into account these gradients.

The following deformation model is an extension of a stretch-bending model developed in a previous work [11]. This model assumed that the sheet adapts to the punch geometry in an earlier stage mainly by bending followed by a dominant stretching process until the sheet failure is reached. By simplicity, the evolution of the sheet curvature at the earlier stage is neglected, assuming this to be fixed and equal to the punch curvature. The model assumes that the deformation process is controlled by the reduction of sheet thickness.

In order to be able to integrate analytically the radial equilibrium equation, which relates principal stresses through the sheet thickness, only a plane-strain condition ($d\varepsilon_2 = 0$) is assumed hereafter. The effect on sheet failure of the through-thickness stress is also taken into account.

Figure 3 shows the model variables in a sheet element located at the dome of the punch that is subjected to stretch-bending. The undeformed sheet dimensions are l_0 and the initial thickness is t_0. In the deformed configuration, t is the current thickness, and the punch radius R matches the radius of curvature of the inner surface.

In this configuration, θ is the bent angle, r is the radius of curvature of a given layer, $z = r - (R + t/2)$ is the position of the layer measured from the middle surface, and r_u is the radius of curvature of the unstretched surface. We note that all layers on the thickness are stretched ($r_u < R$) when stretching dominates over bending. The current length of a generic layer, which initially had an undeformed length of $l_0 = r_u\,\theta$, is $l = r\,\theta$.

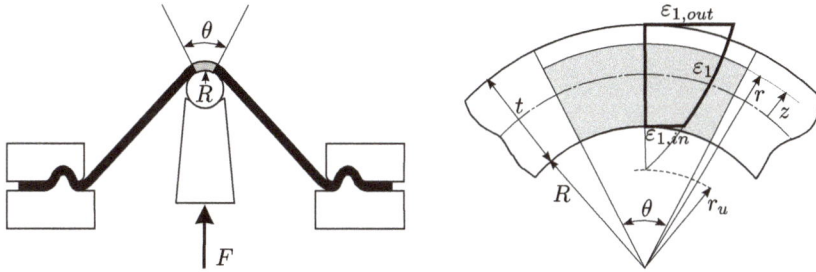

Figure 3. Variables in a deformed sheet element.

The principal true strain distributions through the sheet thickness can be written as

$$\varepsilon_1 = \ln \frac{r}{r_u}, \qquad \varepsilon_2 = 0, \qquad \varepsilon_3 = -\varepsilon_1 \tag{7}$$

where r_u can be obtained from the material incompressibility condition of the sheet element:

$$l_0 \cdot t_0 = (R + t/2)\,\theta \cdot t \tag{8}$$

leading to

$$r_u = (R + t/2) \cdot t/t_0 \tag{9}$$

Substituting Equation (9) into Equation (7), the major strain can be expressed as a function of the variables t and r, and the parameters t_0 and R, as follows:

$$\varepsilon_1 = \ln \frac{r}{R + t/2} - \ln \frac{t}{t_0} \tag{10}$$

We note that the major strain at the outer and inner sides of the sheet ($\varepsilon_{1,out}$ and $\varepsilon_{1,in}$, as shown in Figure 3) are given by Equation (10) by substituting $r = R + t$ and $r = R$, respectively.

The distribution of principal stresses in the sheet thickness are obtained from the radial equilibrium equation [17]:

$$r\frac{d\sigma_3}{dr} = \sigma_1 - \sigma_3 \tag{11}$$

Given that $d\varepsilon_1 = dr/r$ from Equation (7), it is convenient to change the variable r for ε_1 in Equation (11) and write the radial equilibrium equation as

$$\frac{d\sigma_3}{d\varepsilon_1} = \sigma_1 - \sigma_3 \tag{12}$$

The anisotropic non-quadratic yield criterion proposed by Hosford is assumed. This provides simplified expressions for the equivalent stress and strain under plane-strain conditions (see Appendix B):

$$\sigma_{eq} = \frac{\sigma_1 - \sigma_3}{C} \tag{13}$$

$$\varepsilon_{eq} = C\varepsilon_1 \tag{14}$$

where C is a function of the anisotropic parameters of the material. For instance, if the longitudinal or circumferential direction (axis 1) is aligned with the rolling direction of the sheet, C is given by

$$C = \left(\frac{(1+r_0)\left(1+r_{90}^{\frac{1}{a-1}}\right)^{a-1}}{r_0 + \left(1+r_{90}^{\frac{1}{a-1}}\right)^{a-1}} \right)^{\frac{1}{a}}$$

(15)

A material behaviour following the Hollomon power law is assumed:

$$\sigma_Y = K \varepsilon_{eq}^n$$

(16)

Thus, setting $\sigma_{eq} = \sigma_Y$ and substituting Equations (13), (14) and (16) into Equation (12), the radial equilibrium condition becomes

$$\frac{d\sigma_3}{d\varepsilon_1} = K' \varepsilon_1^n$$

(17)

where $K' = K \cdot C^{n+1}$.

Integrating the differential Equation (17), being $\sigma_{3,out} = 0$ at the outer surface of the sheet, the distribution of σ_3 through the sheet thickness is obtained as

$$\sigma_3 = -K' \frac{\varepsilon_{1,out}^{n+1} - \varepsilon_1^{n+1}}{n+1}$$

(18)

The circumferential stress gradient through the sheet thickness can be now determined from Equations (13), (14), (16) and (18) as

$$\sigma_1 = K' \left(\varepsilon_1^n - \frac{\varepsilon_{1,out}^{n+1} - \varepsilon_1^{n+1}}{n+1} \right)$$

(19)

and its derivative is given by

$$\frac{d\sigma_1}{d\varepsilon_1} = K' \left(n \varepsilon_1^{n-1} + \varepsilon_1^n \right)$$

(20)

On the other hand, the yield stress is found from Equations (12)–(14) and (17) as

$$\sigma_Y = \frac{1}{C} \frac{d\sigma_3}{d\varepsilon_1} = \frac{K'}{C} \varepsilon_1^n$$

(21)

Differentiating Equations (13) and (14) and combining with Equations (17) and (20), one obtains

$$\frac{d\sigma_Y}{d\varepsilon_{eq}} = \frac{1}{C^2} \left(\frac{d\sigma_1}{d\varepsilon_1} - \frac{d\sigma_3}{d\varepsilon_1} \right) = \frac{K'}{C^2} n \varepsilon_1^{n-1}$$

(22)

Finally, dividing the last two expressions, the non-dimensional strain-hardening function in stretch-bending for a certain layer on the thickness is given by

$$\frac{1}{\sigma_Y} \frac{d\sigma_Y}{d\varepsilon_{eq}} = \frac{n}{C \varepsilon_1}$$

(23)

We note that this expression is equivalent to Equation (3) for in-plane stretching.

4. Maximum Force Criteria in Stretch-Bending

In this section, MFCs are generalised to stretch-bending conditions using the proposed deformation model. The analysis focuses on reformulating the Hill and Aretz necking criteria for metal sheets presented above.

The axial force (per unit of width) can be calculated as:

$$f_1 = \int_R^{R+t} \sigma_1 \, dr = \overline{\sigma}_1 \cdot t \tag{24}$$

where $\overline{\sigma}_1$ is the average major stress on thickness. Using the average strains $\overline{\varepsilon}_1$ and $\overline{\varepsilon}_3 = -\overline{\varepsilon}_1$ to characterise the sheet deformation under plane-strain conditions, the derivative of the axial force is given by

$$\frac{df_1}{dt} = \overline{\sigma}_1 + t\frac{d\overline{\sigma}_1}{dt} = \overline{\sigma}_1 + \frac{d\overline{\sigma}_1}{d\overline{\varepsilon}_3} = \overline{\sigma}_1 - \frac{d\overline{\sigma}_1}{d\overline{\varepsilon}_1} \tag{25}$$

Thus, the Hill necking criterion ($df_1 = 0$) yields

$$\frac{1}{\overline{\sigma}_1}\frac{d\overline{\sigma}_1}{d\overline{\varepsilon}_1} = 1 \tag{26}$$

which in the actual formulation becomes

$$\frac{1}{\overline{\sigma}_Y}\frac{d\overline{\sigma}_Y}{d\overline{\varepsilon}_{eq}} = \frac{1}{C} \tag{27}$$

As can be seen, the right-hand side of Equation (27) is equal to the Hill criterion for in-plane stretching under plane-strain conditions, given by setting $\beta = 0$ in Equation (3). We note that $C = \rho$ in this situation. However, the left-hand side needs to be averaged over the sheet thickness.

Similarly, the modification of the Hill criterion proposed by Aretz can be expressed as

$$\frac{1}{\overline{\sigma}_Y}\frac{d\overline{\sigma}_Y}{d\overline{\varepsilon}_{eq}} = \frac{n}{C \cdot FLD_0} \tag{28}$$

by introducing the correction factor n/FLD_0.

Hereafter, the above necking models will be referred to as Hill-based MFC-SB (Equation (27)) and Aretz-based MFC-SB (Equation (28)), where SB stands for stretch-bending. Both criteria are numerically evaluated using the deformation model described in Section 3.

Computational Implementation

In order to provide a self-contained document, some aspects of the computational implementation are summarised in this section.

It is useful to use dimensionless parameters and variables in the deformation model. Thus, those previously defined in Figure 3 become

$$\tau = \frac{t}{t_0}, \qquad \zeta = \frac{z}{t}, \qquad \kappa = \frac{R+t/2}{t} = \frac{\tau}{t_0/R} + \frac{1}{2} \tag{29}$$

where τ is the thickness reduction, ζ is the non-dimensional position of a layer measured from the middle surface, κ is the relative curvature of the middle surface of the sheet, and t_0/R is the bending ratio. Using these variables, the major strain given by Equation (10) is expressed as

$$\varepsilon_1 = \frac{1+\kappa\zeta}{\tau} \tag{30}$$

The Riemann integral is used to evaluate the average value of variables σ_Y and ε_{eq} through the sheet thickness, that is,

$$X = \{\sigma_Y, \varepsilon_{eq}\}, \qquad \overline{X} = \int_{-1/2}^{1/2} X \, d\zeta = \frac{1}{N}\sum_{j=0}^{N} X, \qquad \text{with } \zeta = -\frac{1}{2} + \frac{j}{N} \tag{31}$$

where the function X is given by Equations (14) and (21), respectively.

As previously stated, the deformation model neglects the bending process that occurs in an earlier stage of the stretch-bending operation and assumes that thickness reduction τ is the only variable. Thus, the iterative calculation procedure consists of finding τ until a necking criterion is reached, for a given value of the bending ratio t_0/R. To avoid negative values of variables, the initial step assumes that major strain is positive through the whole thickness. Because of the strain gradient, this condition is always satisfied when $\varepsilon_{1,in} \geq 0$. In this limit, that is, $\varepsilon_{1,in} = 0$, this leads to a thickness reduction:

$$\tau_0 = \frac{R}{t_0}\sqrt{\frac{R}{2\,t_0} - 1} \tag{32}$$

At the end of every thickness decrement ($\Delta\tau < 0$), the non-dimensional strain-hardening function is calculated as

$$\lambda^{(i+1)} = \frac{1}{\overline{\sigma}_Y^{(i+1)}}\frac{\Delta\overline{\sigma}_Y^{(i+1)}}{\Delta\overline{\varepsilon}_{eq}^{(i+1)}} = \frac{1}{\overline{\sigma}_Y^{(i+1)}}\frac{\overline{\sigma}_Y^{(i+1)} - \overline{\sigma}_Y^{(i)}}{\overline{\varepsilon}_{eq}^{(i+1)} - \overline{\varepsilon}_{eq}^{(i)}}, \qquad i = \{0,1,2...\} \tag{33}$$

where $\overline{\varepsilon}_{eq}^{(i+1)}$ and $\overline{\sigma}_Y^{(i+1)}$ are determined by Equation (31). Accordingly, the iterative procedure is established as follows:

1. Compute $\overline{\varepsilon}_{eq}^{(0)}$ and $\overline{\sigma}_Y^{(0)}$ at initial stage, where $\tau^{(0)} = \tau_0$.
2. Compute the non-dimensional strain-hardening function $\lambda^{(i+1)}$ for $\tau^{(i+1)} = \tau^{(i)} + \Delta\tau$.
3. If $\lambda^{(i+1)} \leq 1/Z$, then necking is attained; else repeat step 2.

Depending on the selected Hill- or Aretz-based MFC-SB criterion, Z is C or $C \cdot FLD_0/n$, respectively. Typical values for discrete parameters are $N = 100$ and $\Delta\tau = -0.001$.

5. Critical-Distance Rule for Necking

In a recent research work [11,18,19], the present authors developed a mesoscopic approach to predict failure in stretch-bent sheets. The proposed necking model combines the concepts of CSR and critical distance to predict the failure of the sheet. The model assumes that necking is controlled by the development of damage in a certain material volume located at the inner side of the sheet. The size of the critical volume is assumed to be a material constant, which can be related with the microstructure of the material.

According to the above idea, the less-stressed material at the inner zone is responsible for containing the plastic instability of the entire sheet thickness. In previous works, the sheet failure was characterised in terms of principal strains [19] or principal stresses, which was more appropriate to analyse non-proportional strain paths [11]. In both cases, the experimental data were successfully analysed. Following the MFC concepts, in the present work, the use of the non-dimensional strain-hardening function (left-hand side of Equation (1)) is proposed to assess necking initiation under stretch-bending conditions.

Figure 4 shows schematically the evolution of the strain-hardening function through the sheet thickness at the onset of necking. To rationalise the present model, the sheet can be assumed to be formed by the superposition of layers or fibres in the thickness, all having the mechanical behaviour of the base material. For a given layer located at a distance d measured from the inner side of the sheet, a *local stability index* can be defined as

$$lsi(d) = \frac{\dfrac{1}{\sigma_Y}\dfrac{d\sigma_Y}{d\varepsilon_{eq}}\bigg|_d}{1/Z} \tag{34}$$

where $1/Z$ depends on the failure criteria; for example, in the case of using the Aretz approach, $1/Z$ is $n/(C \cdot FLD_0)$.

Thus, layers exhibiting *lsi* less than unity are assumed to be layers that are not able to resist the plastic instability of the sheet thickness. We note, from Figure 4, that this condition is first reached at the upper side and propagates downwards in thickness during the forming process.

Otherwise, fibres having a *lsi* value greater than unity are considered to be fibres that contain the plastic instability of the entire thickness. As can be seen, this material extends from the bottom layer to a depth d_{cr}, called here the critical distance, at which *lsi* takes the unit value. This material volume prevents the sheet from necking.

According to the above description, the failure by necking of the sheet will occur when, at a certain critical distance d_{cr} from the inner side of the sheet, the local stability index becomes equal to the unit value. As is discussed in the next section, the value of d_{cr} is influenced only slightly by the bending ratio in stretch-bending; thus it can be considered in practice a material property to be determined experimentally.

This criterion is called here the critical-distance rule (CDR) by analogy to those previously proposed in the literature, such as the MPR, CSR and convex-side rule (CxSR).

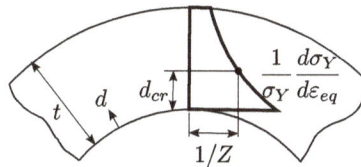

Figure 4. Critical-distance rule.

6. Practical Application and Discussion

This section evaluates the capability of the failure criteria described above to predict localised necking under stretch-bending conditions. The experimental results analysed were found from the literature and have already been used in a previous work [11].

Briefly, the experiments involved stretch-bending tests under plane-strain conditions using cylindrical punches of different radii. The materials were 1008 AK steel, 70/30 brass, and 6010 aluminium, from experimental work conducted by Tharrett and Stoughton [9,20], and 7075-O aluminium, from the research carried out by Martínez-Donaire et al. [16]. All specimens failed in the zone in contact with the punch, under simultaneous bending and streching conditions. The material properties are reported in Table 1.

Table 1. Mechanical properties and material constants.

	1008 AK Steel	70/30 Brass	AA6010	AA7075-O
t_0 (mm)	1.04	0.81	0.89	1.6
σ_0 (MPa)	187.0	112.5	202.0	102.3
K (MPa)	556.8	809.1	543.9	400.3
n	0.24	0.50	0.24	0.25
r_0	1.740	0.870	1.590	0.812
r_{90}	1.800	0.730	1.760	1.317
a	6	8	8	8
FLD_0	0.358	0.358	0.166	0.251

Figure 5 presents the experimental results provided in the mentioned references. Figure 5a–c depicts the major strains measured on the convex side of the sheet $\varepsilon_{1,out}$ as a function of the *current* bending ratio t/R. Instead, Figure 5d shows $\varepsilon_{1,out}$ versus the *initial* bending ratio t_0/R. The cases for which a visible neck was observed are represented as solid circles. Otherwise, open circles are used. The values of FLD_0 (see Table 1) are represented as solid stars.

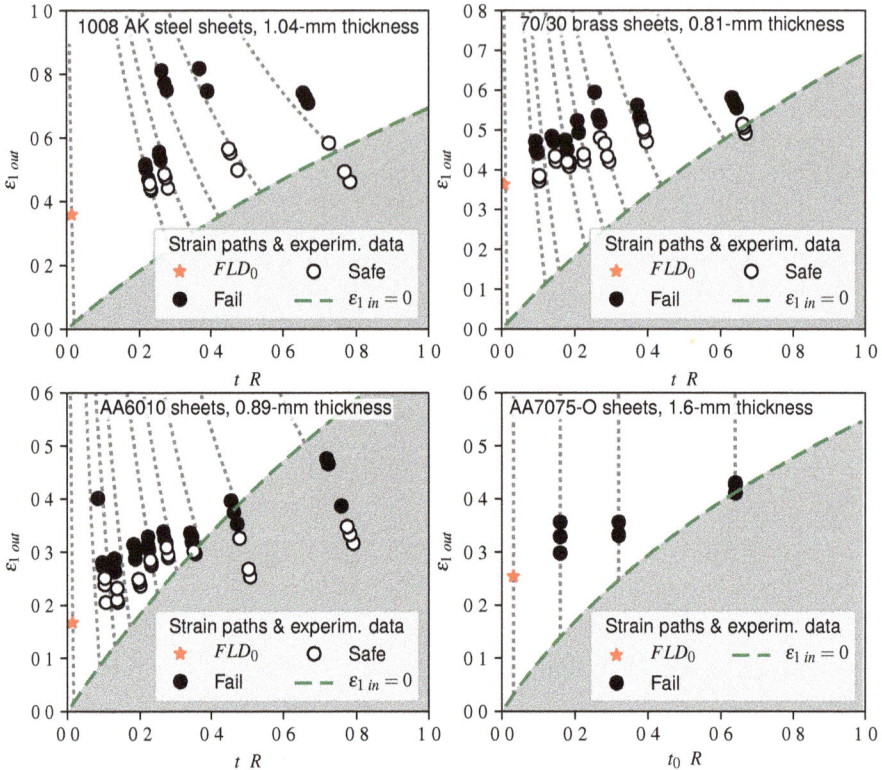

Figure 5. Influence of punch radius R on the formability of stretch-bent 1008 AK steel, 70/30 brass, AA6010 and AA7075-O sheets.

Figure 5 also presents the strain paths predicted by the proposed stretch-bending deformation model for the different punch radii R. As can be observed, these agree reasonably well with the experimental results. The major differences are obtained by the largest values of t/R, and they may be attributed to a potential indentation of the forming tool into the sheet thickness. In this situation, the transverse shear stress cannot be neglected, and the plane-section assumption in the deformed sheet element is no longer valid.

The shaded areas in Figure 5 and subsequent figures represent stretch-bending conditions for which the material in the inner side of the sheet was shortened, that is, $\varepsilon_{1,in} \leq 0$. This situation inhibits the onset of the plastic instability, giving way to the eventual development of failure by ductile fracture at the outer side of the sheet. It should be noted that for 70/30 brass and 6010 aluminium sheets, the authors observed the initiation of cracks before necking for the largest values of t/R. As can be observed, this observation agreed very well with the model predictions.

The value of the strains at the outer sheet surface predicted by Hill- and Aretz-based MFC-SB criteria at the onset of necking are represented in Figure 6. As can be seen, the Hill-based MFC-SB criterion given by Equation (27) underestimated the experimental results for 1008 AK steel sheets, whereas it overestimated these for 70/30 brass and 6010 aluminium sheets. Instead, the predictions of the Aretz-based MFC-SB (28) agreed very well with the experimental results for all materials analysed, except in general for the smallest punch radii.

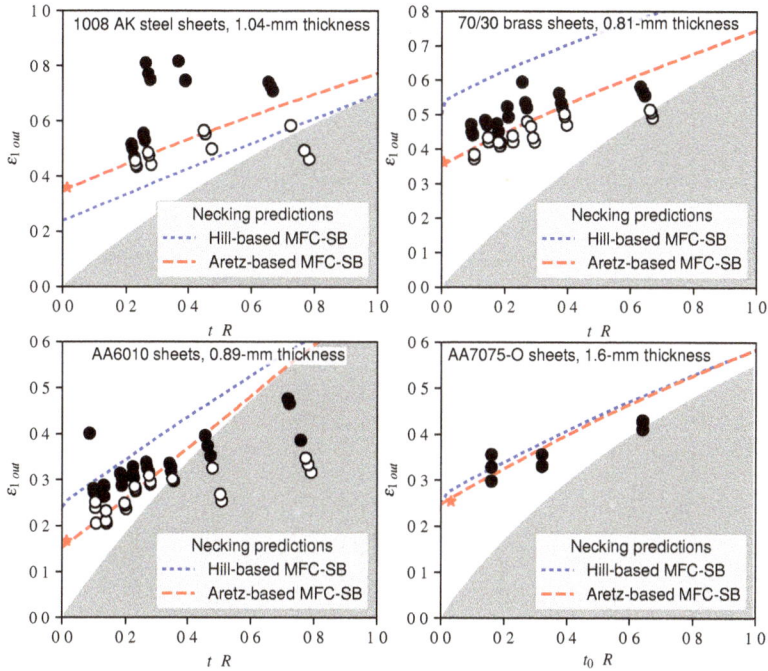

Figure 6. Necking predictions of the proposed maximum force criterion in stretch-bending (MFC-SB) for stretch-bent 1008 AK steel, 70/30 brass, AA6010 and AA7075-O sheets.

It should be noted that both Hill- and Aretz-based MFC-SB criteria showed the same trend with the bending ratio. The basic difference was that the Aretz-based approach had been calibrated to meet the experimental FLD_0 value.

To assess the proposed CDR criterion, the critical distance d_{cr} on the sheet thickness was estimated for the different materials. Figure 7a reproduces graphically the calculation procedure to obtain d_{cr} for 1008 AK steel sheets. Figure 7a (left) shows the gradient through the sheet thickness of the local stability index based on the Aretz correction.

As can be seen, the distance d at which the *lsi* value becomes equal to 1 was smaller in the failed specimen that in the successful specimen. This clearly indicates that the material volume resisting the necking of the sheet was smaller in the former than in the latter. Figure 7a (right) depicts the non-dimensional value d/t for a local stability index equal to 1 versus the bending ratio t_0/R for all tested specimens of 1008 AK steel. The almost horizontal line dividing successful tests from failures determines the value of the critical distance, which here is $d_{cr}/t_0 \approx 0.3$ for the steel sheets. We note that the slight slope of the line is due to the material thickening as the sheet curvature increased.

In practice, it is enough to choose a few significant experimental data in the range of intermediate or high values of t_0/R to determine the critical distance. The following values were obtained for the different materials: $d_{cr}/t_0 \approx 0.3$ for 1008 AK steel and AA6010, $d_{cr}/t_0 \approx 0.4$ for 70/30 brass and $d_{cr}/t_0 \approx 0.5$ for AA7075-O sheets.

Figure 7b shows the necking predictions of the CDR criterion along with the Aretz-based MFC-SB discussed previously. In general, the CDR criterion improved the predictions over the whole range of t_0/R values. As can be seen, for low values of t_0/R, the slight improvement led to excellent predictions of experimental data of steel, brass and AA6010 sheets. However, for higher values of t_0/R, the enhancements of predictions was clearly more pronounced, particularly in aluminium sheets.

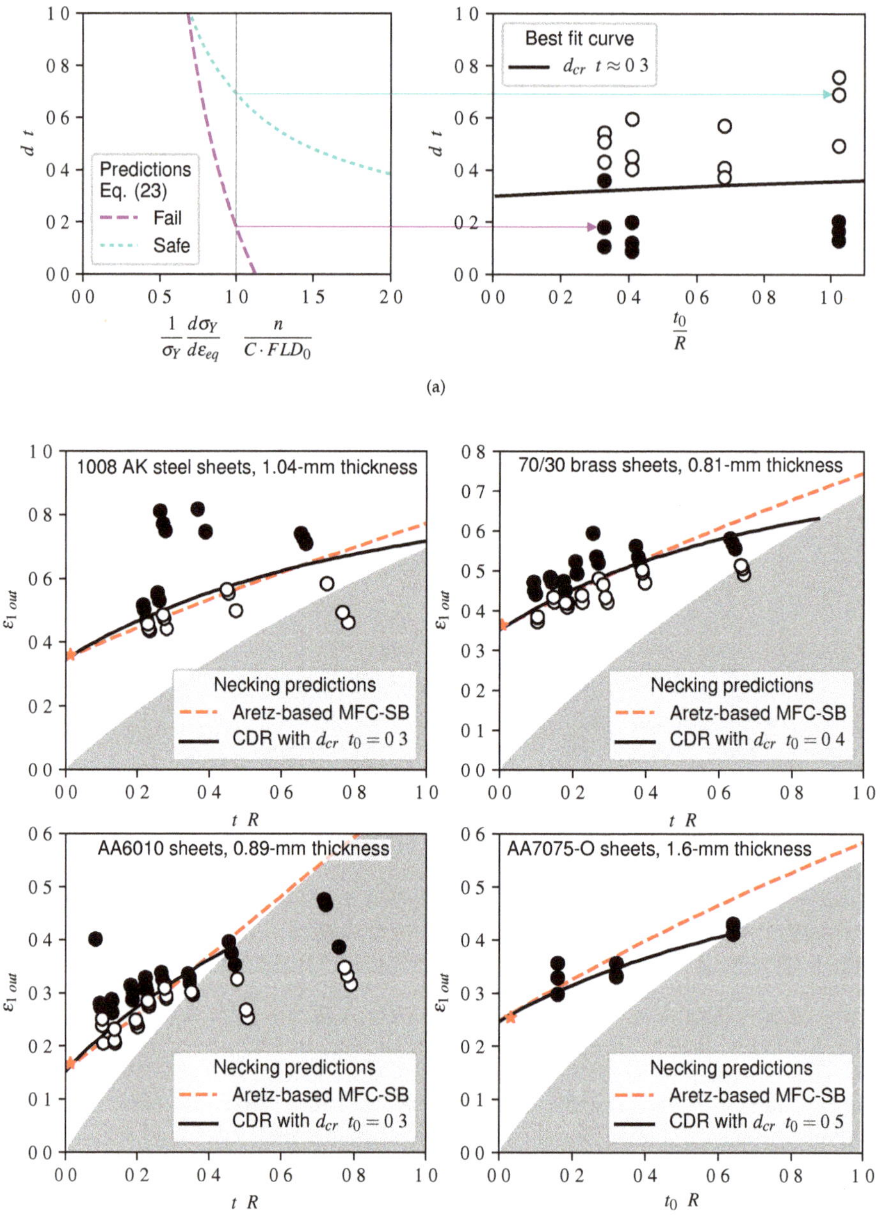

Figure 7. Practical application of the proposed critical-distance rule (CDR) criterion. (**a**) Graphical representation for the calibration of the CDR (Equation (34)) for 1008 AK steel sheets: (left) predictions of the non-dimensional strain-hardening characteristic through the sheet thickness by Equation (23), illustrated for the experimental data of a failed and a successful specimen; (right) graphical determination of the critical distance as an almost horizontal line that separates failed from successful tests. (**b**) Necking predictions of proposed Aretz-based maximum force criterion in stretch bending (MFC-SB; Equation (28)) and CDR criterion (Equation (34)) for stretch-bent 1008 AK steel, 70/30 brass, AA6010 and AA7075-O sheets.

7. Conclusions

The maximum force principle has been analysed to predict necking under stretch-bending conditions. Two kinds of failure criteria have been proposed, the first based on the generalisation of traditional MFCs to stretch-bending and the other based on the concept of damage in a critical material volume. The following conclusions can be drawn from this study:

- The strain-hardening function $(d\sigma_Y/d\varepsilon)/\sigma_Y$ used to propose both types of failure criteria seems to control necking in both stretching and stretch-bending processes.
- The good results obtained by the proposed criteria are largely due to the calibration of the failure models from the experimental FLD in the absence of bending. Although the Aretz proposal for the modification of the Hill necking criterion under plane-strain conditions has been used in this work, the procedure can be generalised to the whole range of strain conditions in the FLD.
- The necking predictions of the proposed Aretz-based MFC-SB agree reasonably well with the experimental data. The lack of precision for high bending ratios seems to be related to the predictions of the deformation model rather than to the failure model itself.
- The necking predictions of the proposed CDR criterion fit well with the experimental data and improve those of the previous criterion over the whole range of t_0/R values. To characterise the failure, a local stability index and a critical distance, which depends on the material, have been proposed. For the material analysed, the critical distance values range from 0.3 to 0.5. Although more exhaustive research is required to relate the critical distance to the material properties, this criterion can be easily implemented in the finite-element method.

Acknowledgments: The authors wish to thank the Spanish Government for its financial support throughout research project DPI2015-64047-R.

Author Contributions: Domingo Morales-Palma developed the analytical models, performed the simulations, analysed the results and wrote the paper; Andrés J. Martínez-Donaire provided support and contributed to the discussions; Carpóforo Vallellano provided support, analysed the results, contributed to the discussions and reviewed the paper.

Conflicts of Interest: The authors declare no conflict of interest.

Abbreviations

The following abbreviations are used in this manuscript:

MFC	Maximum force criterion (Considère's criterion)
FLD	Forming limit diagram
MFCs	Maximum force criteria (Considère's criterion and related extensions and modifications: Swift, Hill, Hora, etc.)
MPR	Mid-plane rule
CSR	Concave-side rule
MFC-SB	Maximum force criterion generalised to stretch-bending processes
CDR	Critical-distance rule
lsi	Local stability index
CxSR	Convex-side rule

Appendix A. Hosford Yield Criterion under Plane-Stress Condition

Assuming that the directions of principal stress coincide with the symmetry axis, the non-quadratic yield criterion proposed by Hosford for anisotropic materials [21] can be written as

$$r_0(\sigma_2 - \sigma_3)^a + r_{90}(\sigma_3 - \sigma_1)^a + r_0\,r_{90}(\sigma_1 - \sigma_2)^a = r_{90}(1 + r_0)\sigma_{eq}^a \tag{A1}$$

where r_0 and r_{90} are the Lankford coefficients along the rolling (0°) and transverse (90°) directions, respectively. This criterion reduces to the Hill quadratic yield criterion by setting $a = 2$. For isotropic materials ($r_0 = r_{90} = 1$), it reduces to the Mises yield criterion by setting $a = 2$ or $a = 4$, and to the

Tresca criterion by setting $a = 1$ or $a = \infty$. The suggested values for the exponent are $a = 6$ for BCC metals and $a = 8$ for FCC materials. The ratios of the plastic strain increments are found from the flow rule as

$$
\begin{aligned}
d\varepsilon_1 : d\varepsilon_2 : d\varepsilon_3 = &- r_{90}(\sigma_3 - \sigma_1)^{a-1} + r_0 \, r_{90}(\sigma_1 - \sigma_2)^{a-1} \\
&: r_0(\sigma_2 - \sigma_3)^{a-1} - r_0 \, r_{90}(\sigma_1 - \sigma_2)^{a-1} \\
&: -r_0(\sigma_2 - \sigma_3)^{a-1} + r_{90}(\sigma_3 - \sigma_1)^{a-1}
\end{aligned}
\tag{A2}
$$

The equivalent strain increment can be found from the plastic work as

$$
d\varepsilon_{eq} = \frac{\sigma_1 \, d\varepsilon_1 + \sigma_2 \, d\varepsilon_2 + \sigma_3 \, d\varepsilon_3}{\sigma_{eq}}
\tag{A3}
$$

Under plane-stress conditions through the sheet thickness ($\sigma_3 = 0$), it is usual to express the stress and strain increments by using the parameters $\alpha = \sigma_2/\sigma_1$, $\beta = d\varepsilon_2/d\varepsilon_1$, $\varphi = \sigma_{eq}/\sigma_1$, and $\rho = d\varepsilon_{eq}/d\varepsilon_1$. Thus, the equivalent stress and strain increments are given by the following [22]:

$$
\sigma_{eq} = \varphi \, \sigma_1 = \left(\frac{r_{90} + r_0 \, \alpha^a + r_0 \, r_{90}(1 - \alpha)^a}{r_{90}(1 + r_0)} \right)^{\frac{1}{a}} \sigma_1
\tag{A4}
$$

$$
d\varepsilon_{eq} = \rho \, d\varepsilon_1 = \frac{1 + \alpha \beta}{\varphi} d\varepsilon_1
\tag{A5}
$$

From the flow rule,

$$
1 : \beta : -(1 + \beta) = 1 + r_0(1 - \alpha)^{a-1} : \frac{r_0}{r_{90}} \alpha^{a-1} - r_0(1 - \alpha)^{a-1} : -\frac{r_0}{r_{90}} \alpha^{a-1} - 1
\tag{A6}
$$

the following relation between β and α is established:

$$
\beta = \frac{r_0 \, \alpha^{a-1} - r_0 \, r_{90}(1 - \alpha)^{a-1}}{r_{90} + r_0 \, r_{90}(1 - \alpha)^{a-1}}
\tag{A7}
$$

In the case of plane-strain conditions ($\beta = 0$), the α and φ parameters are simplified to

$$
\alpha = \frac{r_{90}^{\frac{1}{a-1}}}{1 + r_{90}^{\frac{1}{a-1}}}
\tag{A8}
$$

$$
\varphi = \frac{1}{\rho} = \left(\frac{r_0 + \left(1 + r_{90}^{\frac{1}{a-1}}\right)^{a-1}}{(1 + r_0)\left(1 + r_{90}^{\frac{1}{a-1}}\right)^{a-1}} \right)^{\frac{1}{a}}
\tag{A9}
$$

As a check, particularising the above expression for Mises plasticity in plane stress, that is, by setting $r_0 = r_{90} = 1$ and $a = 2$, the parameters α, β, φ and ρ yield

$$
\alpha = \frac{2\beta + 1}{\beta + 2}, \quad \beta = \frac{2\alpha - 1}{2 - \alpha}, \quad \varphi = \sqrt{1 - \alpha + \alpha^2}, \quad \rho = \frac{2}{\sqrt{3}}\sqrt{1 + \beta + \beta^2}
\tag{A10}
$$

Appendix B. Hosford Yield Criterion under Plane-Strain Condition

This following formulation is used to analyse the stretch-bending deformation model proposed in this paper. We consider a stretch-bent sheet in the rolling direction under plane-strain conditions ($d\varepsilon_2 = 0$) as represented in Figure A1a. From the flow rule given in Equation (A2), the stress in direction 2 is found as

$$\sigma_2 = \frac{r_{90}^{\frac{1}{a-1}}\sigma_1 + \sigma_3}{r_{90}^{\frac{1}{a-1}} + 1} \tag{A11}$$

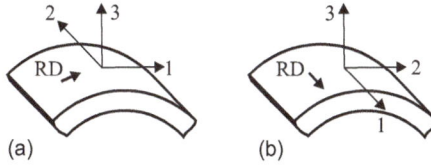

Figure A1. Stretch-bent metal sheet under plane-strain deformation: (a) $d\varepsilon_2 = 0$; (b) $d\varepsilon_1 = 0$.

Substituting σ_2 into Equation (A1), the equivalent stress reduces to

$$\sigma_{eq} = \frac{\sigma_1 - \sigma_3}{C} \tag{A12}$$

where C is found from the anisotropy parameters as

$$C = \left(\frac{(1+r_0)\left(1 + r_{90}^{\frac{1}{a-1}}\right)^{a-1}}{r_0 + \left(1 + r_{90}^{\frac{1}{a-1}}\right)^{a-1}} \right)^{\frac{1}{a}} \tag{A13}$$

We note that the above expression matches $\rho = 1$ in Equation (A9) for plane-stress conditions. Setting $d\varepsilon_2 = 0$ in Equation (A3), the equivalent strain increment is expressed as

$$d\varepsilon_{eq} = \frac{(\sigma_1 - \sigma_3)\,d\varepsilon_1}{\sigma_{eq}} = C\,d\varepsilon_1 \tag{A14}$$

Similarly, we consider a sheet stretch-bent in the transverse direction under plane-strain conditions ($d\varepsilon_1 = 0$), as represented in Figure A1b. Following the same procedure as before, the corresponding Equations (A12) and (A13) are now given by

$$\frac{\sigma_2 - \sigma_3}{\sigma_{eq}} = \left(\frac{\left(1 + r_0^{\frac{1}{a-1}}\right)^{a-1}(r_{90} + r_{90}/r_0)}{\left(1 + r_0^{\frac{1}{a-1}}\right)^{a-1} + r_{90}} \right)^{\frac{1}{a}} \tag{A15}$$

As a check, by particularising Equations (A13) and (A15) for Hill plasticity in plane stress, that is, by setting $a = 2$, one obtains the following [23]:

$$\frac{\sigma_1 - \sigma_3}{\sigma_{eq}} = \sqrt{\frac{(1+r_0)(1+r_{90})}{1+r_0+r_{90}}} \tag{A16}$$

$$\frac{\sigma_2 - \sigma_3}{\sigma_{eq}} = \sqrt{\frac{(1+r_0)(r_{90}+r_{90}/r_0)}{1+r_0+r_{90}}} \tag{A17}$$

References

1. Swift, H.W. Plastic instability under plane strain. *J. Mech. Phys. Solids* **1952**, *1*, 1–18.
2. Hill, R. On discontinuous plastic states, with special reference to localized necking in thin sheets. *J. Mech. Phys. Solids* **1952**, *1*, 19–30.
3. Hora, P.; Tong, L.; Reissner, J. A prediction method for ductile sheet metal failure using FE-simulation. In Proceedings of the NUMISHEET, Dearborn, MI, USA, 29 September–3 October 1996; pp. 252–256.
4. Bressan, J.D.; Williams, J.A. The use of a shear instability criterion to predict local necking in sheet metal deformation. *Int. J. Mech. Sci.* **1983**, *25*, 155–168.
5. Hora, P.; Tong, L.; Berisha, B. Modified maximum force criterion, a model for the theoretical prediction of forming limit curves. *Int. J. Mater. Form.* **2013**, *6*, 267–279.
6. Brunet, M.; Morestin, F. Experimental and analytical necking studies of anisotropic sheet metals. *J. Mater. Process. Technol.* **2001**, *112*, 214–226.
7. Aretz, H. An extension of Hill's localized necking model. *Int. J. Eng. Sci.* **2010**, *56*, 609–618.
8. Marciniak, Z.; Kuczyński, K. Limit strains in the processes of stretch-forming sheet metal. *Int. J. Mech. Sci.* **1967**, *9*, 609–620.
9. Tharrett, M.R.; Stoughton, T.B. *Stretch-Bend Forming Limits of 1008 AK Steel*; SAE Paper 2003-01-1157; Society of Automotive Engineers: Troy, MI, USA, 2003.
10. Stoughton, T.B.; Yoon, J.W. A new approach for failure criterion for sheet metals. *Int. J. Plast.* **2011**, *27*, 440–459.
11. Morales-Palma, D.; Vallellano, C.; García-Lomas, F.J. Assessment of the effect of the through-thickness strain/stress gradient on the formability of stretch-bend metal sheets. *Mater. Des.* **2013**, *50*, 798–809.
12. Marciniak, Z.; Duncan, J.L.; Hu, S.J. *Mechanics of Sheet Metal Forming*, 2nd ed.; Butterworth-Heinemann: Oxford, UK, 2002.
13. Aretz, H. Numerical analysis of diffuse and localized necking in orthotropic sheet metals. *Int. J. Plast.* **2007**, *23*, 798–840.
14. Abed-Meraim, F.; Balan, T.; Altmeyer, G. Investigation and comparative analysis of plastic instability criteria: Application to forming limit diagrams. *Int. J. Adv. Manuf. Technol.* **2014**, *71*, 1247–1262.
15. Stoughton, T.B.; Zhu, X. Review of theoretical models of the strain-based FLD and their relevance to the stress-based FLD. *Int. J. Plast.* **2004**, *40*, 1463–1486.
16. Martínez-Donaire, A.J.; Vallellano, C.; Morales, D.; García-Lomas, F.J. Experimental and numerical analysis of the failure of AA7075-O stretch-bend sheets. *Steel Res. Int.* **2012**, 251–254.
17. Hill, R. *The Mathematical Theory of Plasticity*; Clarendon Press: Oxford, UK, 1950.
18. Morales, D.; Martínez-Donaire, A.J.; Vallellano, C.; García-Lomas, F.J. Bending effect in the failure of stretch-bend metal sheets. *Int. J. Mater. Form.* **2009**, *2* (Suppl. S1), 813–816.
19. Vallellano, C.; Morales, D.; Martínez-Donaire, A.J.; García-Lomas, F.J. On the use of Concave-Side Rule and Critical-Distance Methods to predict the influence of bending on sheet-metal formability. *Int. J. Mater. Form.* **2010**, *3* (Suppl. S1), 1167–1170.
20. Tharrett, M.R.; Stoughton, T.B. Stretch-Bend Forming Limits of 1008 AK Steel, 70/30 Brass, and 6010 Aluminum. In *Dislocations, Plasticity and Metal Forming, Proceedings of the 10th International Symposium on Plasticity and Its Current Applications, Quebec, Canada, 7–11 July 2003*; Neat Press: Jonesboro, AR, USA, 2003; pp. 199–201.
21. Hosford, W.F. *Mechanical Behavior of Materials*; Cambridge University Press: Cambridge, UK, 2005.
22. Hosford, W.F. Comments on anisotropic yield criteria. *Int. J. Mech. Sci.* **1985**, *27*, 423–427.
23. Tan, Z.; Persson, B.; Magnusson, C. Plastic bending of anisotropic sheet metals. *Int. J. Mech. Sci.* **1995**, *37*, 405–421.

metals

MDPI

Article

Anisotropic Hardening Behaviour and Springback of Advanced High-Strength Steels

Jaebong Jung [1], Sungwook Jun [1], Hyun-Seok Lee [2], Byung-Min Kim [1], Myoung-Gyu Lee [3] and Ji Hoon Kim [1,*]

[1] School of Mechanical Engineering, Pusan National University, Busan 46241, Korea; sylar999@pusan.ac.kr (J.J.); sungwook@pusan.ac.kr (S.J.); bmkim@pusan.ac.kr (B.-M.K.)
[2] NARA Mold & Die Co., Ltd., Changwon 51555, Korea; hslee@naramnd.com
[3] Department of Materials Science and Engineering, Korea University, Seoul 02841, Korea; myounglee@korea.ac.kr
*   Correspondence: kimjh@pusan.ac.kr; Tel.: +82-51-510-3031

Received: 26 September 2017; Accepted: 12 October 2017; Published: 6 November 2017

Abstract: Advanced high-strength steels (AHSSs) exhibit large, and sometimes anisotropic, springback recovery after forming. Accurate description of the anisotropic elasto-plastic behaviour of sheet metals is critical for predicting their anisotropic springback behaviour. For some materials, the initial anisotropy is maintained while hardening progresses. However, for other materials, anisotropy changes with hardening. In this work, to account for the evolution of anisotropy of a dual-phase steel, an elastoplastic material constitutive model is developed. In particular, the combined isotropic–kinematic hardening model was modified. Tensile loading–unloading, uniaxial and biaxial tension, and tension–compression tests were conducted along the rolling, diagonal, and transverse directions to measure the anisotropic properties, and the parameters of the proposed constitutive model were determined. For validation, the proposed model was applied to a U-bending process, and the measured springback angles were compared to the predicted ones.

Keywords: anisotropy; combined isotropic–kinematic hardening; dual-phase steel

---

1. Introduction

Weight reduction, in order to improve fuel efficiency and meet CO_2 regulations for addressing global warming while maintaining safety regulations, is an important issue in automotive manufacturing [1,2]. In this work, we investigate the application of advanced high-strength steels (AHSSs), with good strength and formability, in automotive parts. The demand for automotive parts made of AHSSs is based on their excellent impact resistance, which is an asset for the reinforcement of the car body structure, and which depends on their high strength [3–7]. However, along with their high strength and low thickness, the AHSSs exhibit anisotropic properties and large springback recovery after forming [8,9]. The higher their strength and the thinner the sheet metals, the greater the tendency for anisotropy and springback to occur [10,11]. For some metals, the anisotropy of a material follows a tendency determined at the initial yielding, which does not change as hardening progresses. However, in several cases of materials in which the anisotropic tendency changes according to the hardening progress, there is a restriction to express the phenomenon of anisotropic springback problem with a yield function and the conventional hardening rules. Therefore, the accurate description of the anisotropic elasto-plastic behaviour of sheet metals is critical for predicting their anisotropic springback behaviour.

To predict springback accurately, it is important to use sophisticated elastic material models [12–15]. A constant elastic modulus is widely used, but it is not suitable for representing anisotropic and nonlinear unloading behaviour. Lems [16] investigated a change in the elastic modulus following the plastic strain at low temperature and its recovery. Morestin and Boivin [17] proved that the variation in the elastic modulus following a plastic strain allows better numerical analysis of the elasto-plastic phenomenon in the case of the springback problem. Steels, for example, have a reduction of up to 20–30% of their initial elastic modulus (E_0), while for aluminium alloys, in general, the reduction is 20%. The chord modulus model [18] represents a changing elastic modulus behaviour according to the hardening progress by reducing the elastic modulus with the increase in equivalent plastic strain. This model can improve the accuracy in springback prediction, because it is effective in expressing the reduction phenomenon and is computationally efficient. Its disadvantage is that a nonlinear stress–strain response cannot be captured. As a result, the stress–strain description is accurate only when fully loaded. In the quasi-plastic-elastic (QPE) model proposed by Sun and Wagoner [19], the elastic modulus is maintained as the initial value until reaching a certain level of stress based on the QPE rules, and then, the elastic modulus, according to the strain amount, is nonlinearly reduced until reaching the next plastic behaviour.

The springback predictions are also sensitive to yield functions for materials having highly anisotropic properties [20–22]. The Yld2000-2d yield function [23] is widely used in sheet metal forming simulations. The Bauschinger effect is usually described by introducing back stresses, as in the Armstrong-Frederick hardening rule [24]. To account for the nonlinear hardening and changing anisotropic tendency, variable parameters and tensors have been introduced into the back stress evolution rule [25,26]. Recently, the homogeneous yield function-based anisotropic hardening (HAH) model [27] was proposed, wherein the plastic behaviour of a metallic material subjected to multiple or continuous strain path changes is described using a collapse of the yield function. A fluctuating term of the HAH model, along with phenomenological or dislocation density-based hardening equations, can depict an anisotropic hardening according to the change in various strain paths [28–30].

The dual-phase (DP) steels studied in this work are generally subjected to intercritical annealing of cold rolled strips followed by quenching and as a result, they have a microstructure consisting of a ferritic matrix and a martensitic islands [31]. The quenching converts the ferrite-mixed austenite phase to martensite, leading to a ferrite–martensite two-phase system. The complex microstructure of DP steels may cause complicated behaviour upon changes in loading paths.

In this work, to account for the evolution of anisotropy in a DP steel, an elastoplastic material constitutive model is developed. In particular, the combined isotropic–kinematic hardening model was modified. Tensile loading–unloading, uniaxial and biaxial tension, and tension–compression tests were conducted along the rolling (RD), diagonal (DD), and transverse (TD) directions to measure the anisotropic properties, and the parameters of the proposed constitutive model were determined. For validation, the proposed model was applied to a U-bending process, and the measured and predicted springback angles were compared.

2. Constitutive Equations

2.1. Elasticity

A stress-strain relationship of a material under a non-yielding condition may be expressed by the generalised Hooke's law [32]:

$$\sigma = \mathbf{C}\varepsilon^e \tag{1}$$

where ε^e is elastic strain, σ is Cauchy stress, and $\mathbf{C}$ is the stiffness matrix. Under the plane stress condition and with the isotropic elasticity, Equation (1) may be rewritten in the matrix form as [32]:

$$\begin{bmatrix} \sigma_{11} \\ \sigma_{22} \\ \sigma_{12} \end{bmatrix} = \frac{1}{1-v^2} \begin{bmatrix} E & vE & 0 \\ vE & E & 0 \\ 0 & 0 & G(1-v^2) \end{bmatrix} \begin{bmatrix} \varepsilon^e_{11} \\ \varepsilon^e_{22} \\ \varepsilon^e_{12} \end{bmatrix} \tag{2}$$

where E is the Young's modulus, G is the shear modulus, and v is the Poisson's ratio. The Young's modulus can be express by various elastic material models, as shown in Figure 1. The constant modulus model underestimates the elastic recovery, whereas the chord modulus model overestimates it. The QPE model [19] can depict the elastic unloading behaviour by introducing a QPE function and its nonlinear evolution. In the QPE model, the elastic modulus is given by:

$$E = E_0 - E_1 \left[1 - \exp\left(-b \int ||d\varepsilon - d\varepsilon^p|| \right) \right]$$ (3)

where E_0 is the initial elastic modulus and E_1 and b are material parameters determined by loading–unloading tests. $d\varepsilon$ and $d\varepsilon^p$ are, respectively, increment of total strain and plastic strain. In the QPE model, the QPE surface is defined where the constant elastic modulus is used:

$$f_1 = \phi_1(\boldsymbol{\sigma} - \boldsymbol{\alpha}^*) - R = 0$$ (4)

where f_1, $\boldsymbol{\alpha}^*$, and R are the QPE function, its center, and the size of the QPE function, respectively. The size of the QPE function is assumed constant in this work. In the QPE mode, the stress is located outside of f_1 and inside of f.

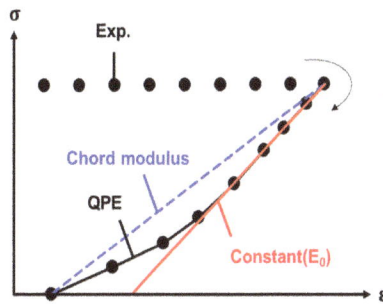

Figure 1. Schematic comparison of the elastic material models during unloading.

2.2. Yield Criterion

The Yld2000-2d yield function [23], f, composed of two functions, ϕ' and ϕ'', was used to account for the anisotropic yielding

$$f(\boldsymbol{\sigma}) = \frac{\phi' + \phi''}{2} = \bar{\sigma}^m$$ (5)

where $\boldsymbol{\sigma}$ is the stress, $\bar{\sigma}$ is the effective stress, and m is the yield function exponent. The details of the yield function are given in Appendix A.

2.3. Hardening Law

The back stress, $\boldsymbol{\alpha}$, is introduced to account for the Bauschinger effect in the Armstrong-Frederick hardening model [24]:

$$f(\boldsymbol{\sigma} - \boldsymbol{\alpha}) - \bar{\sigma}_{iso}^m = 0$$ (6)

where the yield function size of the isotropic hardening law of Voce type [33], $\bar{\sigma}_{iso}$, is given by:

$$\bar{\sigma}_{iso} = \bar{\sigma}^0 + Q(1 - \exp(-b\bar{\varepsilon}))$$ (7)

where $\bar{\sigma}^0$, Q, and b are material parameters. The yield function is an m-th order homogeneous function, whose size is decided by the equivalent plastic strain $\bar{\varepsilon}$. Yielding occurs when the yield function equals its size. In the Armstrong–Frederick hardening model, the increment in the back stress is composed of two terms:

$$d\alpha = d\alpha_1 - d\alpha_2 = C\frac{(\sigma - \alpha)}{\bar{\sigma}_{iso}}d\bar{\varepsilon} - \gamma\alpha d\bar{\varepsilon} \tag{8}$$

where C and γ are material parameters. The second term of the furthest right side of Equation (8) is called a recall term, which is introduced by expressing the transient behaviour of the gradually disappearing memory of the material when subjected to a rapid change in the stress. The hardening rate increases if C of the first term increases and non-linear hardening appears when γ increases under a reverse loading condition.

Chung et al. [25] proposed a back stress evolution rule, where C and γ are functions of the equivalent plastic strain, for expressing general hardening behaviour. In this paper, C and γ were expressed as:

$$C(\bar{\varepsilon}) = C_1 + C_2\exp(-C_3\bar{\varepsilon}) \tag{9}$$

$$\gamma(\bar{\varepsilon}) = \gamma_1 + \gamma_2\exp(-\gamma_3\bar{\varepsilon}) \tag{10}$$

where C_1, C_2, C_3, γ_1, γ_2, and γ_3 are material parameters.

To account for the anisotropic evolution of back stress [26], the back stress evolution rule was modified as:

$$d\alpha = d\alpha_1 - d\alpha_2 = \Gamma_1 C\frac{(\sigma - \alpha)}{\bar{\sigma}_{iso}}d\bar{\varepsilon} - \Gamma_2\gamma\alpha d\bar{\varepsilon} \tag{11}$$

where Γ_1 and Γ_2 are 3×3 diagonal tensors given by:

$$\Gamma_1 = \begin{bmatrix} 1 & 0 & 0 \\ 0 & \Gamma_{1,22} & 0 \\ 0 & 0 & \Gamma_{1,12} \end{bmatrix} \tag{12}$$

$$\Gamma_2 = \begin{bmatrix} 1 & 0 & 0 \\ 0 & \Gamma_{2,22} & 0 \\ 0 & 0 & \Gamma_{2,12} \end{bmatrix} \tag{13}$$

This model is denoted as 'the hardening model with anisotropic evolution'. If the back stress evolves isotropically ($\Gamma_{1,22} = \Gamma_{1,12} = \Gamma_{2,22} = \Gamma_{2,12} = 1$), it is denoted as 'the hardening model with the isotropic evolution'.

3. Experiments

A dual-phase steel with tensile strength of 980 MPa and thickness of 1.1 mm (DP980) was used for the mechanical and U-bending tests. The uniaxial tension, uniaxial tensile loading–unloading, and biaxial tension tests were conducted to determine the material parameters of the QPE model and the Yld2000-2d yield function. The hardening parameters of the isotropic and anisotropic evolution models were obtained from the tension–compression tests. The U-bending test was conducted to evaluate the effect of the anisotropic evolution on springback.

3.1. Uniaxial Tensile and Loading–Unloading Tests

The uniaxial tensile and loading–unloading tests were carried out using a universal testing machine. The ASTM E8 standard size specimens were manufactured by wire electrodischarge machining. The effect of the coupon manufacturing methods on tensile properties were discussed elsewhere [34]. The uniaxial tensile tests were conducted for the RD, DD, and TD, as shown in Figure 2a. The loading–unloading tests were conducted for pre-strains of 3, 5, and 7% in the RD, as shown in Figure 2b. The material properties were obtained following the standard test methods ASTM E8 (Standard Test Method for Tension Testing of Metallic Materials) and E517 (Standard Test Method for Plastic Strain Ratio r for Sheet Metal), as listed in Table 1.

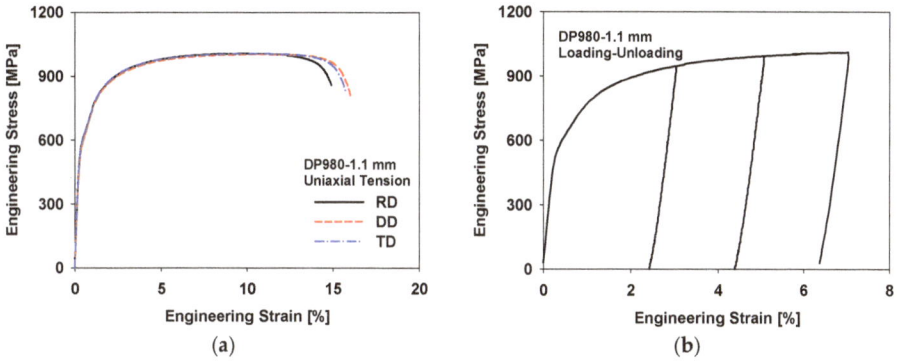

Figure 2. Engineering stress–strain curves of (**a**) uniaxial tensile and (**b**) loading–unloading tests.

Table 1. Uniaxial tensile properties of DP980.

Direction	Young's Modulus [GPa]	Yield Strength [MPa]	r-Value	Tensile Strength [MPa]	Elonagtion [%]
RD	202.2	618.1	0.757	1014.1	15.6
DD	206.1	610.7	0.851	1015.6	16.4
TD	208.8	626.7	0.856	1021.9	14.0

The parameters of the QPE model were derived from the loading–unloading data, as listed in Table 2.

Table 2. Parameters of the QPE model.

E_0 [GPa]	E_1 [GPa]	b	R [MPa]
205.8	105.7	345	300

3.2. Biaxial Tensile Test

The balanced biaxial (BB) tensile test was carried out using a biaxial tensile tester [35] with a cruciform specimen, as shown in Figure 3. A constant loading rate of 9.8 kN/min was used. The plastic strain ratio in balanced biaxial tension r_b was obtained. The results of the balanced biaxial tension are shown in Figure 4 and listed in Table 3.

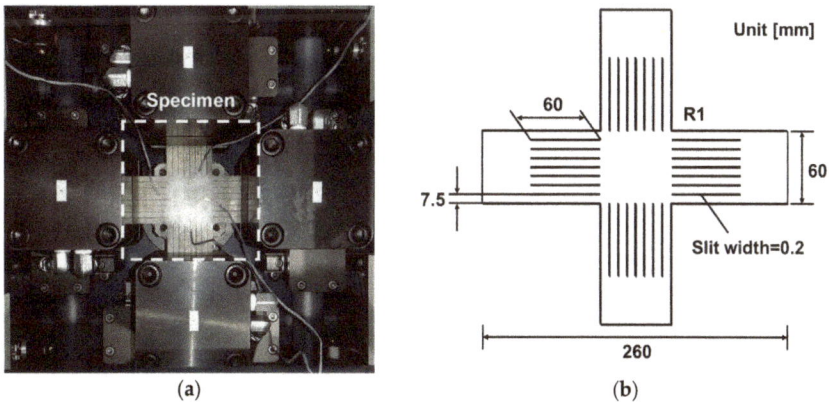

Figure 3. Biaxial tensile test: (**a**) biaxial tensile tester; (**b**) cruciform specimen.

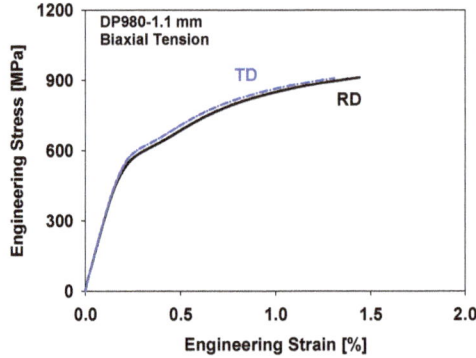

Figure 4. Engineering stress–strain curves of the biaxial tension.

Table 3. Balanced biaxial tensile properties of DP980.

Elastic Modulus [GPa]	Yield Strength [MPa]	r_b
297.0	601.9	0.915

The Yld2000-2d parameters were obtained using the data in Tables 1 and 3, as listed in Table 4. The variation of the yield strength and the comparison of the initial yield criteria of the von-Mises and Yld2000-2d yield functions are shown in Figure 5.

Table 4. Parameters of the anisotropic Yld2000-2d yield function.

m	α_1	α_2	α_3	α_4	α_5	α_6	α_7	α_8
6	0.978	0.978	1.049	1.008	1.026	1.044	0.998	1.023

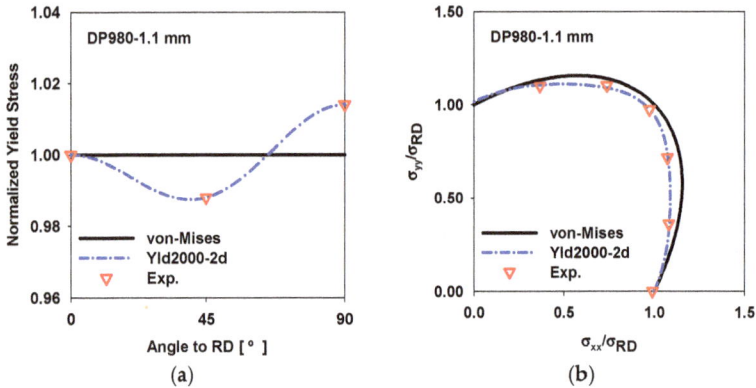

Figure 5. (a) Variation of yield strength and (b) initial yield criteria.

3.3. Tension-Compression Test

In order to investigate the anisotropic evolution of hardening, tension–compression tests were carried out in the three material directions using a tension–compression tester, as shown in Figure 6. In the tension-compression test, the sheet specimen is placed between the comb-shaped anti-buckling fixtures. Buckling of the sheet metals, which may occur during compression, is suppressed by the hydraulic force acting perpendicular to the sheet plane through the comb-shaped fixtures. Details of the tension-compression test can be found elsewhere [36]. In the tension-compression tests, the tension by a strain of 7% was followed by the compression to a strain of −7%. The true stress-accumulated plastic strain curves can be divided into three stages, as shown in Figure 7. In stage A, where hardening hardly progresses, the anisotropic tendency of yielding matches the initial yield strength trend in Figure 5a. The curves of RD and TD become similar in stage B, and the flow stress of RD becomes larger than that of TD in stage C where the initial anisotropic tendency is reversed. The change in anisotropic tendency is illustrated in Figure 8.

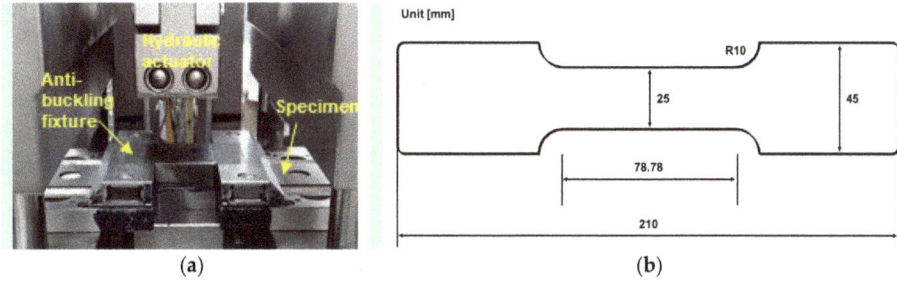

Figure 6. The tension–compression test: (**a**) tension–compression tester; (**b**) specimen.

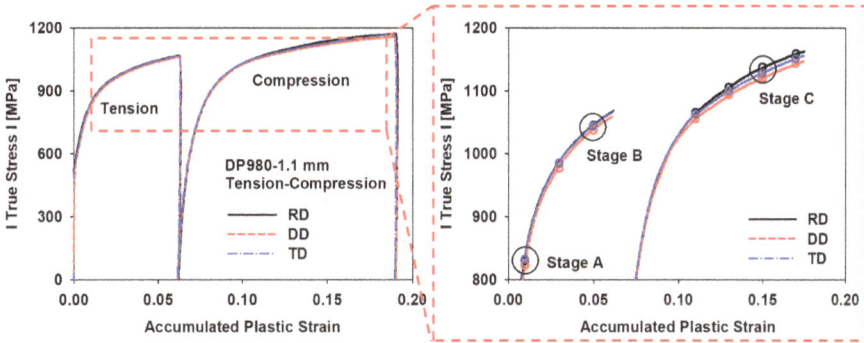

Figure 7. Absolute true stress–accumulated plastic strain curves of the tension-compression tests.

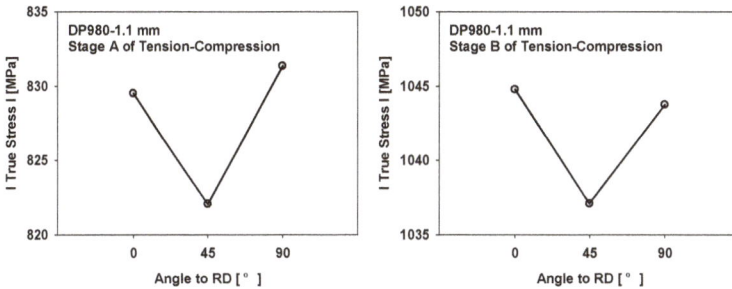

Figure 8. *Cont.*

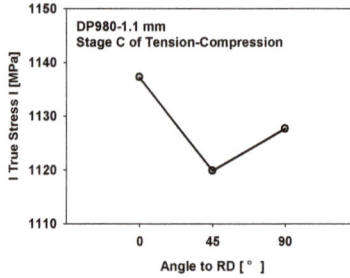

Figure 8. Variation of anisotropy according to the hardening progress.

The hardening model parameters were obtained from the tension–compression data by optimisation using the *fminsearch* function of the MATLAB program, as listed in Table 5. The MATLAB function finds parameters that minimise the error function value using the Nelder–Mead method [37]. After calculating a residual by performing a finite element (FE) analysis using an estimated value, the *fminsearch* function calculates the updated parameter values and repeats the FE analysis. This process is repeated until no further reduction in residual is achieved. In the case of the hardening model with isotropic hardening, it was optimised only for the rolling direction data.

Table 5. Parameters of hardening models with isotropic and anisotropic evolution.

Hardening Model with Isotropic Evolution								
$\bar{\sigma}^0$ [MPa]	Q [MPa]	b	C_1 [MPa]	C_1 [MPa]	C_1	γ_1	γ_2	γ_3
583.9	409.7	3.861	17,896	26,144	9.537	34.74	80.27	6.551
Hardening Model with Anisotropic Evolution								
$\bar{\sigma}^0$ [MPa]	Q [MPa]	b	C_1 [MPa]	C_1 [MPa]	C_1	γ_1	γ_2	γ_3
583.9	409.7	3.861	17,896	26,144	9.537	34.74	80.27	6.551
$\Gamma_{1,22}$	$\Gamma_{1,12}$	$\Gamma_{2,22}$	$\Gamma_{2,12}$					
0.948	0.975	1.014	0.940					

In Figure 9, the tension-compression curves in the three material directions were compared. In the case of the isotropic evolution, it follows the initial trend of anisotropy determined by the yield function. However, in the case of the anisotropic evolution, the anisotropic tendency is changed during reverse loading. The errors of fit of the tension–compression curves predicted by the isotropic and anisotropic evolution models are compared in Figure 10. The error of fit in the RD value is similar because the parameters of the isotropic model were obtained from the RD data. However, the errors in the DD and TD are smaller for the anisotropic models.

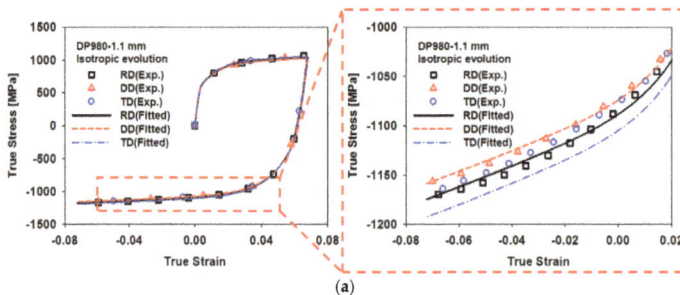

(a)

Figure 9. *Cont.*

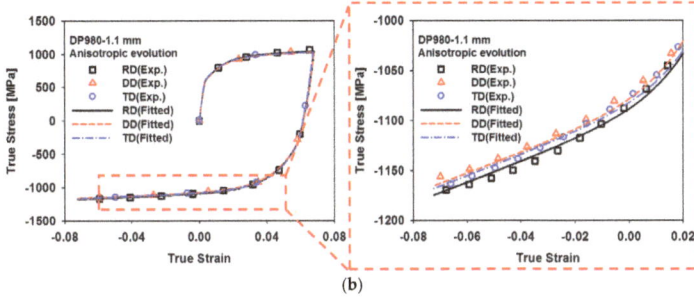

Figure 9. True stress–strain curves under tension–compression calculated by the hardening models with (**a**) isotropic evolution and (**b**) anisotropic evolution.

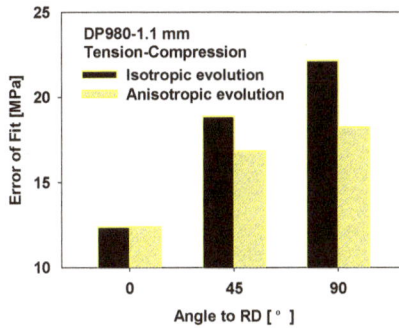

Figure 10. Errors of fit of the isotropic and anisotropic evolution models.

4. Application

U-bending tests were conducted using specimens with size of $300 \times 30 \times 1.1$ mm^3 for the three material directions. The experiments were performed using a servo-press. The dimensions of the test are shown in Figure 11a. The holding force was linearly increased from 5.65 kN to 7.92 kN from the beginning to the end of the punch stroke. The specimens before and after forming and springback are shown in Figure 12a.

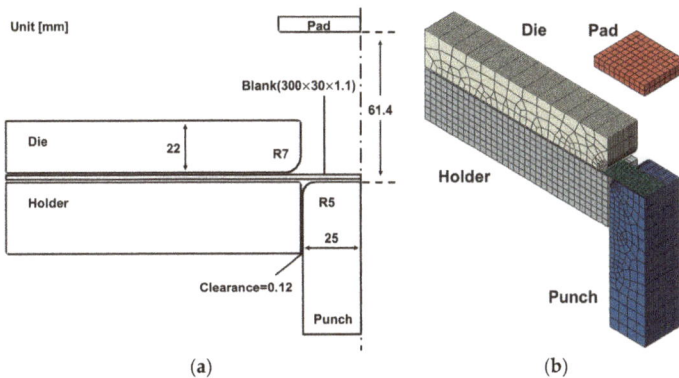

Figure 11. U-bending test: (**a**) dimensions and (**b**) finite element model.

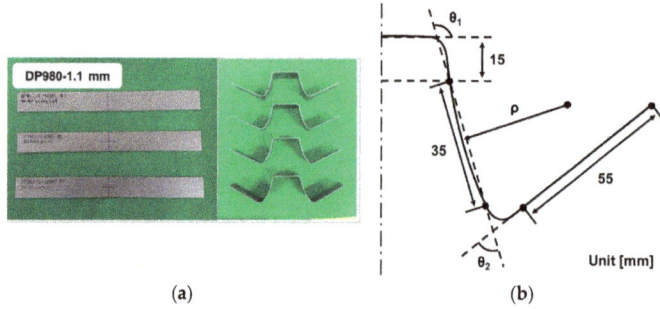

Figure 12. (a) Specimens of the U-bending test and (b) measures of springback (ρ, θ_1, and θ_2).

Simulations of the U-bending process were carried out using the commercial finite element software Abaqus/Standard, as shown in Figure 11b. Four-node shell elements with reduced integration (S4R) were used with an element size of 0.5 mm. The number of integration points through the thickness was seven. The coefficient of friction was assumed as 0.12. The deformation of the tools may affect the U-bending test results significantly [38]. However, for simplicity, the punch, die, and holder were assumed as rigid bodies in this work.

The measured and calculated shapes of the specimens after springback are compared for the three material directions in Figure 13. The predicted shapes showed good agreements with the measurements in RD, while there are some differences in DD and TD. The predictions with the isotropic and anisotropic evolution showed little differences for RD because the material parameters of the isotropic evolution model were fitted for RD. However, the predictions with the isotropic and anisotropic evolution showed differences in TD, although the amount of the difference is small.

In order to evaluate the springback quantitatively, three measures of springback (ρ, θ_1, and θ_2), defined in Figure 12b, were taken from the specimen geometry, as shown in Figure 14. The predictions with the isotropic and anisotropic evolution showed little differences in the U-bending shape of the RD and DD specimens. For the TD specimens, the predictions with the anisotropic evolution showed better agreements with the measurements. This is because the anisotropic hardening evolution rule reduces the error when representing the stress-strain behaviour in directions other than RD. In the case of the isotropic evolution model, the initial yielding anisotropic tendency of the yield function is maintained in the springback result of U-bending. However, the initial anisotropic tendency was changed during deformation, as shown in Figure 8. In the case of the anisotropic evolution model, the evolving anisotropy tendency is represented, which is more similar to the measurements.

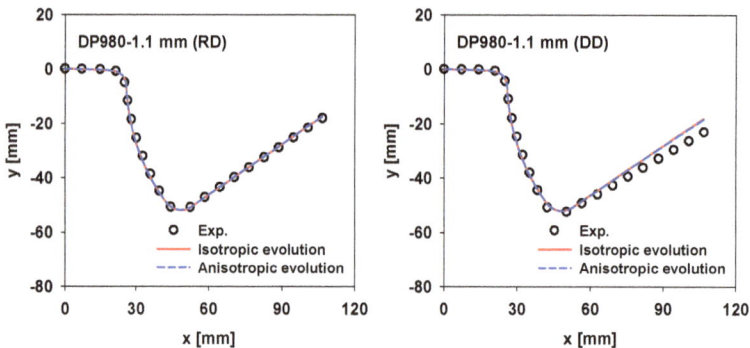

Figure 13. *Cont.*

90

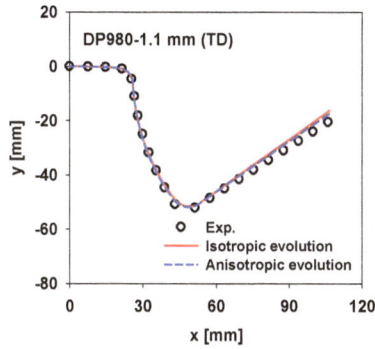

Figure 13. Intermediate shape after the springback for verification of difference between hardening models with isotropic and anisotropic evolution.

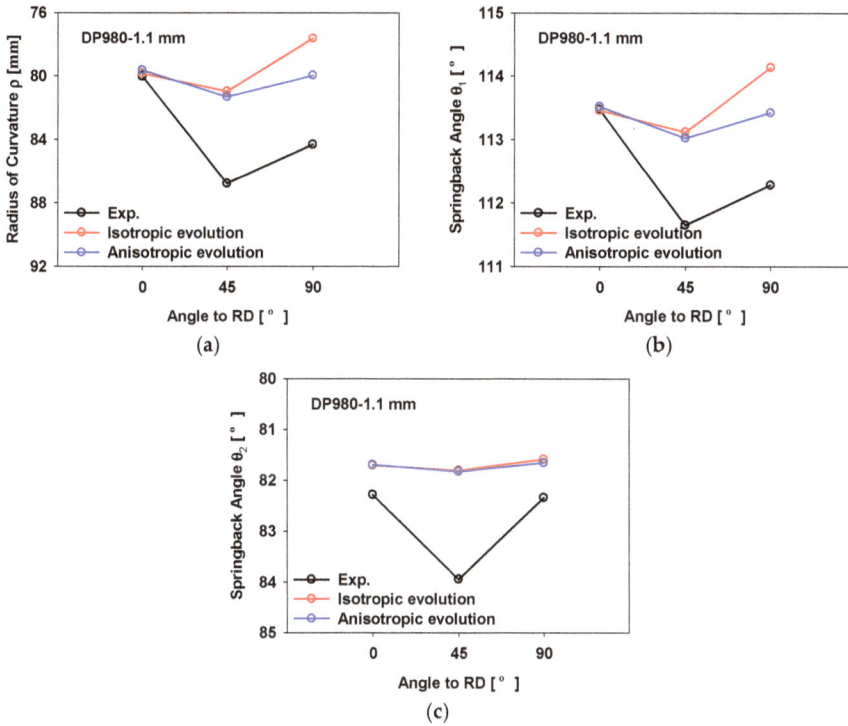

(a)

(b)

(c)

Figure 14. Comparison of the measured and calculated springback measures: (**a**) curvature ρ; (**b**) springback angle θ_1; (**c**) springback angle θ_2.

5. Conclusions

To account for the evolution of anisotropy of a dual-phase steel, an elastoplastic material constitutive model is developed by modifying the combined isotropic-kinematic hardening model. The tensile loading–unloading, uniaxial and biaxial tension, and tension–compression tests were conducted along the rolling, diagonal, and transverse directions to measure the anisotropic properties, and the parameters of the proposed constitutive model were determined. For validation, the proposed model was applied to a U-bending process, and the measured and predicted springback angles were compared. The following conclusions were drawn:

- The dual-phase steel studied in this work exhibited a change in the anisotropic properties during deformation.
- The hardening model with the anisotropic evolution successfully captured the evolution of the angular variation of anisotropic properties.
- Using the hardening model with anisotropic evolution, it was possible to accurately predict the springback in different directions.

Acknowledgments: This work was supported by the Small and Medium Business Administration of Korea (SMBA) grant funded by the Korean government (MOTIE) (No. S2315965) and the National Research Foundation of Korea (NRF) grant funded by the Korean government (MSIP) (No. 2012R1A5A1048294).

Author Contributions: Jaebong Jung and Ji Hoon Kim conceived and designed the experiments; Jaebong Jung and Sungwook Jun performed the experiments; Jaebong Jung, Sungwook Jun, Ji Hoon Kim, Byoung Min Kim, and Myoung-Gyu Lee analysed the data; Hyun-Seok Lee contributed with materials/analysis tools; Jaebong Jung and Ji Hoon Kim wrote the paper.

Conflicts of Interest: The authors declare no conflict of interest.

Appendix A

The Yld2000-2d yield function [6] has eight parameters (α_{1-8}) that require eight mechanical properties (σ_0, σ_{45}, σ_{90}, σ_b, r_0, r_{45}, r_{90}, r_b). The m value is suggested as six for BCC and eight for FCC metal. In Equation (4), the yield function is given by a combination of two functions:

$$\phi' = \left| X_1' - X_2' \right|^m \tag{A1}$$

$$\phi'' = \left| 2X_2'' + X_1'' \right|^m + \left| 2X_1'' + X_2'' \right|^m \tag{A2}$$

where the X_i' and X_i'' are principal values of the tensor $\mathbf{X}'$ and $\mathbf{X}''$, respectively, which are expressed by the product of the anisotropic components (C_{ij} or L_{ij}) and stress components (s_{ij} or σ_{ij}).

References

1. Keeler, S.; Kimchi, M. *Advanced High-Strength Steels Application Guidelines Version 5.0*; WorldAutoSteel: Middletown, OH, USA, 2014.
2. Cheah, L.W. Cars on a Diet: The Material and Energy Impacts of Passenger Vehicle Weight Reduction in the U.S. Ph.D. Thesis, Massachusetts Institute of Technology, Cambridge, MA, USA, September 2010.
3. Amigo, F.J.; Camacho, A.M. Reduction of induced central damage in cold extrusion of dual-phase steel DP980 using double-pass dies. *Metals* **2017**, *7*, 335. [CrossRef]
4. Xue, X.; Pereira, A.B.; Amorim, J.; Liao, J. Effects of pulsed Nd: YAG laser welding parameters on penetration and microstructure characterization of a DP1000 steel butt joint. *Metals* **2017**, *7*, 292. [CrossRef]
5. Evin, E.; Tomas, M. The influence of laser welding on the mechanical properties of dual phase and TRIP steels. *Metals* **2017**, *7*, 239. [CrossRef]
6. Moeini, G.; Ramazani, A.; Myslicki, S.; Sundararaghavan, V.; Konke, C. Low cycle fatigue behaviour of DP steels: Micromechanical modelling vs. validation. *Metals* **2017**, *7*, 265. [CrossRef]
7. Emre, H.E.; Kacar, R. Resistance spot weldability of galvanize coated and uncoated TRIP steels. *Metals* **2016**, *6*, 299. [CrossRef]
8. Hassan, H.U.; Traphoner, H.; Guner, A.; Tekkaya, A.E. Accurate springback prediction in deep drawing using pre-strain based multiple cyclic stress-strain curves in finite element simulation. *Int. J. Mech. Sci.* **2016**, *110*, 229–241. [CrossRef]
9. Lee, H.S.; Kim, J.H.; Kang, G.S.; Ko, D.C.; Kim, B.M. Development of seat side frame by sheet forming of DP980 with die compensation. *Int. J. Precis. Eng. Manuf.* **2017**, *18*, 115–120. [CrossRef]
10. Abvabi, A.; Rolfe, B.; Hodgson, P.D.; Weiss, M. The influence of residual stress on a roll forming process. *Int. J. Mech. Sci.* **2015**, *101–102*, 124–136. [CrossRef]

11. Liu, X.; Cao, J.; Chai, X.; Liu, J.; Zhao, R.; Kong, N. Investigation of forming parameters on springback for ultra high strength steel considering Young's modulus variation in cold roll forming. *J. Manuf. Process.* **2017**, *29*, 289–297. [CrossRef]

12. Lee, J.; Lee, J.Y.; Barlat, F.; Wagoner, R.H.; Chung, K.; Lee, M.G. Extension of quasi-plastic-elastic approach to incorporate complex plastic flow behaviour–application to springback of advanced high-strength steels. *Int. J. Plast.* **2013**, *45*, 140–159. [CrossRef]

13. Chen, Z.; Bong, H.J.; Li, D.; Wagoner, R.H. The elastic-plastic transition of metals. *Int. J. Plast.* **2016**, *83*, 178–201. [CrossRef]

14. Torkabadi, A.; Perdahcioglu, E.S.; Meinders, V.T.; Boogaard, V.D. On the nonlinear anelastic behaviour of AHSS. *Int. J. Solids Struct.* **2017**, in press. [CrossRef]

15. Lee, J.Y.; Lee, M.G.; Barlat, F.; Bae, G. Piecewise linear approximation of nonlinear unloading-reloading behaviours using a multi-surface approach. *Int. J. Plast.* **2017**, *93*, 112–136. [CrossRef]

16. Lems, W. The change of Young's modulus of copper and silver after deformation at low temperature and its recovery. *Physica* **1962**, *28*, 445–452. [CrossRef]

17. Morestin, F.; Boivin, M. On the necessity of taking into account the variation in the Young modulus with plastic strain in elastic-plastic software. *Nucl. Eng. Des.* **1996**, *162*, 107–116. [CrossRef]

18. Yoshida, F.; Uemori, T.; Fujiwara, K. Elastic-plastic behaviour of steel sheets under in-plane cyclic tension-compression at large strain. *Int. J. Plast.* **2002**, *18*, 633–659. [CrossRef]

19. Sun, L.; Wagoner, R.H. Complex unloading behaviour: Nature of the deformation and its consistent constitutive representation. *Int. J. Plast.* **2011**, *27*, 1126–1144. [CrossRef]

20. Cardoso, R.P.R.; Yoon, J.W. Stress integration method for a nonlinear kinematic/isotropic hardening model and its characterization based on polycrystal plasticity. *Int. J. Plast.* **2009**, *25*, 1684–1710. [CrossRef]

21. Cao, J.; Lee, W.; Cheng, H.S.; Seniw, M.; Wang, H.P.; Chung, K. Experimental and numerical investigation of combined isotropic-kinematic hardening behaviour of sheet metals. *Int. J. Plast.* **2009**, *25*, 942–972. [CrossRef]

22. Sumikawa, S.; Ishiwatari, A.; Hiramoto, J.; Urabe, T. Improvement of springback prediction accuracy using material model considering elastoplastic anisotropy and Bauschinger effect. *J. Mater. Proc. Technol.* **2016**, *230*, 1–7. [CrossRef]

23. Barlat, F.; Brem, J.C.; Yoon, J.W.; Chung, K.; Dick, R.E.; Lege, D.J.; Pourboghrat, F.; Choi, S.H.; Chu, E. Plane stress yield function for aluminum alloy sheets-part 1: Theory. *Int. J. Plast.* **2003**, *19*, 1297–1319. [CrossRef]

24. Frederick, C.O.; Armstrong, P.J. A mathematical representation of the multiaxial Bauschinger effect. *Mater. High Temp.* **1966**, *24*, 1–26. [CrossRef]

25. Chung, K.; Lee, M.G.; Kim, D.; Kim, C.; Wenner, M.L.; Barlat, F. Spring-back evaluation of automotive sheets based on isotropic-kinematic hardening laws and non-quadratic anisotropic yield functions part 1: Theory and formulation. *Int. J. Plast.* **2005**, *21*, 861–882.

26. Lee, M.G.; Kim, D.; Chung, K.; Youn, J.R.; Kang, T.J. Combined isotropic-kinematic hardening laws with anisotropic back-stress evolution for orthotropic fiber-reinforced composites. *Polym. Polym. Compos.* **2004**, *12*, 225–234.

27. Barlat, F.; Gracio, J.J.; Lee, M.G.; Rauch, E.F.; Vincze, G. An alternative to kinematic hardening in classical plasticity. *Int. J. Plast.* **2011**, *27*, 1309–1327. [CrossRef]

28. Lee, J.Y.; Lee, J.W.; Lee, M.G.; Barlat, F. An application of homogeneous anisotropic hardening to springback prediction in pre-strained U-draw/bending. *Int. J. Solids Struct.* **2012**, *49*, 3562–3572. [CrossRef]

29. Lee, J.Y.; Barlat, F.; Lee, M.G. Constitutive and friction modeling for accurate springback analysis of advanced high strength steel sheets. *Int. J. Plast.* **2015**, *71*, 113–135. [CrossRef]

30. Choi, J.; Lee, J.; Lee, M.G.; Barlat, F. Advanced constitutive modeling of AHSS sheets for application to springback prediction after U-draw double stamping process. *J. Phys.* **2016**, *734*. [CrossRef]

31. Speich, G.R.; Demarest, V.A.; Miller, R.L. Formation of austenite during intercritical annealing of dual-phase steels. *Metall. Mater. Trans. A* **1981**, *12*, 1419–1428. [CrossRef]

32. Shames, I.H. *Introduction to Solid Mechanics*, 2nd ed.; Prentice-Hall, Inc.: Upper Saddle River, NJ, USA, 1989.

33. Voce, E. A practical strain hardening function. *Metallurgica* **1955**, *51*, 219–226.

34. Krahmer, D.M.; Polvorosa, R.; López de Lacalle, L.; Alonso-Pinillos, U.; Riu, F. Alternatives for Specimen Manufacturing in Tensile Testing of Steel Plates. *Exp. Tech.* **2016**, *40*, 1555–1565. [CrossRef]

35. Hanabusa, Y.; Takizawa, H.; Kuwabara, T. Numerical verification of a biaxial tensile test method using a cruciform specimen. *J. Mater. Proc. Technol.* **2013**, *213*, 961–970. [CrossRef]

36. Lee, M.G.; Kim, J.H.; Kim, D.; Seo, O.S.; Nguyen, N.T.; Kim, H.Y. Anisotropic Hardening of Sheet Metals at Elevated Temperature: Tension-Compressions Test Development and Validation. *Exp. Mech.* **2013**, *53*, 1039–1055. [CrossRef]
37. Nelder, J.A.; Mead, R. A simplex method for function minimization. *Comput. J.* **1965**, *7*, 308–313. [CrossRef]
38. Del Pozo, D.; López de Lacalle, L.N.; López, J.M.; Hernández, A. Prediction of press/die deformation for an accurate manufacturing of drawing dies. *Int. J. Adv. Manuf. Technol.* **2008**, *37*, 649–656. [CrossRef]

Article

Radial Basis Functional Model of Multi-Point Dieless Forming Process for Springback Reduction and Compensation

Misganaw Abebe, Jun-Seok Yoon and Beom-Soo Kang *

Department of Aerospace Engineering, Pusan National University, Busan 46241, Korea;
misge98@gmail.com (M.A.); hamjang21c@gmail.com (J.-S.Y.)
* Correspondence: bskang@pusan.ac.kr; Tel.: +82-51-510-2310; Fax: +82-51-512-4491

Received: 29 September 2017; Accepted: 23 November 2017; Published: 27 November 2017

Abstract: Springback in multi-point dieless forming (MDF) is a common problem because of the small deformation and blank holder free boundary condition. Numerical simulations are widely used in sheet metal forming to predict the springback. However, the computational time in using the numerical tools is time costly to find the optimal process parameters value. This study proposes radial basis function (RBF) to replace the numerical simulation model by using statistical analyses that are based on a design of experiment (DOE). Punch holding time, blank thickness, and curvature radius are chosen as effective process parameters for determining the springback. The Latin hypercube DOE method facilitates statistical analyses and the extraction of a prediction model in the experimental process parameter domain. Finite element (FE) simulation model is conducted in the ABAQUS commercial software to generate the springback responses of the training and testing samples. The genetic algorithm is applied to find the optimal value for reducing and compensating the induced springback for the different blank thicknesses using the developed RBF prediction model. Finally, the RBF numerical result is verified by comparing with the FE simulation result of the optimal process parameters and both results show that the springback is almost negligible from the target shape.

Keywords: multi-point dieless forming; springback reduction; springback compensation; radial basis function

1. Introduction

Springback is a common and critical existence in sheet metal forming processes, which is caused by the elastic redistribution of the internal stresses after unloading of the external forces. Since multi-point dieless forming (MDF) is only a multi-curvature bending and small deformation, the tendency of inducing springback is high. Springback prediction using finite element (FE) simulation is familiar for any types of forming processes.

Previous researchers are investigated and proposed some techniques to compensate and reduce springback on MDF using numerical simulation and physical experiment, such as, Li et al. [1] proposed the multi-step forming to reduce the unloading springback on multi-point forming using numerical simulation. Hwang et al. [2] introduced the springback adjustment for multi-point forming of thick plate in shipbuilding using FE analysis and the iterative displacement adjustment algorithm. Shim et al. [3] investigates the tension force on MDF stretch forming process of aluminum alloy sheet to reduce the induced springback using FE simulation. However, using only FE simulation to compensate the machine tool geometry or any design or process parameters is computationally expensive. To overcome this problem, this study proposed surrogate modeling to replace the computational expensive FE simulation as a function of target curvature radius, blank thickness, and punch holding time (stress relief period) after the final displacement of the punch.

The application of surrogate modeling on sheet metal forming were investigated by different researchers, for instance, Liu and Liang [4] proposed a fuzzy genetic algorithm to compensate for springback of multi-curvature forming. Zhang et al. [5] introduced an algorithm to compensate springback by modifying the double curved plate. Kitayama and Yoshioka [6] proposed a springback reduction technique with the control of punch speed and blank holder force through sequential approximate optimization using the radial basis function (RBF) network. Behera et al. [7] proposed a solution to improve the accuracy of single point incremental sheet forming by using multivariate adaptive regression splines as an error prediction tool to generate continuous error response surfaces for individual features and feature combinations. In order to reduce springback effects after forming, Li [8] also deals with the variable of blank holder force using the least square support vector regression by establishing the adaptive metamodeling optimization system. Khada and El-Morsy [9] applied Kriging metamodel to predict the springback in the air bending process with process parameters of blank material and geometry. Hassen et al. [10] investigates the springback for deep drawing process using the parameters of blank holding force and friction coefficient as a function of time. In this study, the sensitivity analyses of the finite element simulation result were investigated by applying a statistical method. The influence of the parameter variation also analyzed using a sequential screening experiment by applying a piecewise constant function. Karaağaç [11] carried experimental investigation to evaluate the process parameters of bending pressure, bending die angle, punch holding time, and the rubber membrane thickness effect on springback in V-bending using the flexforming process, and he also used the fuzzy logic system to estimate the springback effect. In [12,13] also investigated the influence of punch holding time and other process parameters for V-bending dies. In addition, different researchers [14–16] have applied the surrogate model to investigate springback with a consideration of different process parameters. However, these techniques were applied to single die forming, and most of the study used blank holder to reduce the induced springback; so this study investigates springback for blank holder free boundary condition of the MDF process to fill the gap.

This study proposes process optimization along with surrogate modeling with the aim of reducing and compensating springback in the forming of AA3003-H14 aluminum alloy sheet of saddle shape by MDF with consideration of three influencing parameters: curvature radius, blank thickness, and punch holding time. To construct the model, first, the study chose Latin hyper cube design of experiment (DOE) method for training and random sampling for testing. The Latin hypercube DOE method facilitates the statistical analyses and the extraction of a prediction model. FE simulation is applied using ABAQUS commercial software (ABAQUS 6.12, Dassault Systèmes Simulia Corp., Providence, RI, USA) to generate the training and testing sample data responses. After obtaining the training sample responses from the numerical simulation, different prediction model has been applied such as regression model, Ordinary and Blind Kriging, and RBF. The R^2 and roots mean square error (RMSE) numerical validation of the testing data, which has the same number of the training sample shown RBF is more acceptable than the others. Then, the study used the prediction model to apply the genetic algorithm (GA) in MATLAB to determine the global optimal parameter's value. First, the study found the optimal value of punch holding time to reduce the spring back for different blank thickness. Second, the proper curvature radius value is determined to compensate the induced springback for 800 mm target curvature radius at a different blank thickness. Finally, the FE simulation result of the optimization process parameters shows that the springback is almost negligible from the target geometry.

The rest of the paper is organized as follows. Section 2 discusses the data acquisition methods, using FE methods it includes the geometry, material, and numerical simulation model, in addition, the defect quantification method is also discussed. Section 3 describes the process parameters, sampling technique, formulation of the RBF, and finally the numerical model verification is discussed to validate the developed prediction model. Section 4 presents the optimal process parameter result to compensate the induced springback for the given target shape.

2. Finite Element Modeling

2.1. Geometry Model

The study investigates the MDF springback effect on the rectangular saddle shape, which is shown in Figure 1a. The general equation of the saddle geometry is obtained from: $Z(x, y) = \sqrt{R_y^2 - y^2} - \sqrt{R_x^2 - x^2}$ where x, y and z are the Cartesian coordinates, R_x is the radius curvature along the x-axis and R_y is the radius curvature along the y-axis.

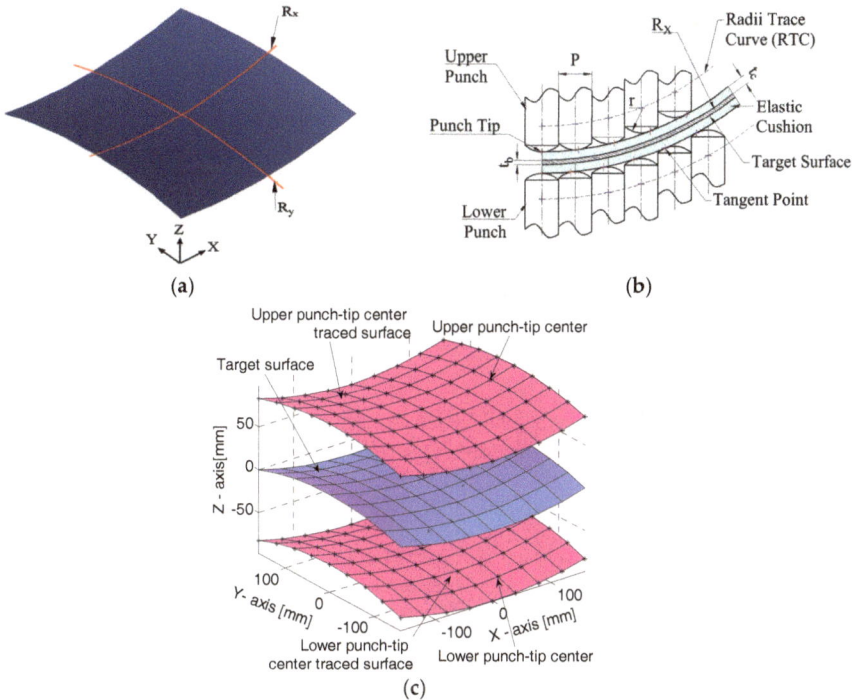

Figure 1. Saddle shape geometry model on multi-point dieless forming (MDF): (**a**) Target shape; (**b**) Punch placement in a two-dimensional view; (**c**) Positions of the lower and upper Radii trace curve (RTC) radii center traced surfaces.

The MDF target surface is generated using the discrete punches element, which has the m number of punches along the x-axis and n number of punches along the y-axis. To construct the target saddle shape in MDF, the lower and the upper punch matrices are placed using Equations (1) and (2), respectively [9].

$$Z_L(x_{ij}, y_{ij}) = \sqrt{\left(R_y^L\right)^2 - y_{ij}^2} - \sqrt{\left(R_x^L\right)^2 - x_{ij}^2}, \tag{1}$$

$$Z_U(x_{ij}, y_{ij}) = \sqrt{\left(R_y^U\right)^2 - y_{ij}^2} - \sqrt{\left(R_x^U\right)^2 - x_{ij}^2}, \tag{2}$$

where $0 < i < m$; $0 < j < n$; m and n are the number of punches along the x- and y-axes, respectively; R_x^L and R_y^L are the curvature radii of the lower punch-tip radius center traced surface along the x- and y-axes, respectively; R_x^U and R_y^U are the curvature radii of the upper punch-tip radius center traced surface along the x- and y-axes, respectively. Figure 1b shows the punch placement in a 2-dimensional view and Figure 1c shows the positions of the lower and upper RTC radii center traced surfaces, which is obtained using Equations (1) and (2), respectively.

2.2. Material Model

AA3003-H14 aluminum alloy sheet metal blank is used to investigate the springback effect on MDF. The material properties are taken from our previous study [17], which is also mentioned in Table 1. Since the MDF process is a small deformation, the study treated the sheet material as an isotropic property, in addition, here, the MDF process is not subjected to cyclic loading so that the study chose isotropic hardening instead of kinematic hardening. Swift strain hardening model is adopted for the material modeling which is $\sigma = K(\varepsilon_0 + \varepsilon)^n$ for $\sigma \geq \sigma_y$, where K is the strength coefficient, ε is the true strain value, ε_0 is the pre-strain constant, and n is the strain hardening exponent.

Table 1. Material properties of AA3003-H14.

Properties	Value	Unit
Young's Modulus (E)	70.1	GPa
Poison ratio (ν)	0.33	N/A
Density (ρ)	2700	kg/m^2
Yield stress (σ_y)	152.2	MPa
Strength Coefficient (K)	192.7	MPa
Strain hardening exponent (n)	0.0394	N/A
Pre-strain constant (ε_0)	2.51×10^{-3}	N/A

For elastic cushion, polyurethane material with a hardness of shore 90A is chosen. Mooney-Rivlin hyperplastic modeling is employed for the simulation model, the model equation is given as:

$$U = C_{10}(\bar{I}_1 - 3) + C_{01}(\bar{I}_2 - 3),\qquad(3)$$

where U is the strain energy per unit of initial volume, $\bar{I}_1$ and $\bar{I}_2$ are the first and the second deviatoric strain invariants, respectively, and C_{10} and C_{01} are the material constants. The martial constants are taken from the previous study [18], which are $C_{01} = 0.6606$ and $C_{10} = -0.0057$.

2.3. Simulation Model

Both explicit and implicit schemes have been used in ABAQUS commercial software to solve a challenging MDF sheet metal forming process that involves a high amount of springback. First, the explicit FE formulation was used to solve the MDF loading process, because the model is highly discontinuous process due to the geometry complexity, material property, and discrete punch element, so the quadratic convergence may be lost, and it may require a larger time increment than the explicit method and several iterations to obtain a solution within the prescribed tolerances.

While the loading process is conducted, a deformable blank sheet was brought into contact with a pair of deformable elastic cushions and the cushion also in contact with the rigid upper and lower discrete punch elements. The contact friction coefficient between the blank and elastic cushion, and elastic cushion and punch is considered as 0.1 [19]. Blank sheet and elastic cushions are meshed using C3D8R solid element and R3D4 shell element is also used for the rigid punch element. To reduce the computational time, as shown in Figure 2, the symmetrical condition is considered. For x-symmetry, the translational displacement in the x-direction, and rotational in y-direction and z-direction were set to zero ($U_x = UR_y = UR_z = 0$), and under y-symmetry, translational displacement in y-direction, and rotation in x-direction and z-direction were set to zero ($U_y = UR_x = UR_z = 0$). After the completion of the loading process using explicit analysis, the deformed shape, strains, and stresses within the blank elements were transferred into the implicit analysis for unloading process. This was accomplished by creating a database file that updated the geometry and stress-strain history of the implicit elements to match the final explicit solution. During the implicit analysis, the punches and elastic cushions were removed from the model and valid boundary conditions were employed to the sheet blank to restrain rigid body motion. As shown in Figure 3, the displacement in *y*-axis (*U*2) at the center of the nodes are fixed, and the X-symmetric and Z-symmetric are also imposed, as shown in the figure. Finally, the implicit analysis was exploited to determine the succeeding springback displacement in the deformed blank that occurred after the forming load was removed. The FE simulation model was experimentally validated in our previous study [17,18].

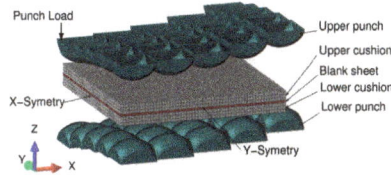

Figure 2. Finite element model for saddle shape.

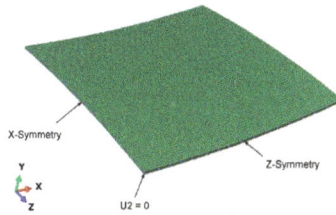

Figure 3. Simulation result before springback and the setup for implicit analysis.

2.4. Defect Quantification

The main objective of the study is investigating the process parameters to compensate for the springback for the given target shape. The study also tried to reduce the induced springback for a different blank thickness. As shown in Figure 4, a numerical simulation was conducted with the process parameters of target curvature radius 800 mm, blank cross-section 320 mm × 320 mm, blank thickness 2 mm, punch width 40 mm, punch-tip radius 30 mm, elastic cushion thickness 16 mm, punch displacement (elastic cushion compressive strain ratio) 6%, and the punch holding time is 0 s. As shown in the figure, the springback is observed from nodal displacement distribution before and after unloading of the punch load. The study used the maximum deviation of the vertical displacement *U*2 from the expected target shape to quantify the springback responses of the numerical simulation result. The equation is given as:

$$f(u) = \max_{1 < i < N} \left(\left| u_t^i - u_d^i \right| \right), \tag{4}$$

where *N* is the number of nodes, u_t^i and u_d^i is the nodal displacement at the *i*th node of the target and deformed shape, respectively.

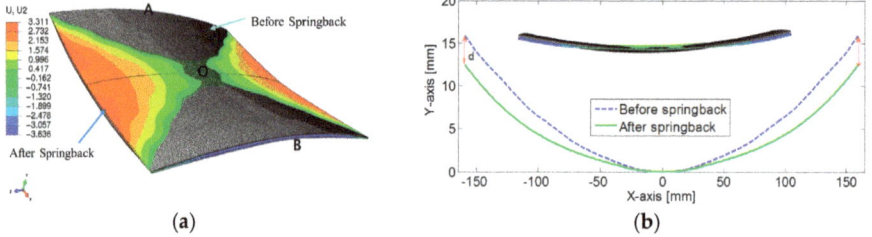

Figure 4. Case of visible springback: (**a**) Nodal displacement distribution; (**b**) Curvature radii at section AB.

3. Surrogate Modeling

3.1. Process Parameter

As mentioned in the previous section, the study has two objectives, the first one is reducing the springback by setting the process parameters and the second one is compensating the induced springback for the expected target shape. Springback is induced because of the elastic recovery (deformation) after the forming process is performed. Previous researchers suggested that creating a restraining force, such as using a blank holder, can decrease the amount of the springback by increasing the sheet tension; however, in our case, there is no blank holder to create such kinds of tension load during the forming process. Here, the study suggested that holding the upper punch load for a certain time can stabilize the induced stress. To check the sensitivity of the process in regards to the punch holding time, the study run a numerical simulations in ABAQUS by varying the punch holding time of 0, 2.5, 5, 10, 20, 30, 40, 50, and 60 s. The other parameters were the same as in Section 2.4. During the quasi-static simulation in ABAQUS, stopping the punch means stopping the simulation, so, to investigate the punch holding time, the study applied a macro speed for the specified time at the end of the forming process and considering it as a punch holding time. As Figure 5 shows, in the first few seconds, the springback result is drastically decreased, however, after that, the springback responses became the same for a few seconds, and getting a slightly increased while the punch holding time is increased. So, to find the optimal punch hold time, the study investigates in between 0 and 20 s.

To compensate the induced springback value, the only option in MDF is changing the curvature radius. Here, the study target curvature radius is 800 mm. To check the upper and lower value, the study conducted two FE simulations. As shown in Figure 6, the study target curvature is between the simulation result of 600 and 800 mm curvature radii after springback. So, the study searches the optimal radius value in between 600 and 800 mm curvature radius for different blank thickness. The considered blank thickness in this study is in between 0.5 and 2 mm. The process parameters are summarized in Table 2.

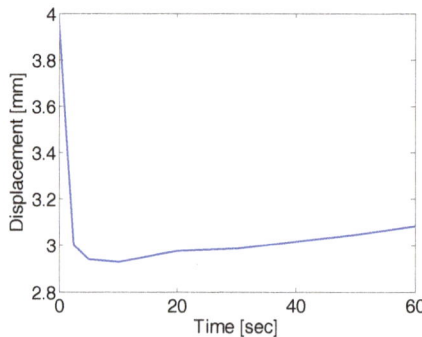

Figure 5. Punch holding time effect on springback.

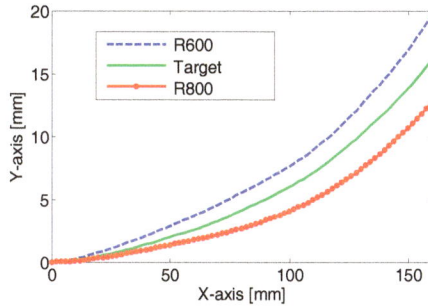

Figure 6. Target curvature radius with the deformed result of *R* = 600 mm and *R* = 800 mm curvature radius after springback.s

Table 2. Process parameters.

Parameters	Lower Bound	Upper Bound	Units
Curvature radius	600	800	mm
Blank thickness	0.5	2	mm
Blank holding time	0	20	s

3.2. Sample Stratagey

To develop the prediction model, the study generates 30 sample data for training using Latin hypercube sampling method. Latin hypercube sampling is one of the best DOE techniques which efficiently sample the large design spaces by maximizing the minimum distance between all of the possible pairs of the sample points in a given sample plan, this shows that a Latin hypercube design method will keep a distance between the samples as large as possible. Figure 7a shows the scatter plots of the training sample data and the views of the plot also shows that the sample is taken only one sample from each row and each column. In addition, 30 random sample data were generated using the MATLAB built function for testing of the developed prediction model. Figure 7b shows the scatter plots of the testing sample data, as compared to the training data, the random data may have a chance to take a sample from the same row and column, which is shown in the front view of the figure. Tables A1 and A2 also show the training and testing raw data, respectively.

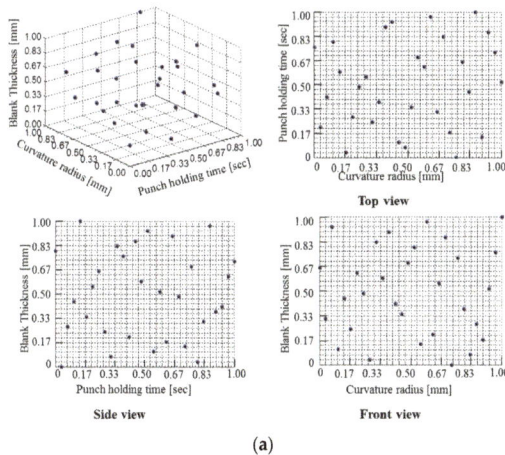

(**a**)

Figure 7. *Cont.*

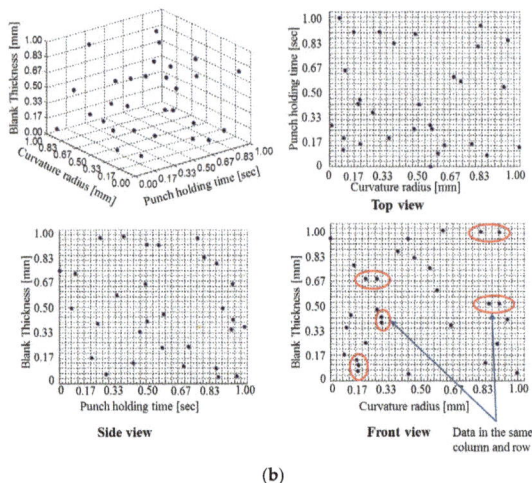

(b)

Figure 7. Training and testing sample data of the model: (**a**) Training data using Latin hypercube sampling; and (**b**) Testing data using random sampling.

3.3. Radial Basis's Function Formulation

RBF was first introduced by Rolland Hardy to fit irregular topographic contours of geographical data [20]. RBF networks are three layer feed-forward networks that are trained using a supervised training algorithm, which is shown in Figure 8, featuring are an input x, hidden units $\mathbf{\Psi}$, weights w, linear output transfer functions, and output $f(x)$. To construct the fitting model using RBF, we first consider the scalar response function f or the yielding responses $y = \{y_1, y_2, \ldots, y_n\}^{\mathrm{T}}$ which is obtained from the simulation or experimental results by employing the input training sample data of $X = \{x_1, x_2, \ldots, x_n\}^{\mathrm{T}}$. Then, we pursue RBF approximation function $\hat{f}(x)$, which is given as

$$\hat{f}(x) = w^{\mathrm{T}}\mathbf{\Psi} = \sum_{i=1}^{n_c} w_i \Psi_i(\|x - c_i\|),$$ (5)

where c_i is the ith of the n_c basis function centers and $\mathbf{\Psi}$ denotes the n_c vector which is contained the values of the basis function $\mathbf{\Psi}$, it is evaluated at the Euclidean distances between the prediction site x and the centers c_i of the basis functions, w_i denotes the weight of the ith basis function. In a previous study, different basis functions are suggested, which is used in RBF, some of the basis functions are stated as follow under categories of fixed and parametric basis functions:

1. Fixed basis functions

    - Linear: $\Psi(x) = (x - c_i)$
    - Cubic: $\Psi(x) = (x - c_i)^3$
    - Thin plate spline: $\Psi(x) = (x - c_i)^2 \ln(x - c_i)$

2. Parametric basis functions

    - Gaussian: $\Psi(x) = e^{-(x-c_i)^2/(2\sigma^2)}$
    - Multiquadratic: $\Psi(x) = ((x - c_i)^2 + \sigma^2)^{1/2}$
    - Inverse multiquadratic: $\Psi(x) = ((x - c_i)^2 + \sigma^2)^{-1/2}$

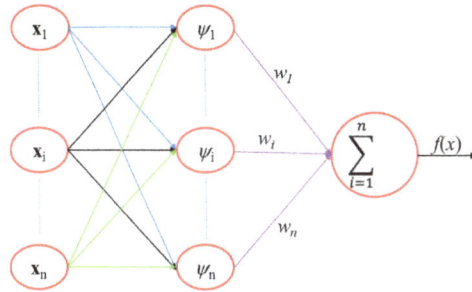

Figure 8. Radial basis function network.

As we have seen from the formulation equation, the weight w parameters are unknown, in addition, for parametric basis functions, the variance (width) σ is also unknown. The estimation of the weight parameters can be found through the interpolation of

$$y(x_j) = \sum_{i=1}^{n_c} w_i \Psi_i(\|x - c_i\|), \; j = 1, 2, \cdots, n. \tag{6}$$

The above equation is linear in terms of the weights w of the basis function; however, the predictor y can express highly nonlinear. To obtain a unique solution, the system in Equation (6) has to be square; this means that n_c should be equal to n. For simplification, the bases coincide with the data points, which means $c_i = x_i$, for $i = 1, 2, \ldots, n$, this leads us to the equation of

$$\Psi w = y, \tag{7}$$

where Ψ is also called as *Gram matrix* and it is defined as $\Psi = \Psi(\|x_j - x_i\|), \; i = j = 1, 2, \ldots, n$. From Equation (7) we can estimate the weight value as

$$w = \Psi^{-1} y, \tag{8}$$

However, if the responses $y = \{y_1, y_2, \ldots, y_n\}^{\mathrm{T}}$ are corrupted by noise, using the above equation may affect the prediction model that to fit the observed data. The noise effect should be considered in the model. To solve this kind of problem, Poggio and Girosi [21] introduced using a regularization parameter λ as model flexibility can control the noise effect on the prediction model. It is also recommended that λ value should be adequately small, such as $\lambda = 1.0 \times 10^{-3}$. This will add to the diagonal matrix of Ψ. So the weight estimation value will be in the form of

$$w = (\Psi^{\mathrm{T}}\Psi + \lambda I)\Psi^{\mathrm{T}} y, \tag{9}$$

where Ψ, and λI are given as follows:

$$w = \begin{bmatrix} \Psi_1(x_1) & \Psi_2(x_1) & \cdots & \Psi_n(x_1) \\ \Psi_1(x_2) & \Psi_2(x_2) & \cdots & \Psi_n(x_2) \\ \vdots & \vdots & \ddots & \vdots \\ \Psi_1(x_n) & \Psi_2(x_n) & \cdots & \Psi_n(x_n) \end{bmatrix}, \tag{10}$$

$$I = \begin{bmatrix} 1 & 0 & \cdots & 0 \\ 0 & 1 & \cdots & 0 \\ \vdots & \vdots & \ddots & \vdots \\ 0 & 0 & \cdots & 1 \end{bmatrix}. \tag{11}$$

To estimate the sigma value the previous studies proposed different methods, for instance, Nakayama, et al. [22] proposed $\sigma = \frac{d_{max}}{\sqrt[m]{nm}}$, where n denotes the number of sampling point, m denotes the number of process parameters, and d_{max} is denotes the maximum distance among the sampling point. Whereas, the accurate estimation of the sigma parameters will allow for us to reduce the generalization (estimated) error of the prediction model, so, this study used the direct search method by investigating the model accuracy using the cross-validation error for each sigma value. Cross-validation is a model verification method, which is used to assess the results of a statistical analysis to simplify the independent data set. Here, the study investigates the model by removing some data from the training sample data and evaluating each sigma value for every removed data and finally we chosen the best sigma value, which has a minimum model cross-validation error. The cross-validation function is given as:

$$E_{crv} = \frac{1}{q}\sum_{i=1}^{q}\left[y(x_i) - \hat{f}(x_i, w)\right], \tag{12}$$

where q is the number of the removed data from the training sample (subsets of the training sample), $y(x_i)$, and $\hat{f}(x_i, w)$ is the true and prediction response value at the ith removed training sample, respectively. If q is equal to the number of training sample data n, the cross-validation error is a nearly balanced estimator of the exact risk. Nevertheless, because of the n subsets being similar to each other the leave-one-out measure variance can be quite high. Hastie et al. [23] proposed a desirable value of q such as $q = 5$ or $q = 10$, it is depends on the total number of sample data. In general, using less number of the training sample subsets q means reducing the cross-validation process computational cost by reducing the total number of the fitted model.

Here, the study used 30 numbers of samples, so we chose q is equal to 5 and the σ value is searched in between 10^{-2} and 10^2. The algorithm is written in MATLAB and the σ value found as 1.6103. In addition, the obtained weight w value also listed in Table 3.

Table 3. The weight value.

No.	Value
1	−0.1325
2	2.7338
3	0.849
4	−0.3378
5	−0.0510
6	−2.4592
7	0.9597
8	−1.9367
9	−0.5711
10	−2.6565
11	−0.6119
12	1.2294
13	3.2205
14	1.7304
15	−1.4591
16	−1.0661
17	−2.6649
18	−3.1082
19	7.7967

Table 3. *Cont.*

No.	Value
20	−2.9087
21	1.2768
22	0.6473
23	3.0969
24	−0.0234
25	−2.3770
26	−1.7703
27	0.3146
28	0.2355
29	−0.1235
30	0.4802

3.4. Numerical Model Verification

For prediction model verification, the roots mean square error (RMSE), and co-efficient of determination (R^2) numerical verification techniques are used in this study. Table 4 compared the model verification of different prediction methods, as we have seen from the model verification results, RBF with multiquadratic basis function have better prediction than the others. Figure 9 shows of the predicted vs. the observed result comparison for RBF with a multiquadratic basis function; the predicted and observed result is quietly fitted to each other. Figure 10 also shows that the residual values are randomly distributed and the residual mean value also almost zero, this verified that the developed model is acceptable to use.

Table 4. Results of model verification for different prediction models.

Model		R^2	RMSE
	Blind Kriging 2nd deg.	0.9837	0.0518
	Blind Kriging 3rd deg.	0.9848	0.0501
	Ordinary Kriging	0.9589	0.4510
	Linear	0.9811	0.0500
	Cubic	0.9854	0.0444
RBF	Thin plate spline	0.8728	0.1297
	Gaussian	0.9868	0.0418
	Multiquadratic	0.9935	0.0294
	Inverse multiquadratic	0.9717	0.0612
	Regression	0.9769	0.0617

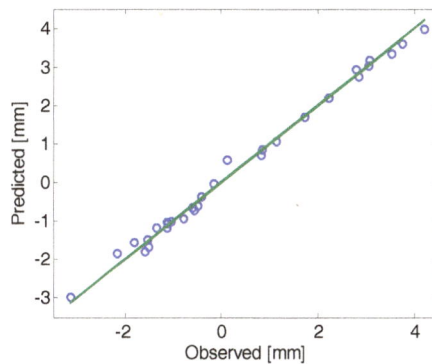

Figure 9. Predicted vs. observed result for RBF with multiquadratic basis function.

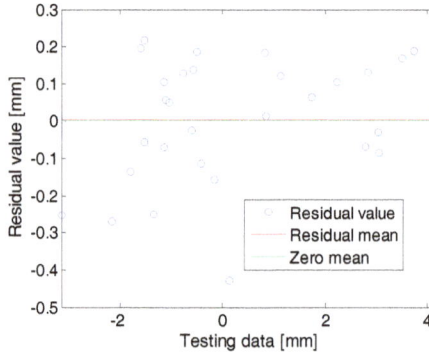

Figure 10. Residual error.

4. Process Optimization

4.1. Optimization Formulation

The objectives of the study are reducing the springback and compensating the induced springback for the expected target shape. As explored in Section 2.4, the springback is examined using the maximum deviation of the vertical displacement of the deformed shape from the expected target shape for both objectives. The formulation function is given as

$$\begin{aligned} \min \ & f(x) \\ \text{s.t. } & x^l \le x_i \le x^u \end{aligned} \tag{13}$$

where $x(x_1, x_2,$ and $x_3)$ is the process parameters, here, the constructed function is the same for both of the objectives. The first objective is to minimize the induced springback for a given curvature radius and blank thickness. So, the curvature radius and blank thickness should be fixed, the only parameter left here is the punch holding time before unloading of the punch load. The second objective is minimizing the maximum curvature radius difference from the target shape to compensate for the expected shape error due to springback. In this case, the punch holding time is fixed to the obtained optimal value of the first objective for the given target curvature radius and blank thickness.

4.2. Result and Discussion

After developing the prediction model using RBF as a function of curvature radius, punch holding time, and blank thickness, the study employed genetic algorithm on MATLAB to determine the global optimal value of the given objective. The algorithm scheme is shown in Figure 11. From the first objective, the optimal value of punch holding time is obtained to reduce the springback for the fixed curvature radius and blank thickness. After obtaining the proper punch holding time, we proceed to find the optimal value of the curvature radius to compensate for the induced springback for the given target shapes, in our case, the target curvature radius was 800 mm and the blank thickness was 2 mm. Table 5 summarizes the obtained process parameter setting to compensate for the induced springback for the expected target shape. Figure 12 shows the curvature radii comparison between the simulation result and the target shape. The RMSE between the simulation result and target shape is 0.0245, this shows us the obtained result is acceptable with the mentioned error.

Table 5. Optimal result.

Curvature Radius (mm)	Blank Thickness (mm)	Punch Holding Time (s)	Deviation from 800 mm Curvature Radius (mm)
724.00	2.00	6.00	0.58

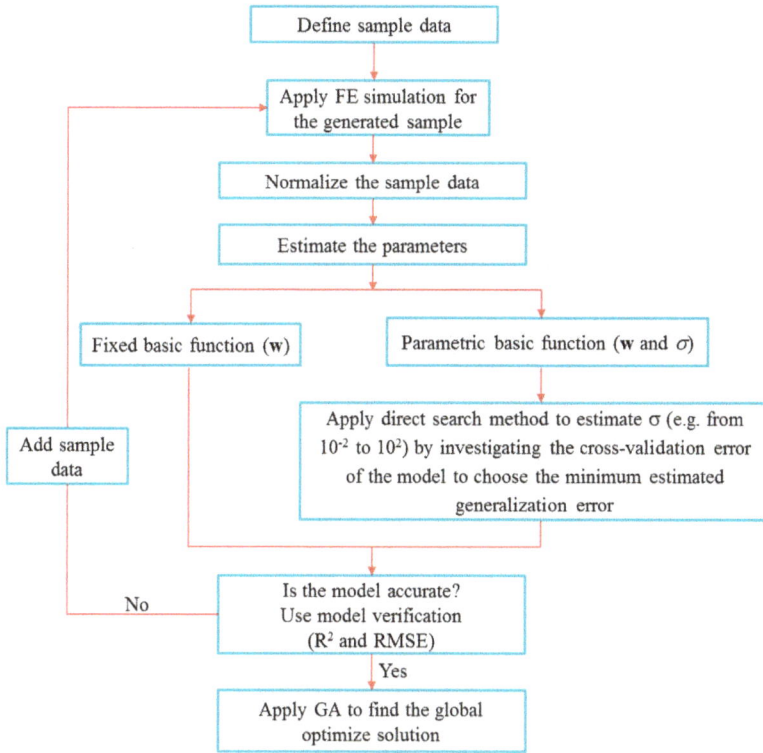

Figure 11. Radial basis optimization algorithm.

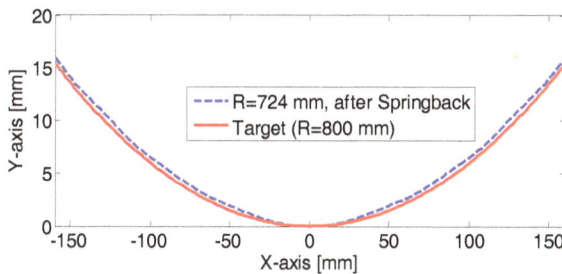

Figure 12. Compensated optimal result.

5. Conclusions

The induced springback in MDF process is investigated using ABAQUS commercial software, and the study tried to reduce the springback by finding the optimal punch holding time. The study also compensates for the induced springback by finding the optimal curvature radius, using the developed surrogate model for different blank thickness. The prediction model was chosen after investigating different types of the prediction model using the model verification of R^2 and RMSE. The study chose RBF with a multi-quadratic base function to develop the model as a function of punch holding time, curvature radius, and blank thickness. Finally, the study found the optimal curvature radius, and the punch holding time for the target curvature and blank thickness with the RMSE of 0.0245 by employing the GA on MATLAB.

Acknowledgments: This work was supported by the National Research Foundation of Korea (NRF) grant funded by the Korea government (MSIP) through the Engineering Research Center (No. 2012R1A5A1048294). Also, this work was supported by the Human Resource Training Program for the Regional Innovation and Creativity through the Ministry of Education and National Research Foundation of Korea (NRF-2015H1C1A1035499).

Author Contributions: Misganaw Abebe and Jun-Seok Yoon conceived and designed the experimental data; Misganaw Abebe and Beom-Soo Kang analyzed the data; Misganaw Abebe wrote the algorithm on MATLAB, Misganaw Abebe and Jun-Seok Yoon wrote the paper and edited by Beom-Soo Kang.

Conflicts of Interest: The authors declare no conflict interest.

Appendix A

Table A1. Training sample data.

Case	Punch Holding Time, x_1 (s)	Curvature Radius, x_2 (mm)	Thickness, x_3 (mm)
1	14.482	766	0.862
2	13.104	738	1.741
3	8.966	779	0.655
4	17.932	772	1.172
5	10.344	731	1.379
6	2.068	641	1.069
7	4.138	745	0.552
8	18.620	628	0.914
9	11.034	662	1.793
10	8.276	614	1.276
11	4.828	703	1.535
12	2.758	683	1.948
13	15.172	800	1.483
14	13.794	634	0.604
15	16.552	710	1.017
16	20.000	697	1.690
17	17.242	600	2.000
18	12.414	607	1.638
19	9.656	786	1.897
20	0.000	648	1.586
21	19.31	655	0.500
22	0.690	759	1.431
23	6.896	724	1.845
24	11.724	676	0.810
25	15.862	669	1.328
26	1.380	717	1.121
27	5.518	690	0.707
28	3.448	793	0.965
29	6.206	752	1.224
30	7.586	621	0.759

Table A2. Testing sample data.

Case	Punch Holding Time, x_1 (s)	Curvature Radius, x_2 (mm)	Thickness, x_3 (mm)
1	1.518	779	1.126
2	1.080	608	0.575
3	10.616	800	1.854
4	15.584	645	1.917
5	18.680	637	1.236
6	2.598	626	1.234
7	11.376	783	1.007
8	9.388	720	1.850
9	0.238	747	1.054
10	6.742	641	0.667
11	3.244	714	1.670
12	15.886	618	1.085
13	6.224	763	0.863
14	10.570	747	1.106
15	3.312	771	0.645
16	12.040	773	0.698
17	5.260	626	1.913
18	13.082	684	1.934
19	13.784	690	1.363
20	14.964	771	0.590
21	9.010	628	0.852
22	1.676	676	1.030
23	4.580	731	1.732
24	18.266	697	1.573
25	3.048	720	0.565
26	16.516	786	0.754
27	10.766	752	1.474
28	19.922	775	1.598
29	1.564	763	1.472
30	8.854	752	1.176

References

1. Li, L.; Seo, Y.H.; Heo, S.C.; Kang, B.S.; Kim, J. Numerical simulations on reducing the unloading springback with multi-step multi-point forming technology. *Int. J. Adv. Manuf. Technol.* **2010**, *48*, 45–61. [CrossRef]
2. Hwang, S.Y.; Lee, J.H.; Yang, Y.S.; Yoo, M.J. Springback adjustment for multi-point forming of thick plates in shipbuilding. *Comput. Aided Des.* **2010**, *42*, 1001–1012. [CrossRef]
3. Shim, D.S.; Baek, G.Y.; Shin, G.Y.; Yoon, H.S.; Lee, K.Y.; Kim, K.H. Investigation of tension force in stretch forming of doubly curved aluminum alloy (Al5083) sheet. *Int. J. Precis. Eng. Manuf.* **2016**, *17*, 433–444. [CrossRef]
4. Liu, W.; Liang, Z. Springback compensation for multi-curvature part based on multi-objective optimization of fuzzy genetic algorithm. In Proceedings of the IEEE Chinese Control and Decision Conference, Guilin, China, 17–19 June 2009; pp. 3659–3664.
5. Zhang, Q.F.; Cai, Z.Y.; Zhang, Y.; Li, M.Z. Springback compensation method for doubly curved plate in multi-point forming. *Mater. Des.* **2013**, *47*, 377–385. [CrossRef]
6. Kitayama, S.; Yoshioka, H. Springback reduction with control of punch speed and blank holder force via sequential approximate optimization with radial basis function network. *Int. J. Mech. Mater. Des.* **2014**, *10*, 109–119. [CrossRef]
7. Behera, A.K.; Verbert, J.; Lauwers, B.; Duflou, J.R. Tool path compensation strategies for single point incremental sheet forming using multivariate adaptive regression splines. *Comput. Aided Des.* **2013**, *45*, 575–590. [CrossRef]
8. Li, E. Reduction of springback by intelligent sampling-based LSSVR metamodel-based optimization. *Int. J. Mater. Form.* **2013**, *6*, 103–114. [CrossRef]

9. Khadra, F.A.; El-Morsy, A.W. Prediction of Springback in the Air Bending Process Using a Kriging Metamodel. *Eng. Technol. Appl. Sci. Res.* **2016**, *6*, 1200.
10. ul Hassan, H.; Fruth, J.; Güner, A.; Mennecart, T.; Tekkaya, A.E. Finite element simulations for sheet metal forming process with functional input for the minimization of springback. In Proceedings of the IDDRG Conference, Zurich, Switzerland, 2–5 June 2013; pp. 393–398.
11. Karaağaç, İ. The evaluation of process parameters on springback in V-bending using the flexforming process. *Mater. Res.* **2017**, *20*, 1291–1299. [CrossRef]
12. Tekaslan, Ö.; Gerger, N.; Şeker, U. Determination of spring-back of stainless steel sheet metal in "V" bending dies. *Mater. Des.* **2008**, *29*, 1043–1050. [CrossRef]
13. Choudhury, I.A.; Ghomi, V. Springback reduction of aluminum sheet in V-bending dies. *Proc. Inst. Mech. Eng. Part B* **2014**, *228*, 917–926. [CrossRef]
14. Jakumeit, J.; Herdy, M.; Nitsche, M. Parameter optimization of the sheet metal forming process using an iterative parallel Kriging algorithm. *Struct. Multidiscip. Optim.* **2005**, *29*, 498–507. [CrossRef]
15. Naceur, H.; Guo, Y.Q.; Ben-Elechi, S. Response surface methodology for design of sheet forming parameters to control springback effects. *Comput. Struct.* **2006**, *84*, 1651–1663. [CrossRef]
16. Wei, L.; Yang, Y.; Xing, Z.; Zhao, L. Springback control of sheet metal forming based on the response-surface method and multi-objective genetic algorithm. *Mater. Sci. Eng. A* **2009**, *499*, 325–328. [CrossRef]
17. Abebe, M.; Lee, K.; Kang, B.S. Surrogate-based multi-point forming process optimization for dimpling and wrinkling reduction. *Int. J. Adv. Manuf. Technol.* **2016**, *85*, 391–403. [CrossRef]
18. Heo, S.C.; Seo, Y.H.; Yoon, J.S.; Song, W.J.; Kang, B.S. Effect of design variables on forming accuracy in thick plate flexible forming process. In Proceedings of the 14th International conference on Metal Forming, AGH University of Science and Technology, Krakow, Poland, 16–19 September 2012; pp. 1403–1406.
19. Abebe, M.; Park, J.W.; Kang, B.S. Reliability-based robust process optimization of multi-point dieless forming for product defect reduction. *Int. J. Adv. Manuf. Technol.* **2017**, *89*, 1223–1234. [CrossRef]
20. Hardy, R.L. Multiquadric equations of topography and other irregular surfaces. *J. Geophys. Res.* **1971**, *76*, 1905–1915. [CrossRef]
21. Poggio, T.; Girosi, F. Regularization algorithms for learning that are equivalent to multilayer networks. *Science* **1990**, *247*, 978–982. [CrossRef] [PubMed]
22. Nakayama, H.; Masao, A.; Rie, S. Simulation-based optimization using computational intelligence. *Optim. Eng.* **2002**, *3*, 201–214. [CrossRef]
23. Hastie, T.; Tibshirani, R.; Friedman, J. *The Elements of Statistical Learning*; Springer: New York, NY, USA, 2001.

metals   MDPI

Article

Revisiting Formability and Failure of AISI304 Sheets in SPIF: Experimental Approach and Numerical Validation

Gabriel Centeno [1],*, Andrés Jesús Martínez-Donaire [1], Isabel Bagudanch [2], Domingo Morales-Palma [1], María Luisa Garcia-Romeu [2] and Carpóforo Vallellano [1]

[1] Department of Mechanical and Manufacturing Engineering, School of Engineering, University of Seville, 41092 Seville, Spain; ajmd@us.es (A.J.M.-D.); dmpalma@us.es (D.M.-P.); carpofor@us.es (C.V.)
[2] Department of Mechanical Engineering and Industrial Construction, University of Girona, 17071 Girona, Spain; isabel.bagudanch@udg.edu (I.B.); mluisa.gromeu@udg.edu (M.L.G.-R.)
* Correspondence: gaceba@us.es; Tel.: +34-954-485965

Received: 11 October 2017; Accepted: 24 November 2017; Published: 28 November 2017

Abstract: Single Point Incremental Forming (SPIF) is a flexible and economic manufacturing process with a strong potential for manufacturing small and medium batches of highly customized parts. Formability and failure in SPIF have been intensively discussed in recent years, especially because this process allows stable plastic deformation well above the conventional forming limits, as this enhanced formability is only achievable within a certain range of process parameters depending on the material type. This paper analyzes formability and failure of AISI304-H111 sheets deformed by SPIF compared to conventional testing conditions (including Nakazima and stretch-bending tests). With this purpose, experimental tests in SPIF and stretch-bending were carried out and a numerical model of SPIF is performed. The results allow the authors to establish the following contributions regarding SPIF: (i) the setting of the limits of the formability enhancement when small tool diameters are used, (ii) the evolution of the crack when failure is attained and (iii) the determination of the conditions upon which necking is suppressed, leading directly to ductile fracture in SPIF.

Keywords: formability; failure; sheet metal forming; Single-Point Incremental Forming (SPIF)

---

1. Introduction

Incremental Sheet Forming (ISF) processes accomplish the current requirements for rapid, adaptive, economic and environmentally friendly manufacturing. It is especially viable for small batches of parts made of sheet and does not need expensive dedicated machines or equipment. Indeed, it is a relative novel process that has been in the spotlight of the metal-forming community for the last two decades. Although the incremental sheet-forming technology is linked to the process of spinning, the current ISF process has its origins in the late 1960s related to the pioneer works of Leszak [1] and Berghahn and Murray [2], both in 1967. Nevertheless, following the analysis of the historical review by Emmens et al. [3] only the latter can be regarded as an actual version of modern ISF. This investigation reveals that those 2 initial patents were not the work leading to the present developments, but the Bachelor Thesis of Mason in 1978 [4] presented to the scientific community later in 1984 [5], which would be the real origin of the current state of the art in ISF.

Single-Point Incremental Forming (SPIF) is the simplest type within ISF processes, which make use of a simple setup not requiring any partial or full die. As shown in Figure 1, the SPIF technology consists of a hemispherical end-forming tool driven by a CNC machine that follows progressively a pre-established trajectory, deforming a peripherally clamped sheet blank into a final component without the use of any specific forming die.

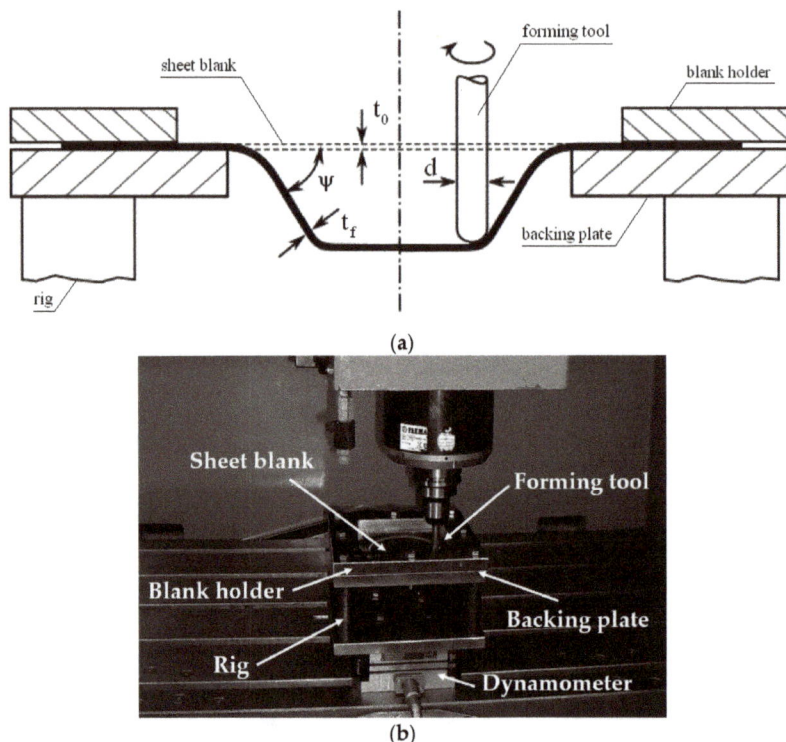

Figure 1. (**a**) Schematic representation of the SPIF process and (**b**) experimental setup utilized.

Formability and failure of sheet metal deformed by SPIF is usually analyzed within Forming Limit Diagrams or FLDs, which include the limit strains at the onset of local necking, represented by the Forming Limit Curve (FLC), as well as at ductile fracture, characterized by the Fracture Forming Line (FFL). The current methods for the evaluation of these limit strains at the onset of necking and fracture have been recently discussed in [6]. In this regard, high ductility metal sheets deformed by conventional forming processes usually start failing at the onset of necking, i.e., the material deforms continuously within this neck under an unstable deformation process, following approximately a near plane strain state, until the ductile fracture takes place. On the contrary, metal sheets deformed by SPIF (or any other ISF variety) within a certain range of process parameters suffer a stable straining above the FLC that may lead directly to ductile fracture. The stabilization mechanisms providing the enhanced formability observed in ISF are presented and discussed in the review paper by Emmens and van den Boogaard [7].

With this background, Silva et al. [8] carried out tests that revealed the possible existence of both deformation mechanisms, either fracture with a previous necking or failure by direct ductile fracture, depending on the ratio between the initial thickness of the sheet and the radius of the tool (t_0/R). For large tool diameters, failure by necking could still occur, whereas for small tool diameters, fracture in absence of necking would be promoted, and formability should then be represented by the FFL. A more recent study [9] demonstrated that in both of the previous cases, i.e., failure controlled either by necking or by ductile fracture, fracture strains are always within a scatter band of the FFL.

In this regard, the authors studied in a previous work [10] the effect of the localized bending induced by the forming tool, evaluated through the above mentioned t_0/R ratio, in the stabilization of plastic deformation above the FLC during ISF. It was observed that for higher tool diameter (20 mm), the failure mode was due to necking followed by ductile fracture. However for the lowest tool diameters considered (10 mm), failure occurred by fracture in the absence of necking.

Furthermore, considering that for a certain range of process parameters corresponding to high t_0/R ratios failure will occur without previous necking in SPIF, Isik et al. [11] proposed a new methodology to determine the maximum strains at fracture directly from the in-plane strain measurements without evaluating the gauge length strains, which simplifies the procedure for obtaining the FFL. What is required is a series of tests on parts with a variable wall angle: truncated conical parts (plane strain conditions) and truncated pyramidal parts (plane strain conditions in the walls of the pyramid and biaxial stretching at the corners). Isik et al. also introduced the concept of Shear Fracture Forming Limit line or SFFL, corresponding to mode II of the fracture mechanics (in-plane shear), which can also be excited under certain loading conditions in ISF. In-plane torsion tests and plane shear tests are required to represent this new forming limit. In order to simplify and facilitate the determination of SFFL, a new geometry manufactured by ISF has been recently proposed [12]. This proposal involves using a truncated lobe conical shape with varying wall angle and measuring the in-plane strains at fracture, thus avoiding the need to measure gauge length strains, which is required with typical test specimens (in-plane torsion and plane shear tests).

In this scientific framework for SPIF, this paper allows the authors presenting the following contributions to the current state of the art in ISF regarding the SPIF process applied to AISI304-H111 sheets: (i) the setting of the limits of the formability enhancement when small tool diameters are used, (ii) the evolution of the crack when failure is attained and (iii) the conditions, validated by the FEA, upon which necking is suppressed, leading directly to ductile fracture in SPIF.

After contextualizing this study within the state of the art in SPIF, Section 2 presents the material and experimental methods utilized, Section 3 focuses on the numerical modelling of the SPIF process carried out and Section 4 discusses the experimental and numerical results obtained. Finally, the contributions of the paper are exposed in Section 5 "Conclusions".

2. Materials and Experimental Methods

This section starts presenting the mechanical characterization of the AISI304-H111 metal sheets obtained from uniaxial tensile tests and providing a power law containing the plastic behavior of the material to be used in the numerical simulations.

In Section 2.2, the forming limits of the material are given by means of conventional Nakazima tests, which combined by a series of stretch-bending tests carried out with a set of cylindrical punches provide the FLD of the sheet metal including bending effects.

Finally, the experimental plan and the experimental methods used in the case of the SPIF tests are presented in Section 2.3.

2.1. Mechanical Characterization

The material analyzed is stainless steel AISI 304-H111 sheet metal of 0.8 mm thickness. The mechanical properties obtained from the tensile tests are summarized in Table 1 [10]. As pointed out in the cited previous work of the authors [10], the plastic behavior of the material fits a Swift's power law as shown in Equation (1).

$$\bar{\sigma} = K\left(\varepsilon_0 + \bar{\varepsilon}^P\right) \tag{1}$$

where E is the elastic modulus, $\sigma_{y0.2}$ is the yield stress, UTS the ultimate tensile stress, K, n and ε_0 are constants of the Swift's power law depending on the material, $\bar{\varepsilon}^P$ is the equivalent plastic strain and $\bar{\sigma}$ the equivalent stress.

Table 1. Mechanical properties from tensile tests and Swift's power law parameters.

E (GPa)	$\sigma_{y0.2}$ (MPa)	UTS (MPa)	K (GPa)	n	ε_0
207	503	669	1.55	0.594	0.055

2.2. Forming Limit Diagram

A series of Nakazima tests were carried out using a hemispherical punch of 100 mm diameter using specimens corresponding to uniaxial, close to plane strain and biaxial strain conditions. The conventional forming limits represented by the FLC and the FFL were obtained. In addition, stretch-bending tests using cylindrical punches of Φ20, Φ10 and Φ6 mm were performed with the aim of evaluating the effect of the bending induced by the tool radius in postponing the onset of necking due to the significant through-thickness strain gradient induced by the curvature of the punches. These former cases led to strain paths in between plane and uniaxial strain. At least 3 replicates of every test were carried out in order to provide statistical meaning to the results obtained.

The tests were performed in a universal sheet metal testing machine Erichsen 142-20 under the testing conditions of the standard ISO 12004-2:2008 [13]. The punch velocity was set to 1 mm/s and the lubricant at the interface punch-sheet was Vaseline + PTFE + Vaseline. The system ARAMIS®, based on digital image correlation (DIC), was used at a rate of 12 frames per second to evaluate the onset of necking by using a methodology proposed by the authors [14].

Once the onset of necking is attained, the major strain (ε_1) of the points distributed around the failure zone develops unstably close to plane strain conditions until the fracture strains is reached, being this behavior characteristic in the necking-controlled failure observed in all of the Nakazima and stretch-bending tests. According to this, the procedure for constructing the FFL starts by measuring the thickness at fracture at several places along the crack in order to obtain the average thickness strain, which is evaluated using the measurements at both sides of the crack for every tested specimen. The average minor strain is evaluated along the fracture line at the last image recorded by ARAMIS® before the crack appearance. The major strain is then calculated by volume constancy as expressed in Equation (2).

$$\bar{\varepsilon}_{1,f} = -\left(\bar{\varepsilon}_{2,f} - \bar{\varepsilon}_{3,f}\right) \tag{2}$$

where $\bar{\varepsilon}_{2,f}$ and $\bar{\varepsilon}_{3,f}$ are the average minor and thickness strains evaluated in a series of points along the crack line. In addition, it must be pointed out that some tested specimens were cut perpendicularly to the crack and the thickness was measured from a profile view in order to validate the previous thickness measurements along the crack. This methodology for determining the strains at fracture is based on the work of Atkins [15] and has been successfully used by the authors in recent research work for measuring fracture strains in forming of sheet metal [10], polymeric sheets [16,17], or even other processes such as tube-end forming [18].

Figure 2 depicts the forming limit diagram of the AISI 304-H111 sheets including bending effects. The evaluation of the FLC and the FFL was performed by using the methodologies exposed above and only taking into account the Nakazima tests. The average necking and fracture strains are provided for the 3 strain paths considered. As can be seen, the FLC presents the expected V-shape whereas the FFL falls to the first quadrant of principal strain following the straight line $\varepsilon_1 + 0.69\varepsilon_2 = 1.08$ with a slope not far from the theoretical value of "-1" proposed by Atkins in [15]. The kink of the strain paths from the onset of necking towards fracture is almost vertical in the first quadrant whereas this transition in the second quadrant suffers from a slight leftward slope, according again to the cited work of Atkins. As can be seen, the bending effect is represented by means of the average necking and fracture strains (not used for obtaining the FLC and FFL) in the stretch-bending tests. Notice that although the significant enhancement of formability attained above the FLC due to this bending effect (which is further discussed in [14]), the fracture strains are placed on the FFL region.

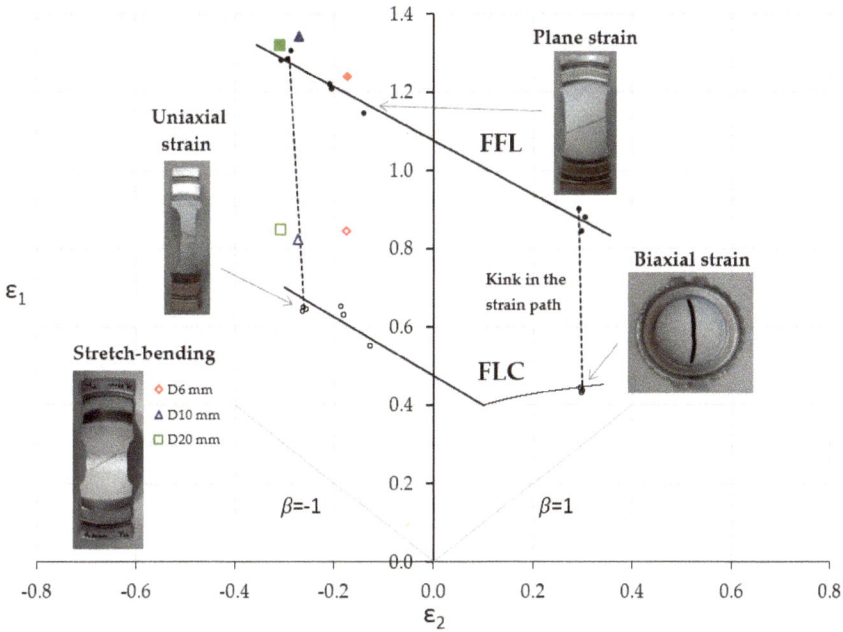

Figure 2. FLD of AISI 304-H111 sheets including bending effects.

2.3. Single Point Incremental Forming Tests

A Computer Numeric Control (CNC) 3-axis milling machine Kondia® HS1000 equipped with the experimental setup shown in Figure 1b was used for carrying out the SPIF tests. As shown in Figure 3a the testing geometry was a conical frustum with circular generatrix of radius 40 mm, initial diameter of the truncated cone 70 mm, and initial drawing angle 20°. Tool diameters of 20, 10 and 6 mm were utilized. The step down was set alternatively to 0.2 mm and 0.5 mm/pass. The step down movement was in the same place during the test, following the forming tool alternatively in-plane clockwise or counterclockwise trajectories for consecutive step downs (see Figure 3b) in order to avoid torsion effects in the final part (see Figure 3c). The tool rotation was free. Special metal-forming lubricant Houghton TD-52 was used with the aim of minimizing friction.

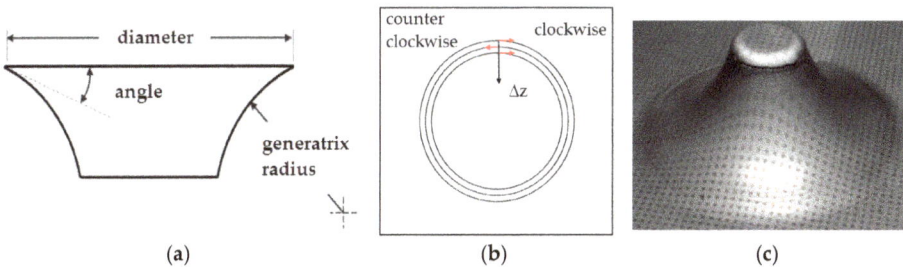

Figure 3. (**a**) Truncated coned geometry, (**b**) tool trajectory and (**c**) final part after testing.

Table 2 shows the SPIF tests carried out within the experimental plan designed, which was already presented in [10]. Once again, three replicates of every SPIF test were carried to provide statistical meaning to the results obtained. The table provides the final depth recorded in the instant in which the failure took place and as well the proportional final forming angle calculated from the predicted trajectories to form the final testing part geometry.

Table 2. Experimental plan of SPIF tests.

Tool Diameter Φ (mm)	Step Down Δz (mm/pass)	Final Depth Z_f (mm)	Final Forming Angle α_f (°)
20	0.2	23.8/23.8/23.8	69.8/69.8/69.8
	0.5	24.5/24.0/24.0	70.9/70.1/70.1
10	0.2	28.0/28.2/28.2	76.1/76.4/76.4
	0.5	27.5/28.0/28.0	70.9/70.1/70.1
6	0.2	28.2/**28.0**/28.8	76.4/76.1/77.3
	0.5	**28.0/28.5/28.0**	76.1/76.9/76.1

The final strain state of the testing specimens deformed by SPIF was evaluated off-line by using the 3D deformation digital measurement system ARGUS® based on circle grid analysis. To this aim a grid pattern of 1 mm diameter was electro-chemically edged on the sheet blank prior to the tests. Figure 4a depicts the grid on the final part deformed by SPIF using a hemispherical forming tool of 20 mm diameter, whereas Figure 4b depicts the contour of the major principal strain evaluated by ARGUS® (notice the similar part orientation in Figure 4a,b). The zone of maximum major strain was located at the vicinity of the crack corresponding to the interpolation of the strains throughout it performed by ARGUS®. Fracture strains in SPIF were obtained following the methodology that was previously explained in Section 2.2.

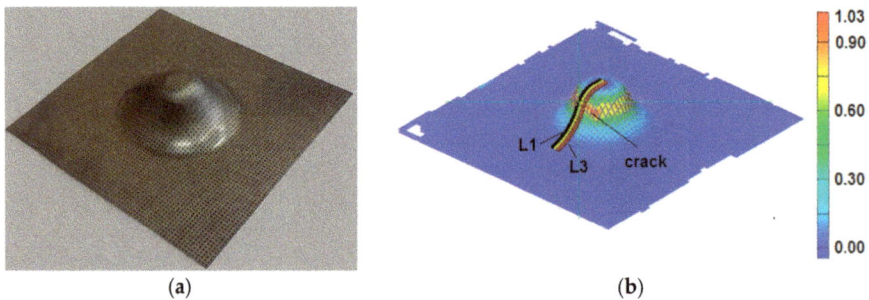

(a)   (b)

Figure 4. (a) Final grid and (b) contour of true major principal strain of a tested part by SPIF.

In order to evaluate the principal strains on the outer surface of the final tested parts deformed by SPIF within the FLD of the material, several sections are selected in every case, such as the section L1 to L3 shown in Figure 4b. The values of the interpolation of principal strains provided by ARGUS® just on the crack line are not taken into account in the FLD, as far as the strains at fracture are calculated using the procedure explained above.

3. Numerical Modelling

This section presents the numerical model carried out using DEFORM™-3D with the aim of analyzing virtually for providing information about the failure prediction and the mechanisms involved in the enhancement of formability attained in SPIF.

DEFORM™-3D is a commercial Finite Element Analysis (FEA) tool based on flow formulation and with implicit computation. Is it a powerful numerical tool that has been mainly used recently for modelling manufacturing processes such as cutting [19] and bulk forming of metals [20], but it has been also used for simulating sheet metal-forming processes [21,22].

The numerical model in DEFORM™-3D was developed using 3D tetrahedrons, having the initial mesh 50,000 elements. As can be seen in Figure 5a, three circular meshing zones were considered, having the intermediate annulus a smaller size of elements in order to provide accurate results within the tool-sheet contact region corresponding to higher values of major strain. Automatic remeshing was used, allowing DEFORM™-3D to adapt the mesh size in the zones attaining the highest strain values. The punch is considered to be a rigid body and follows the real trajectory of the experiments. The elements corresponding to the area of sheet metal in contact with the backing plate are considered to be clamped, as shown in Figure 5b. The sheet metal behaves as an elastic-plastic rate-independent material with kinematic hardening. The elastic-plastic behavior is supposed to be isotropic following the Swift's power law presented in Section 2.1. Due to the high computational cost and taking into account the negligible influence of the step down in formability triggered in [10], the step down was set to 0.5 mm/pass in order to reduce the simulation steps. The simulations were performed until a reference final depth related to the tool depths at failure presented in Table 2, which were chosen to be 24 mm for the case of a tool diameter of Φ20 mm and 28 mm for the case of Φ10 mm respectively.

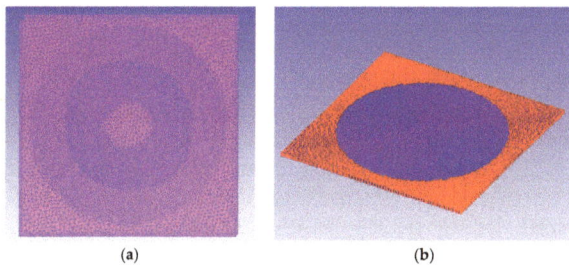

(a)   (b)

Figure 5. (a) Meshing zones considered and (b) boundary conditions (clamped condition in "red").

Figure 6a,b depicts the deformed shape for the simulation of the SPIF process considering a step down of 0.5 mm/pass using a tool of Φ20 and Φ10 mm respectively. Figure 6c shows the contour of major principal strains corresponding to the case of a tool of Φ20 mm (shown in Figure 6a). Notice that in this case, the maximum value of the major strain (ε_1) provided by the FEA was 0.879.

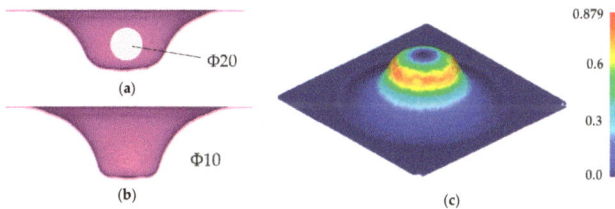

Figure 6. (a) Deformed shape provided by DEFORM™-3D of the final testing part in SPIF using a tool of Φ20 diameter and (b) Φ10 mm. (c) contour of principal strain for the case of Φ20 mm.

Finally, it is worth mentioning that the overall central processing unit (CPU) time for a typical analysis shown in Figure 6c, which include 6 remeshing processes, was approximately 85 h on a laptop using one Intel i7-6500U CPU (2.60 GHz) processor.

4. Results and Discussion

In this section, the experimental and numerical results regarding formability and failure of AISI304-H111 sheets deformed in SPIF are presented. In Section 4.1, the limits of the formability enhancement for small tool diameters are evaluated within the FLD of the material. A fractography analysis is used to clarify the mode of failure attained, the location where the crack initiates and how it evolves once failure is reached. Section 4.2 discusses the numerical results obtained, presenting the conditions upon which necking is suppressed, leading directly to ductile fracture in SPIF.

4.1. Experimental Results

Figure 7 depicts the principal strain state at the outer surface of the final testing part deformed by SPIF using tool diameters of Φ20, Φ10 and Φ6 mm represented within the FLD of the AISI 304 metal sheets for a step down of 0.5 mm/pass (as discussed in [10], variations of step down in the range of 0.2–0.5 mm did not have a relevant influence in formability). Although the three cases show an important enhancement of formability well above the FLC, the increase of formability as well as the mode of failure differs for the different punch radii. In this regard, the transition between the last points of formability provided by ARGUS® and the principal strains at fracture evaluated with the procedure exposed in Section 2.2, is represented in Figure 7 in dotted line. As it was discussed in [10], in the case of a forming tool of Φ20 mm diameter the failure mechanism was postponed necking followed by ductile fracture, whereas in the case of Φ10 mm, the fractography showed a series of grooves, which has been related to an incipient necking [23]. In this sense, the strain state attained in the case of Φ6 mm (see Figure 7) almost coincides with the obtained in the case of Φ10 mm. However, in order to evaluate the mode of failure, a fractography analysis is performed. Besides, it must be noticed that the fracture strains in SPIF are slightly above the scatter band of ±10% with respect to the FFL (which considered only the fracture strains attained in the Nakazima tests). This experimental fact also found in other works [8,10], implies that for some materials Nakazima tests might not be suitable for evaluating the FFL to be used in SPIF. Indeed, the level of triaxiality in stretching (e.g., Nakazima) is much higher than in ISF and then, the ability to reach the fracture limit will depend on the sensitivity of the material to the triaxiality state for fracture.

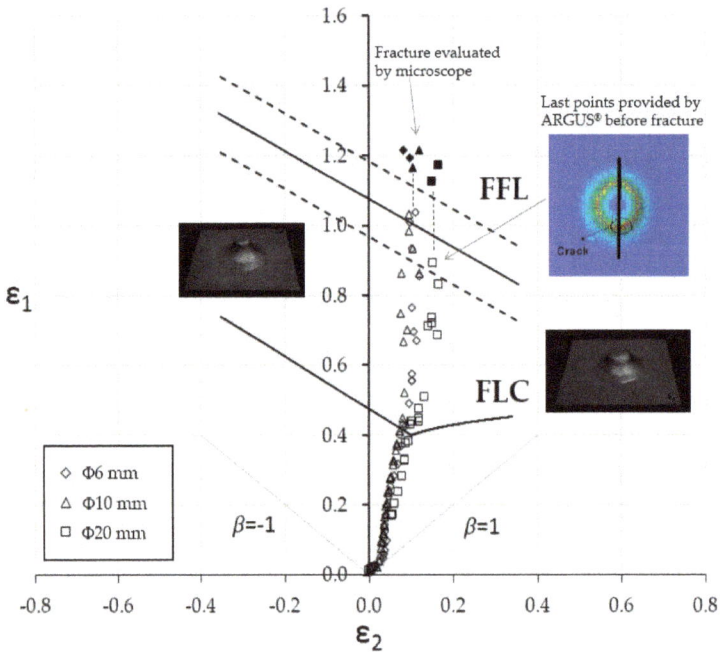

Figure 7. Principal strain state of the final testing part deformed by SPIF using tool diameters of Φ20, Φ10 and Φ6 mm represented within the FLD of the material.

In order to evaluate the failure mode and set the limit strain conditions at failure regarding the enhancement of formability attained in SPIF related to the t_0/R ratio, the fracture zone was analyzed by microscope for the case of the smallest tool diameter of $\Phi6$ mm. Figure 8 depicts the fractography at two sections of the final truncated cone. Section A-A′ corresponds to the zone in which it seems that the crack is about to initiate whereas section B-B′ is placed within the crack line close to its end. On section A-A′, a very incipient necking can be observed. On the contrary, section B-B′ depicts a ductile fracture corresponding to a monotonous decrease of thickness until fracture. In both sections, the indentation produced by the tool can be easily observed. This indentation produced in certain SPIF conditions, mostly for small tool diameters, has been also evaluated using numerical tools [24].

This fractographic analysis led to the conclusion of the failure mode attained. In this sense, it seems that the part could be deformed plastically above the FLC presenting the cited monotonous decrease of thickness until ductile fracture took place, being only possible to register a much postponed incipient necking. Indeed, there was a competition between the bending effect represented by the t_0/R ratio, which is the key factor for the increased formability in SPIF, and the tool indentation, that made this specific process to reach a threshold for the enhancement of formability for a tool diameter within the range of 10 mm to 6 mm.

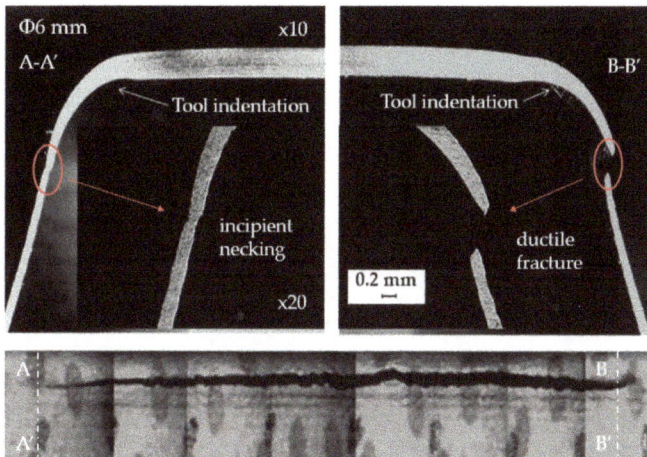

Figure 8. Fractography of the fracture zone at 2 sections for a tool diameter of $\Phi6$ mm. Section A-A′ corresponds to the crack initiation whereas section B-B′ is placed close to its end.

Moreover, related to the determination of the failure mode, it is important to establish how the crack initiates and develops in the SPIF process. With this aim, the online measurement of the forming force allowed triggering the actual depth in which failure was attained for every test (information shown in Table 2). Indeed, at the precise instant in which failure is reached, there is a drop down in the vertical force evolution that serves to stop the process and calibrate the final depth.

Some representative tests corresponding to a tool diameter of $\Phi6$ mm (marked in bold in Table 2) were selected to show the initiation and evolution of the crack. Figure 9 depicts the developed surfaces containing the crack corresponding to the 3 tests of a step down 0.5 mm/pass (Test 1 to 3 in Table 2) reaching final depths of 28.0, 28.5 and 28.0 mm and the second test (Test 2) corresponding to a step down of 0.2 mm/pass. Assuming the alternative movement of the tool for successive passes shown in Figure 3b, it is easy to understand the direction of the rightward crack evolution in 3 of the cases, and the leftward crack evolution in the case of reaching failure at a final depth of 28.5 mm (Test 2 corresponding to a step down of 0.5 mm/pass). This observation of the crack allows concluding that the crack initiates at a certain location where the limiting principal strains are reached (or the FFL in the case of total absence of necking) and it develops following the tool movements until the SPIF test is stopped. Notice the break with a certain angle of the crack straight line at the end location (marked with a circle in Test 2 corresponding to a step down of 0.5 mm/pass).

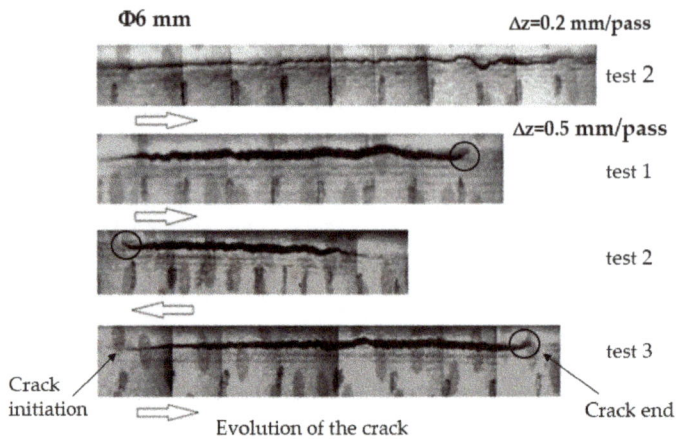

Figure 9. Evolution of the crack on the final testing part deformed by SPIF using a tool of Φ6 mm diameter for some representative tests marked in bold in Table 2.

As far as the authors are aware, this observation of the crack initiation and development in SPIF has not been previously explained in the state of the art of ISF. Indeed, this is an important fact in order to choose the correct location for carrying out a fractographic analysis of the crack and determining the mode of failure as well as the average strains at fracture, as it was previously carried out in this section using the exposed methodology.

4.2. Numerical Results

The numerical values of the principal strain states in SPIF predicted from the FEA carried out using DEFORM™-3D are compared with the experimental results presented in Figure 7 (step down of 0.5 mm/pass) within the FLD of the material for the tool diameters of Φ20 and Φ10 mm, i.e., including the transition from a theoretically necking controlled failure (Φ20 mm) to a failure mode of direct ductile fracture (Φ10 mm), as it was further discussed from the fractographic analysis carried out by the authors in [10].

In this regard, Figure 10 presents the numerical prediction of principal strain state at an average final depth corresponding to failure versus the experimental principal strains provided by ARGUS® as well as the strains at fracture evaluated as with the procedure exposed above, for a SPIF test using a tool of Φ10 mm diameter and a step down of 0.5 mm/pass. As can be seen, the numerical prediction is in good agreement with the experimental results. Besides, the numerical results obtained using DEFORM™-3D allow obtaining a direct transition from stable plastic deformation towards ductile fracture in the absence of necking. Indeed, due to the discrete pattern of the initial grid of circle (of 1 mm diameter and separation of circle centers of 2 mm) that, once the sheet is deformed into a final part, is analyzed using ARGUS®, there exists a gap between the highest value of major principal strain evaluated in the vicinity of the crack by ARGUS® and the fracture strains obtained at the very location of the crack. This lack of information has been covered by the numerical analysis carried out, allowing the authors to confirm the absence of necking in SPIF by FEA using the commercial software DEFORM™-3D.

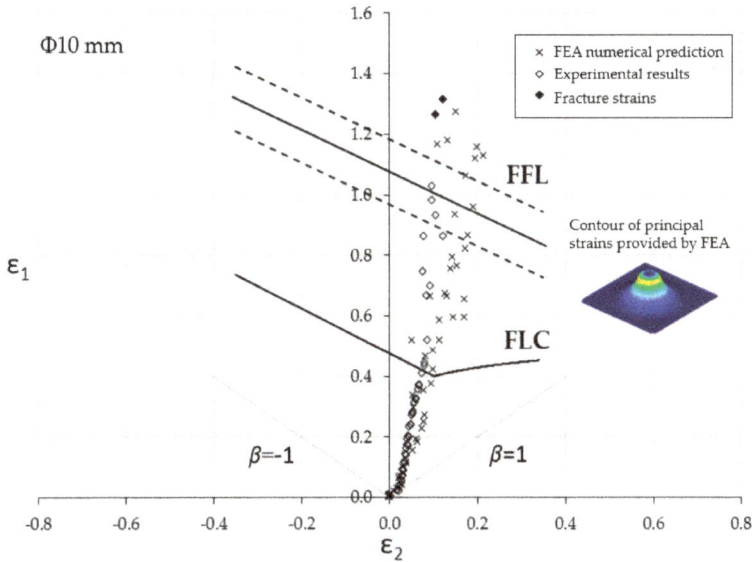

Figure 10. Numerical predictions versus experimental values of principal strains within the FLD of the material in the final testing part deformed by SPIF using a tool of Φ10 mm diameter.

On the contrary, the mode of failure observed in SPIF of AISI 304 sheets using a tool of Φ20 mm diameter was a necking-controlled failure, characterized by a postponed onset of necking (well above the FLC of the material) leading unstably towards ductile fracture (see [10] for further details). In this regard, Figure 11 presents the numerical predictions versus the experimental strains for this case. As can be seen, the numerical prediction is in good agreement with the experimental results, showing both results a gap of formability until the fracture strains, showing the dotted line the transition from the cited onset of postponed localized necking towards fracture.

Finally, the FEA allowed using ductile damage criteria in order to predict failure. In ISF processes, as it was discussed above, failure by direct ductile fracture might be attained within a range of process parameters depending on the material to be deformed. Besides, in most of usual testing geometries such as truncated cones and pyramids, fracture occurs under in-plane tension (see Figure 12a) corresponding to the mode I of fracture mechanics (see the recent unified vision of Martins et al. [25]). In these cases, the non-coupled damage criterion of McClintock [26] based on void growth applies as follows in Equation (3).

$$D_{crit} = \int_0^{\bar{\varepsilon}} \frac{\sigma_H}{\bar{\sigma}} d\bar{\varepsilon} \tag{3}$$

where the ratio of the hydrostatic (σ_H) to the equivalent stress ($\bar{\sigma}$) represents the stress triaxiality and the critical damage can be calculated under Hill's anisotropic plasticity criterion with plane stress conditions as expressed in Equation (4).

$$D_{crit} = \frac{(1+r)}{3}\left(\varepsilon_{1f} + \varepsilon_{2f}\right) \tag{4}$$

where the average anisotropy coefficient (r) is considered to be equal to 1 due to the simplification of considering an isotropic material in the FEA. Considering the average major and minor fracture strains calculated in SPIF this critical damage can be evaluated as $D_{crit} = 0.8$.

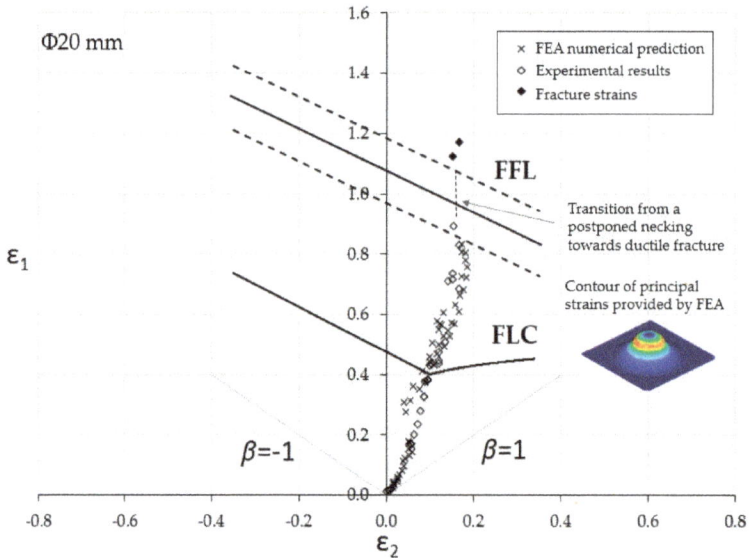

Figure 11. Numerical predictions versus experimental values of principal strains within the FLD of the material in the final testing part deformed by SPIF using a tool of Φ20 mm diameter.

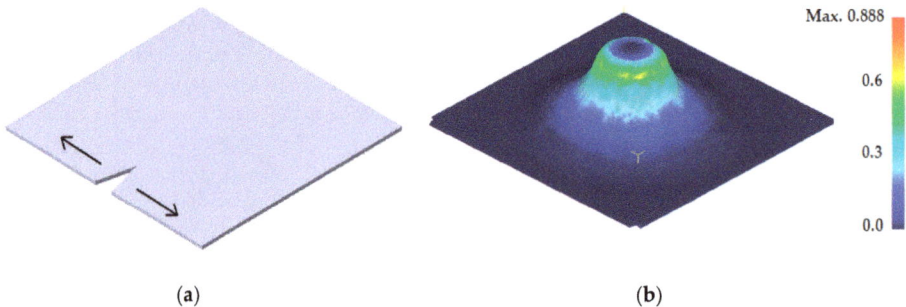

(a)   (b)

Figure 12. (a) Crack produced under In-plane tension corresponding to the Mode I and **(b)** Contour of accumulated damage using McClintock damage criterion for Φ10 mm and step down 0.5 mm/pass.

Analyzing the accumulated damage in the case of a tool of Φ10 mm diameter and a step down of 0.5 mm/pass, the case in which a direct transition from stable plastic deformation towards ductile fracture in the absence of necking was attained, the capacity of the numerical model for predicting failure by fracture in SPIF can be evaluated. Indeed, Figure 12b depicts the accumulated ductile damage for a process stage corresponding to the final depth at failure.

As can be seen, the maximum value predicted for the accumulated damage was 0.888, which is a reasonable prediction on the safe side, i.e., the predicted failure by fracture would have been attained in a previous stage of deformation corresponding to a smaller value of final depth.

To sum up, it must be notice that the relatively simple numerical model of the SPIF process presented in this research work allows evaluating formability and failure in incremental forming, providing fair predictions of the conditions upon which necking in SPIF is postponed or even suppressed. However, it is worth mentioning that there are very recent numerical works in ISF making use of more sophisticated modelling (considering material anisotropy, non-quadratic yield criteria, dynamic adaptive meshing, mixed hardening models, etc.) with the aim of reproducing more specific characteristic of the experimental procedure [24,27] or providing more accurate predictions for the onset of failure [28].

5. Conclusions

This research work revisits formability and failure of AISI304 sheets deformed by single point incremental forming. Formability in SPIF has been compared to conventional testing conditions, including Nakazima and stretch-bending. With this purpose, experimentation in SPIF and stretch-bending has been carried out and a Finite Element modelling of the SPIF process has been performed. The results obtained contributed to the current state of the art in SPIF as follows:

- The limit strains in SPIF have been experimentally evaluated through the competition between the bending effect represented by the t_0/R ratio and the tool indentation, setting a threshold for the enhancement of formability for a tool diameter within the range of 10 mm to 6 mm.
- The crack initiation and development in SPIF has been triggered, this analysis being crucial for determining the correct location to analyze the mode of failure and fracture strains.
- The FEA performed allowed the numerical establishment of the conditions upon which necking in SPIF was suppressed, the mode of failure being direct ductile fracture, and proved to be an useful tool providing fair and safe failure predictions by using the McClintock damage criterion.

Acknowledgments: The authors would like to express their gratitude for the funding received from the University of Girona (MPCUdG2016/036) and the Spanish Ministry of Education through the major grants DPI2015-64047-R and DPI2016-77156-R. The third author acknowledges gratefully the support from the Spanish doctoral grant FPU12/05402. The collaborations of BsC Álvaro Fernández Díaz and BsC José Manuel Carmona Romero are also greatly acknowledged.

Author Contributions: Gabriel Centeno, Isabel Bagudanch and María Luisa Garcia-Romeu designed the experimental plan and conducted the SPIF tests. Andrés Jesús Martinez-Donaire and Carpóforo Vallellano carried out the stretch-Bending and Nakazima tests. Domingo Morales-Palma and Gabriel Centeno performed the numerical simulations. All the authors contributed in the interpretation of the results obtained. The paper was written by Gabriel Centeno.

Conflicts of Interest: The authors declare no conflict of interest.

References

1. Leszak, E. Apparatus and Process for Incremental Dieless Forming. U.S. Patent 3,342,051, 19 September 1967.
2. Berghahn, W.G.; Murray, G.F. Method of Dieless forming Surfaces of Revolution. U.S. Patent 3,316,745, 2 May 1967.
3. Emmens, W.C.; Van den Boogaard, A.H. The technology of incremental sheet forming—A brief review of the history. *J. Mater. Process. Technol.* **2010**, *210*, 981–997. [CrossRef]
4. Mason, B. Sheet Metal Forming for Small Batches. Bachelor's Thesis, University of Nottingham, Nottingham, UK, 1978.
5. Mason, B.; Appleton, E. Sheet metal forming for small batches using sacrificial tooling. In Proceedings of the 3rd International Conference on Rotary Metalworking, Kyoto, Japan, 8–10 September 1984; pp. 495–511.
6. Centeno, G.; Martínez-Donaire, A.J.; Morales-Palma, D.; Vallellano, C.; Martins, P.A.F. Novel experimental techniques for the determination of the forming limits at necking and fracture. In *Materials Forming and Machining*; Davim, P., Ed.; Woodhead Publishing: Cambridge, UK, 2016; pp. 1–24.
7. Emmens, W.C.; Van den Boogaard, A.H. An overview of stabilizing deformation mechanisms in incremental sheet forming. *J. Mater. Process. Technol.* **2009**, *209*, 3688–3695. [CrossRef]
8. Silva, M.B.; Nielsen, P.S.; Bay, N.; Martins, P.A.F. Failure mechanisms in single point incremental forming of metals. *Int. J. Adv. Manuf. Technol.* **2011**, *56*, 893–903. [CrossRef]
9. Madeira, T.; Silva, C.M.A.; Silva, M.B.; Martins, P.A.F. Failure in single point incremental forming. *Int. J. Adv. Manuf. Technol.* **2015**, *80*, 1471–1479. [CrossRef]

10. Centeno, G.; Bagudanch, I.; Martínez-Donaire, A.J.; Garcia-Romeu, M.L.; Vallellano, C. Critical analysis of necking and fracture limit strains and forming forces in single-point incremental forming. *Mater. Des.* **2014**, *63*, 20–29. [CrossRef]
11. Isik, A.K.; Silva, M.B.; Tekkaya, A.E.; Martins, P.A.F. Formability limits by fracture in sheet metal forming. *J. Mater. Process. Technol.* **2014**, *214*, 1557–1565. [CrossRef]
12. Soeiro, J.M.C.; Silva, C.M.A.; Silva, M.B.; Martins, P.A.F. Revisiting the formability limits by fracture in sheet metal forming. *J. Mater. Process. Technol.* **2009**, *217*, 184–192. [CrossRef]
13. International Standard ISO 12004-2:2008. *Metallic Materials-Sheet and Strip-Determination of Forming Limit Curves, Part 2: Determination of Forming Limit Curves in the Laboratory*; International Organisation for Standardization: Geneva, Switzerland, 2008.
14. Martínez-Donaire, A.J.; García-Lomas, F.J.; Vallellano, C. New approaches to detect the onset of localised necking in sheets under through-thickness strain gradients. *Mater. Des.* **2014**, *57*, 135–145. [CrossRef]
15. Atkins, A.G. Fracture in forming. *J. Mater. Process. Technol.* **1996**, *56*, 609–618. [CrossRef]
16. Centeno, G.; Morales-Palma, D.; Gonzalez-Perez-Somarriba, B.; Bagudanch, I.; Egea-Guerrero, J.J.; Gonzalez-Perez, L.M.; Garcia-Romeu, M.L.; Vallellano, C. A functional methodology on the manufacturing of customized polymeric cranial prostheses from CAT using SPIF. *Rapid Prototyp. J.* **2017**, *23*, 771–780. [CrossRef]
17. Bagudanch, I.; Centeno, G.; Garcia-Romeu, M.L.; Vallellano, C. Revisiting formability and failure of polymeric sheets deformed by Single Point Incremental Forming. *Polym. Degrad. Stab.* **2017**, *144*, 366–377. [CrossRef]
18. Centeno, G.; Silva, M.B.; Alves, L.M.; Vallellano, C.; Martins, P.A.F. Towards the characterization of fracture in thin-walled tube forming. *Int. J. Mech. Sci.* **2016**, *119*, 12–22. [CrossRef]
19. Arısoy, Y.M.; Guo, C.; Kaftanoğlu, B.; Özel, T. Investigations on microstructural changes in machining of Inconel 100 alloy using face turning experiments and 3D finite element simulations. *Int. J. Mech. Sci.* **2016**, *107*, 80–92. [CrossRef]
20. Amigo, F.J.; Camacho, A.M. Reduction of induced central damage in cold extrusion of dual-phase steel DP800 using double-pass dies. *Metals* **2017**, *7*, 335. [CrossRef]
21. Palaniswamy, H.; Ngaile, G.; Altan, T. Finite element simulation of magnesium alloy sheet forming at elevated temperatures. *J. Mater. Process. Technol.* **2004**, *146*, 56–60. [CrossRef]
22. Lee, Y.S.; Kwon, Y.N.; Kang, S.H.; Kim, S.W.; Lee, J.H. Forming limit of AZ31 alloy sheet and strain rate on warm sheet metal forming. *J. Mater. Process. Technol.* **2008**, *201*, 431–435. [CrossRef]
23. Wilson, D.V.; Mirshams, A.R.; Roberts, W.T. An experimental study of the effect of sheet thickness and grain size on limit-strains in biaxial stretching. *Int. J. Mech. Sci.* **1983**, *25*, 859–870. [CrossRef]
24. Neto, D.M.; Martins, J.M.P.; Oliveira, M.C.; Menezes, L.F.; Alves, J.L. Evaluation of strain and stress states in the single point incremental forming process. *Int. J. Adv. Manuf. Technol.* **2016**, *85*, 521–534. [CrossRef]
25. Martins, P.A.F.; Bay, N.; Tekkaya, A.E.; Atkins, A.G. Characterization of fracture loci in metal forming. *Int. J. Mech. Sci.* **2014**, *83*, 112–123. [CrossRef]
26. McClintock, F.A. A criterion for ductile fracture by the growth of holes. *J. Appl. Mech.* **1968**, *35*, 363–371. [CrossRef]
27. Esmaeilpoura, R.; Kima, H.; Park, T.; Pourboghratab, F.; Mohammed, B. Comparison of 3D yield functions for finite element simulation of single point incremental forming (SPIF) of aluminum 7075. *Int. J. Mech. Sci.* **2017**, *133*, 544–554. [CrossRef]
28. Mirnia, M.J.; Shamsari, M. Numerical prediction of failure in single point incremental forming using a phenomenological ductile fracture criterion. *J. Mater. Process. Technol.* **2017**, *244*, 17–43. [CrossRef]

metals

MDPI

Article

Effect of Constitutive Equations on Springback Prediction Accuracy in the TRIP1180 Cold Stamping

Ki-Young Seo [1], Jae-Hong Kim [1], Hyun-Seok Lee [2], Ji Hoon Kim [3] and Byung-Min Kim [3,*]

[1] Division of Precision and Manufacturing Systems, Pusan National University, Busandaehak-ro 63beon-gil, Geumjeong-gu, Busan 609-735, Korea; seokiyoung11@pusan.ac.kr (K.-Y.S.); kjh86@pusan.ac.kr (J.-H.K.)
[2] Press Die Research Team, Nara Mold & Die Co. Ltd., Gongdan-ro 675, Seongsan-gu, Changwon-City, Gyeongnam 642-120, Korea; hslee@naramnd.com
[3] School of Mechanical Engineering, Pusan National University, Busandaehak-ro 63beon-gil, Geumjeong-gu, Busan 609-735, Korea; kimjh@pusan.ac.kr
[*] Correspondence: bmkim@pusan.ac.kr; Tel.: +82-51-510-2319; Fax: +82-51-581-3075

Received: 30 October 2017; Accepted: 25 December 2017; Published: 30 December 2017

Abstract: This study aimed to evaluate the effect of constitutive equations on springback prediction accuracy in cold stamping with various deformation modes. This study investigated the ability of two yield functions to describe the yield behavior: Hill'48 and Yld2000-2d. Isotropic and kinematic hardening models based on the Yoshida-Uemori model were adopted to describe the hardening behavior. The chord modulus model was used to calculate the degradation of the elastic modulus that occurred during plastic loading. Various material tests (such as uniaxial tension, tension-compression, loading-unloading, and hydraulic bulging tests) were conducted to determine the material parameters of the models. The parameters thus obtained were implemented in a springback prediction finite element (FE) simulation, and the results were compared to experimental data. The springback prediction accuracy was evaluated using U-bending and T-shape drawing. The constitutive equations wielded significant influence over the springback prediction accuracy. This demonstrates the importance of selecting appropriate constitutive equations that accurately describe the material behaviors in FE simulations.

Keywords: advanced high-strength steel; yield function; hardening model; springback; deformation mode

---

1. Introduction

In recent years, lightweight vehicles have gained attention as fuel efficiency and gas emission regulations become increasingly stringent [1]. As weight reduction has become a key goal, many researchers have devoted significant efforts to selecting the materials for manufacturing automotive parts. Advanced high strength steel (AHSS) has been widely used in the automotive industry for its light weight, crashworthiness, and productivity. However, it is difficult to achieve dimensional accuracy with AHSS because its higher elastic recovery and yield strength cause excessive springback [2]. Fabricating a target product shape with AHSS is challenging for part manufacturers, requiring a considerable amount of time as well as additional costs to modify tools for this springback.

Finite element (FE) simulation may be applied to describe AHSS material behaviors and springback, as this provides a cost-effective and reliable method for predicting springback. The constitutive equations used in FE simulations strongly influence the accuracy of the prediction results. Thus, many researchers have suggested varying constitutive equations. Multiple equations have been used to describe the hardening behavior of AHSS, including the nonlinear kinematic hardening model proposed by Chaboche [3], the kinematic hardening model based on cyclic plasticity suggested by Yoshida, and the distortional hardening model recommended by Barlat. In addition, Hill'48 and Yld2000-2d have been widely used to model the anisotropic behavior of AHSS.

Many published studies have investigated the influence of constitutive equations on simulated prediction results. Lee et al. [4] performed a springback evaluation of automotive sheets based on an isotropic kinematic hardening model and anisotropic yield functions. It was found that the hardening behaviors, including Bauschinger and transient elements, were well represented by the modified Chaboche model. Furthermore, the work-hardening data for dual phase steel (DP steel) was found to better conform to a power law-type hardening law than to the Voce-type law. Zang et al. [5] developed an elasto-plastic constitutive model based on one-surface plasticity. Their results demonstrated that the resulting material model is able to accurately predict springback when materials show a constant offset in permanent softening. Furthermore, Larsson et al. [6] concluded that neither isotropic nor kinematic hardening models were sufficient to describe the plastic-hardening behavior seen in non-linear strain paths. Thus, Larsson employed a combined isotropic-kinematic hardening model to evaluate the effects of springback in steel sheets. Eggertsen et al. [7] predicted the springback using various hardening models and yield functions. The Yoshida-Uemori hardening model has been shown to yield results that fit experimental measurements better than other options. Kim et al. [8] performed die compensation based on the Yld2000-2d yield function and Yoshida-Uemori hardening model. It was concluded that the dimensional accuracy of AHSS products can be achieved efficiently through die compensation using the material models in the multi-stage stamping process. Previous studies have demonstrated that the descriptions of anisotropic behavior, the Bauschinger effect, transient behavior, and the permanent softening effect are important. Successful springback prediction via FE simulation is principally dependent on selecting accurate yield criterion, hardening models, and material coefficients [9]. However, the above studies dealt with simply configured products such as those used in U-bending tests [10] and did not focus on AHSS products with various deformation modes. Therefore, it is necessary to evaluate the effect of the constitutive equations on springback prediction accuracy in AHSS cold stamping with multiple deformation modes.

The objective of this study was to evaluate the effect of constitutive equations on springback prediction accuracy in TRIP1180 cold stamping. In this study, two types of yield function were considered to describe the yield behavior: Hill'48 and the Yld2000-2d. Isotropic and kinematic hardening models based on the Yoshida-Uemori model were also adopted to describe the hardening behavior. The chord modulus model was utilized in the FE simulation alongside the hardening model constants. Various material tests, such as uniaxial tension, tension–compression, loading–unloading, and hydraulic bulging tests were conducted to determine material parameters for the models. The obtained parameters were utilized in the FE simulation to predict springback, and the results were compared with experimental data. In addition, U-bending and T-shape drawing were employed to evaluate the accuracy of the springback predictions.

2. Constitutive Equations for the TRIP1180 Sheet Steel

A TRIP1180 steel sheet with a thickness of 1.0 mm was investigated in this study. The TRIP1180 was used as received. The constitutive equations Hill'48 and Yld2000-2d were used to describe its yield behavior. Moreover, isotropic and kinematic hardening models based on the Yoshida-Uemori model were adopted to express hardening behavior. The chord modulus model was used to describe the degradation of the elastic modulus that occurs during plastic loading. The material constants of TRIP1180 for the constitutive equations were obtained from uniaxial tension, tension-compression, loading-unloading, and hydraulic bulging tests.

2.1. Yield Function

Yield functions define the transition of a material from elastic to plastic behavior in complex stress states. In this study, the Hill'48 and Yld2000-2d yield functions were used to evaluate the anisotropic yield behavior of TRIP1180. In order to determine the material parameters of the yield functions, ASTM E8 standard uniaxial tension tests were performed on specimens using a MTS universal testing machine for different rolling directions ($0°$, $45°$, $90°$), and hydraulic bulge tests were conducted to obtain a stable biaxial stress-strain curve with an Erichsen bulge tester. The biaxial yield stress and biaxial anisotropic plasticity coefficients were derived from the developed curve. A mechanical measurement device was placed on the top of the specimen to allow in-plane elongation and curvature measurements using an extensometer. The membrane stress and thickness strain were calculated using these measurements as described in the literature [11]. The yield stress in balanced biaxial tension (σ_b) was calculated based on the work-equivalence principle by comparing the bulge and uniaxial tension flow curves. The results of these tests are shown in Figure 1.

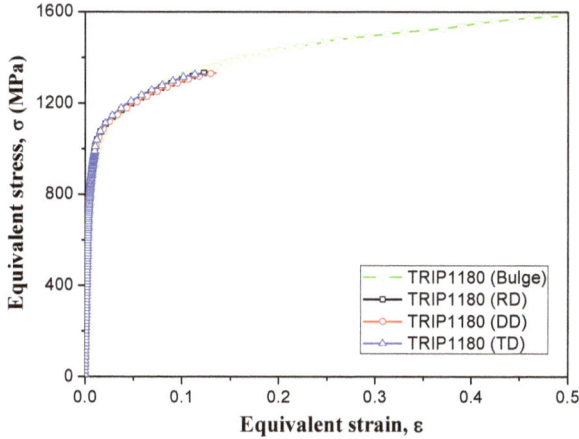

Figure 1. Results of uniaxial tension and bulge tests.

2.1.1. Hill'48 Yield Function

The anisotropic yield criterion proposed by Hill [12] is one of the most widely used yield functions. The Hill'48 yield function is also easy to express, and as such has been widely used to investigate the effect of anisotropy on springback, especially in steel sheets. This function is defined as follows:

$$2f(\sigma) = F(\sigma_{yy} - \sigma_{zz})^2 + G(\sigma_{zz} - \sigma_{xx})^2 + H(\sigma_{xx} - \sigma_{yy})^2 + 2(L\sigma^2_{yz} + M\sigma^2_{zx} + N\sigma^2_{xy}) = 1 \quad (1)$$

Under plane stress conditions ($\sigma_{zz} = \sigma_{yz} = \sigma_{zx} = 0$, $L = M = 0$), the Hill'48 model can be mathematically represented as follows:

$$2f(\sigma) = (G + H)\sigma^2_{xx} + (F + H)\sigma^2_{yy} - 2H\sigma_{xx}\sigma_{yy} + 2N\sigma^2_{xy} = 1 \quad (2)$$

where σ_{xx}, σ_{yy}, and σ_{zz} are the normal stresses in the rolling, transverse, and thickness directions, respectively; σ_{xy}, σ_{yz}, and σ_{zx} are the shear stresses in the xy, yz, and zx planes, respectively; and F, G, H, and N are the anisotropic coefficient parameters. The material parameters of the Hill'48 yield function are principally obtained from Lankford values at angles of $0°$, $45°$, and $90°$ to the rolling direction. The anisotropic parameters F, G, H and N can be formulated in terms of the r-values r_0, r_{45}, r_{90} as follows:

$$F = \frac{r_0}{r_{90}(1 + r_{90})}, G = \frac{1}{(1 + r_0)}, H = \frac{r_0}{(1 + r_0)}, N = \frac{(r_0 + r_{90})(1 + 2r_{45})}{2r_{90}(1 + r_0)} \quad (3)$$

The mechanical properties and material constants thus determined are summarized in Tables 1 and 2.

Table 1. Material properties of TRIP1180 steel.

Test Direction	E_0 (GPa)	YS (MPa)	UTS (MPa)	Elongation (%)	R-Value
Rolling direction (0°)	200.5	861.9	1180	17.2	0.795
Diagonal direction (45°)	200.7	866.6	1175	16.0	0.958
Transverse direction (90°)	206.3	866.2	1182	14.9	0.967

Table 2. Material constants of TRIP1180 for the Hill'48 yield function.

Material	F	G	H	N
TRIP1180	0.4580	0.5571	0.4429	1.480

2.1.2. Yld2000-2d Yield Function

The Yld2000-2d function [13] proposed by Barlat et al. can describe yield behaviors for various deformation modes. This yield function has eight anisotropic coefficients related to the experimental yield stresses (σ_0, σ_{45}, σ_{90}, σ_b) and anisotropic parameters (r_0, r_{45}, r_{90}, r_b). This function can be expressed as shown in Equation (4):

$$f = \frac{\phi' + \phi''}{2}, f = \left| X_1' - X_2' \right|^a + \left| 2X_2'' + X_1'' \right|^a + \left| 2X_1'' + X_2'' \right|^a = 2\bar{\sigma} \tag{4}$$

where $\bar{\sigma}$ is the effective stress and a is a constant related to the crystalline structure of the material, which was set to 6 in this study (for FCC, $a = 8$ and for BCC $a = 6$, thus, for TRIP1180 $a = 6$). In Equation (4), $\phi' = \left| X_1' - X_2' \right|^a$ and $\phi'' = \left| 2X_2'' + X_1'' \right|^a + \left| 2X_1'' + X_2'' \right|^a$ where X_1 and X_2 are the principal values of the matrices, X' and X'', whose components are obtained from the following linear transformations of the Cauchy stress (σ) and deviatoric Cauchy stress (σ'), respectively:

$$X' = C'\sigma' = C'T\sigma = L'\sigma, X'' = C''\sigma' = C''T\sigma = L''\sigma \tag{5}$$

where

$$
\begin{bmatrix} L_{11}' \\ L_{12}' \\ L_{21}' \\ L_{22}' \\ L_{66}' \end{bmatrix} =
\begin{bmatrix} 2/3 & 0 & 0 \\ -1/3 & 0 & 0 \\ 0 & -1/3 & 0 \\ 0 & 2/3 & 0 \\ 0 & 0 & 1 \end{bmatrix}
\begin{bmatrix} \alpha_1 \\ \alpha_2 \\ \alpha_7 \end{bmatrix},
\begin{bmatrix} L_{11}'' \\ L_{12}'' \\ L_{21}'' \\ L_{22}'' \\ L_{66}'' \end{bmatrix} =
\frac{1}{9}
\begin{bmatrix} -2 & 2 & 8 & -2 & 0 \\ 1 & -4 & -4 & 4 & 0 \\ 4 & -4 & -4 & 1 & 0 \\ -2 & 8 & 2 & -2 & 0 \\ 0 & 0 & 0 & 0 & 1 \end{bmatrix}
\begin{bmatrix} \alpha_3 \\ \alpha_4 \\ \alpha_5 \\ \alpha_6 \\ \alpha_8 \end{bmatrix} \tag{6}
$$

The material constants (eight anisotropic coefficients) included in the L' and L'' tensors can be determined according to the rolling direction using the yield stress and anisotropic coefficient (three uniaxial yield stresses, three r-values in the three material directions, and the balanced-biaxial r-values and yield stress: σ_0, σ_{45}, σ_{90}, σ_b, r_0, r_{45}, r_{90} and r_b). This calculation procedure involves solving a system of nonlinear equations. This was performed in the current experiment by using the Newton-Raphson iteration method. The determined anisotropic coefficients are summarized in Table 3.

Table 3. Material constants of TRIP1180 using the Yld2000-2d yield function.

α_1	α_2	α_3	α_4	α_5	α_6	α_7	α_8
0.9471	1.0199	0.9867	0.9925	1.0141	0.9815	0.9910	1.0007

Based on previous experimental results, the yield surface of the von-Mises, Hill'48, and Yld2000-2d models can be plotted alongside experimental results. The yield surface results are shown in Figure 2. It can be seen that the Yld2000-2d model matches well with the experimental results.

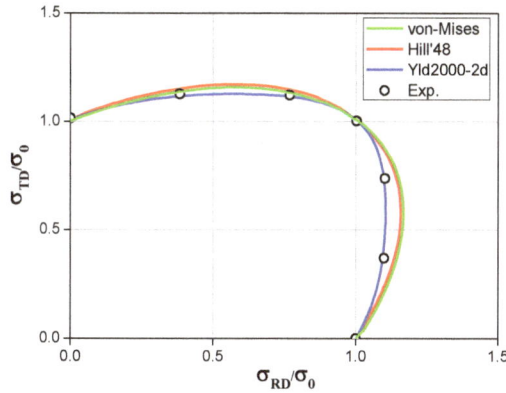

Figure 2. Yield surfaces characterized by different yield functions.

2.2. Hardening Model

In plasticity, the hardening rule is used to describe the behavior of a material during plastic deformation. In this study, the isotropic and Yoshida-Uemori kinematic hardening models were applied to evaluate the hardening behavior of TRIP1180. In order to accurately predict the springback, it is essential to analyze the stress-strain behaviors of sheet metals during tension-compression loading. For this reason, tension-compression tests were performed on a specimen modified from the standard SEP1240 [14] with a gauge length of 50 mm, as depicted in Figure 3a. A vertical load was applied to the uniform elongation portion of center of the specimen to prevent buckling, as shown in Figure 3b. This allowed tension-compression tests to be performed reliably.

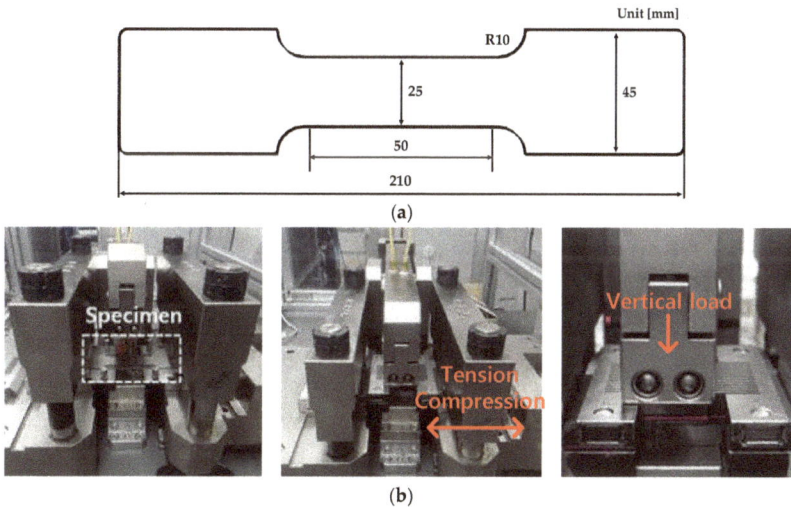

Figure 3. (a) Dimensions of the specimen used for tension–compression test. (b) A schematic view of the tension-compression test.

2.2.1. Isotropic Hardening Model

When expansion of the yield surface is uniform in all directions in the stress space, the hardening behavior is referred to as isotropic. The Swift isotropic hardening model [15] used in this study can successfully describe isotropic behavior under these conditions. The Swift isotropic hardening model is defined as follows:

$$\bar{\sigma} = K(\bar{\varepsilon}_0 + \bar{\varepsilon}_p)^n \tag{7}$$

where $\bar{\sigma}$ is the effective stress and $\bar{\varepsilon}_p$ is the effective strain (total true strain minus recoverable strain). Thus, $\bar{\varepsilon}_p$ represents the residual true strains after elastic unloading. Constants K, n, and ε_0 are material constants related to the hardening behavior. The material parameters of the Swift hardening model were principally obtained via uniaxial tension tests, and the determined material constants are summarized in Table 4.

Table 4. Coefficients of the Swift isotropic hardening model for TRIP1180.

Material	K (MPa)	n	ε_0
TRIP1180	1672.8	0.1044	4.8605×10^{-14}

2.2.2. Yoshida-Uemori Hardening Model

As the material behavior is considerably complex during cyclic loading, a hardening rule should be able to accurately predict deformation behavior during cyclic loading. The Yoshida-Uemori model [16] is one of the most sophisticated models and can reproduce transient Bauschinger effects, permanent softening, and work hardening stagnation during large elasto-plastic deformation.

The Yoshida-Uemori model accounts for both the translation and expansion of the bounding surface, while the active yield surface evolves in a kinematic manner. A schematic of yield surfaces according to the Yoshida-Uemori model was presented in Chongthairungruang et al. [17]. The relative displacement of the two yield surfaces in a bounding surface can be defined as follows:

$$\alpha_* = \alpha - \beta \tag{8}$$

where α represents the current center of the yield surface, β represents the center of the bounding surface, and α_* represents the relative position of the two surfaces. An additional definition for α_* is given in Equation (9), which determines the relative movement of the yield and bounding surfaces:

$$\alpha_* = C\left[\left(\frac{a}{Y}\right)(\sigma - \alpha) - \sqrt{\frac{a}{\alpha_*}}\alpha_*\right]\dot{\bar{\varepsilon}}$$
$$a = B + R - Y \tag{9}$$

where B represents the initial size of the bounding surface, R represents the isotropic hardening component, Y represents the initial yield strength, and C is a material parameter of the kinematic yield surface hardening rule. The isotropic and kinematic hardening behaviors of the bounding surface can be defined as follows:

$$dR = m(R_{sat} - R)\dot{\bar{\varepsilon}} \tag{10}$$

$$d\beta = m\left(\frac{2}{3}bD^p - \beta\right)\dot{\bar{\varepsilon}} \tag{11}$$

where R_{sat} is the saturated value of the isotropic hardening stress R for an infinitely large plastic strain and m is a material parameter controlling the rate of isotropic hardening. D^p is an increment of the plastic deformation rate and b is a material constant. The Yoshida-Uemori model constants were derived via inverse finite element optimization. Inverse optimization was performed using Matlab's fminsearch function, which identifies the constant value that minimizes the error value relative to the tension-compression experiment results using the Nelder-Mead method [18]. The Yoshida-Uemori model constants determined for the TRIP1180 sheets are presented in Table 5.

Table 5. Material constants of TRIP1180 for the Yld2000-2d yield function.

Y	B	R_{sat}	b	m	C_1	C_2	$\varepsilon_{p,ref}$
800	284.9	294.2	88	9.62	366.8	366.8	0.005

A comparison of the experimental stress-strain curves and the calculated results based on the selected hardening models is shown in Figure 4. It can be observed that the Yoshida-Uemori hardening model matches well with the experimental results and captures the Bauschinger effect.

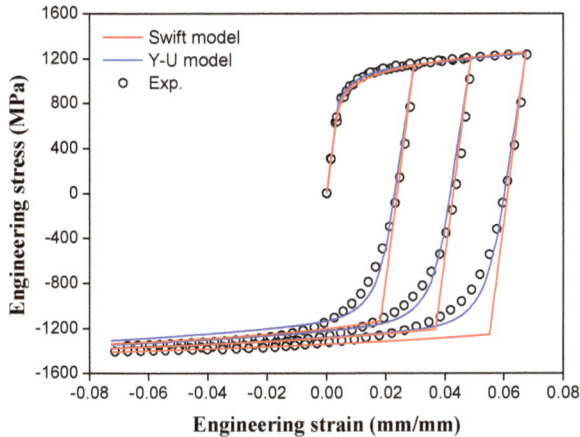

Figure 4. Measured and calculated tension-compression stress-strain curves from multiple hardening models.

2.3. Chord Modulus Model

In order to determine the material constants of the chord modulus model, ASTM E8 standard loading–unloading tests were performed on specimens using a MTS universal testing machine for different rolling directions (0°, 45°, 90°). The elastic modulus changes during plastic deformation are reflected in the chord modulus model. The results of this test are shown in Figure 5. Generally, the elastic modulus of steel sheets decreases as the effective strain increases [19,20]. The initial elastic modulus under uniaxial tension, denoted as E_0, was determined using the linear regression fitting method. The elastic moduli for pre-strained sheet specimens were defined as the slope of a straight line drawn through the two stress–strain end points at a corresponding prescribed plastic strain. Changes in the chord modulus were applied to the FE simulation along with the hardening model constants. This phenomenon was formulated as shown in Equation (12):

$$E = E_0 - (E_0 - E_a)[1 - \exp(-\xi\bar{\varepsilon}_p)] \tag{12}$$

where E represents the unloading elastic model under uniaxial tension, E_a represents the chord modulus obtained under an infinitely large plastic pre-strain, and $\bar{\varepsilon}_p$ and ξ are material parameters determining the rate at which E decreases. The optimized constants are given in Table 6.

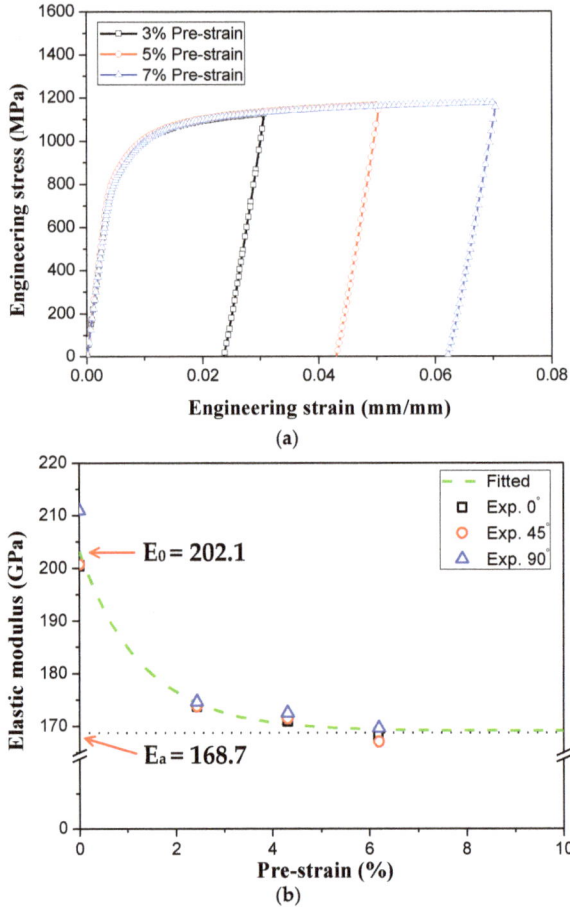

Figure 5. (a) Results of loading-unloading test. (b) Results of chord modulus model.

Table 6. Material constants of TRIP1180 softening behavior.

Material	E_0 (GPa)	E_a (GPa)	ξ
TRIP1180	202.1	168.7	72.9

3. Test Conditions

In this study, FE simulations and experiments were performed for various forming processes including U-bending and T-shape drawing. A commercial program (PamStamp 2G) was used to perform the FE simulation. The experimental and analytical results were compared using various constitutive equations and the material constants determined in Section 2. Both U-bending and T-shape drawing were employed to evaluate the accuracy of springback predictions.

3.1. U-Bending Test

Previous works have confirmed the U-bending test to be a significant verification model for springback prediction [21–23]. The tools used in U-bending are shown in Figure 6, consisting of a punch, blank holder, and die. The dimension of the blank was 300.0 mm × 30.0 mm × 1.0 mm.

The gap between the die and the punch was designed to be 1.1 mm. Testing was conducted using a 200-ton servo press machine. The total punch stroke was 60.0 mm, with a punch speed of 1 mm/s and a blank holding force of 20 kN. Additional tests were performed in various rolling directions (0°, 45°, 90°). Each set of experimental conditions was repeated five times to ensure the reliability of the experiment. After stamping, the final dimensions of the formed specimens were measured along the middle cross section using a laser coordinate measuring machine (a two-dimensional inspection machine), allowing for comparison between the experimental and FE simulation results.

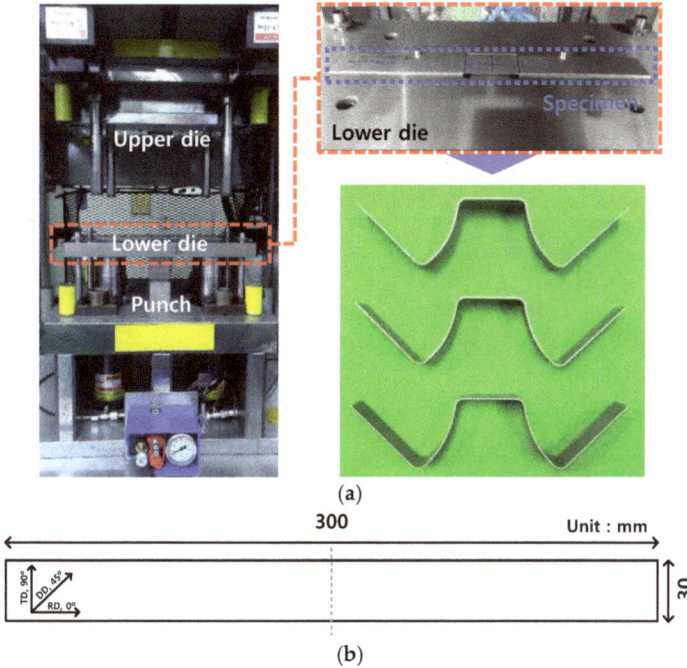

(a)

(b)

Figure 6. (a) Tools for U-bending test. (b) Blank size of U-bending test.

To evaluate the springback behavior observed in the U-bending tests, FE simulations were conducted for forming and springback analysis. The analytical model was designed to mimic the experiment, though only a half model of the tools and blank was simulated as shown in Figure 7, considering the geometric symmetry of the test. The specimen used for FEA was a Belytschko-Lin-Tsay (BLT) shell element of uniform size (1.0 mm × 1.0 mm) with five integration points in the thickness direction. The die was assumed to be a rigid body. The FE simulation conditions were identical to the experimental conditions. The Coulomb friction coefficient between the die and specimen was set to 0.12, a value that assumed an unlubricated condition. In the FE simulation, mass-scaling and mesh-refinement techniques were applied to ensure the efficiency of the analysis. The shapes of the specimens calculated in the FE simulations were compared to those from the experimental results [24].

3.2. T-Shape Drawing Test

In this study, a T-shape drawing test was performed to evaluate the effect of the constitutive equations on the prediction accuracy of springback in complex deformation modes. The blank size used in T-shape drawing and the experimental set-up for the T-shape drawing test, consisting of a punch, blank holder, and die, are shown in Figure 8. The gap between the die and punch was designed to be

1.1 mm, and the test was conducted using a 200-ton servo press machine. The total punch stroke was 22.0 mm, with a punch speed of 20 mm/s and blank holding force of 90 kN. Experiments were repeated five times to ensure reliability. After stamping, the final dimensions of the formed specimens were measured using a 3D optical scanning system (three-dimensional inspection equipment), allowing for comparison between the experimental and FE simulation results.

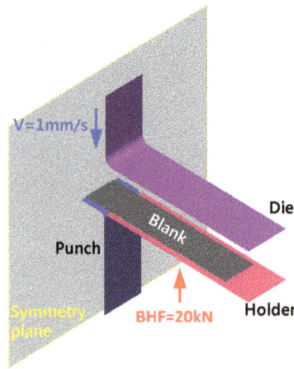

Figure 7. Finite element (FE) model of U-bending.

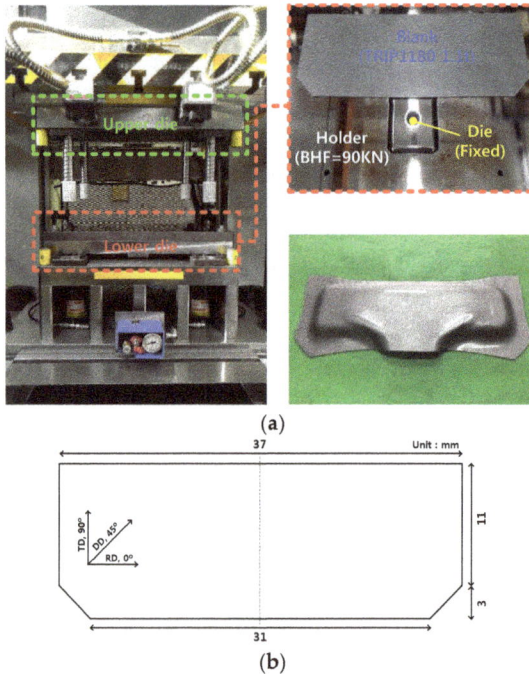

Figure 8. (**a**) Experimental set-up for T-shape drawing test. (**b**) Blank size of T-shape drawing test.

To investigate the springback behavior during T-shape drawing tests, FE simulations were conducted for forming and springback analysis. The analytical model was designed to mimic the experiments, and Figure 9 shows the FE model that was used. A significant number of conditions for the T-shape drawing FE simulation were equivalent to those in the U-bending test simulation and the experimental conditions of the T-shape drawing test. The shapes of the specimens determined via FE simulations were compared with those obtained experimentally. In this study, a commercial reverse-design program (Geomagic Design X), was employed to quantitatively compare the configurations. For this comparison, the experimental results were input as the reference configuration to measure the dimensional errors between the experimental and analytical results.

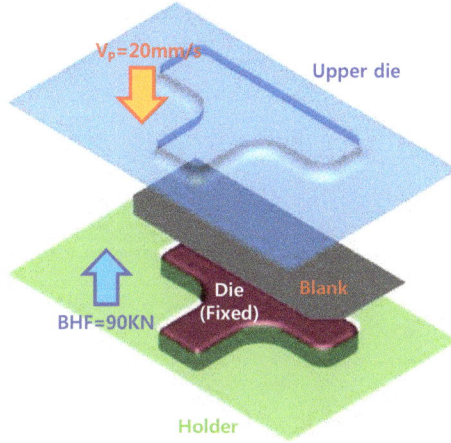

Figure 9. Finite element (FE) model for T-shape drawing test.

4. Results and Discussion

4.1. Springback Prediction for U-Bending Test

In order to determine the springback prediction accuracy dependent on the constitutive equations, the predicted U-bending test results were compared to the experimental results. The result is shown in Figure 10. The combination of the Hill'48 yield function and isotropic hardening model resulted in specimen shapes different from those observed experimentally. The combined Yld2000-2d yield function and Yoshida-Uemori model, however, predicted shapes that were similar to those of the manufactured parts.

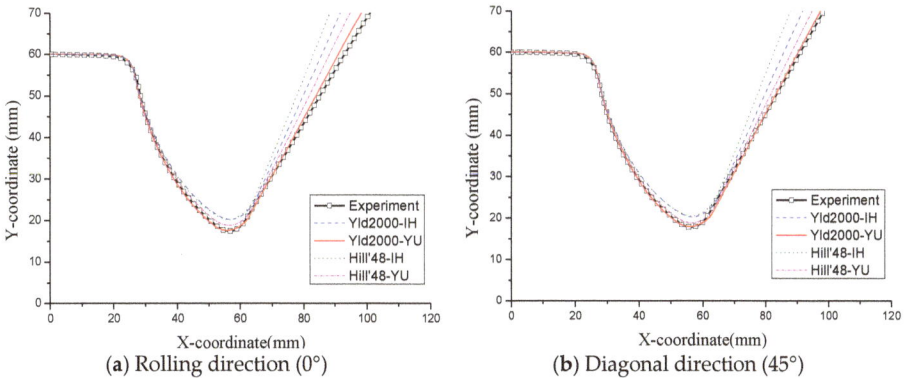

(**a**) Rolling direction (0°)   (**b**) Diagonal direction (45°)

Figure 10. *Cont.*

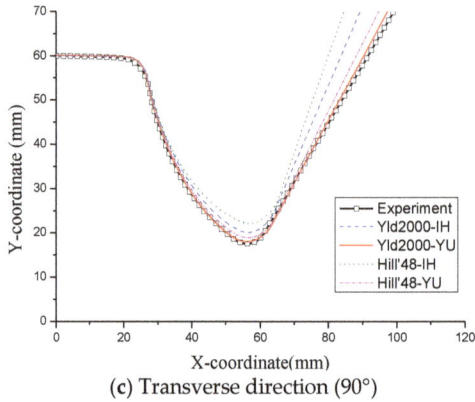

(c) Transverse direction (90°)

Figure 10. Shape comparison between predictions and experiments for U-bending test.

Springback parameters were employed in this study to quantitatively compare the springback. In Figure 11a, the springback parameters of the defined Numisheet'93 benchmark problem are shown [25]. The 2D draw-bending test proposed as a benchmark problem in Numisheet'93 involves two-dimensional blank holders to show both the effects of the material as well as the process parameters. As previously mentioned, when the results of the combined Yld2000-2d yield function and Yoshida-Uemori model were used, the prediction accuracy for springback was excellent in various rolling directions, including 0°, 45°, and 90°, as shown in Figure 11.

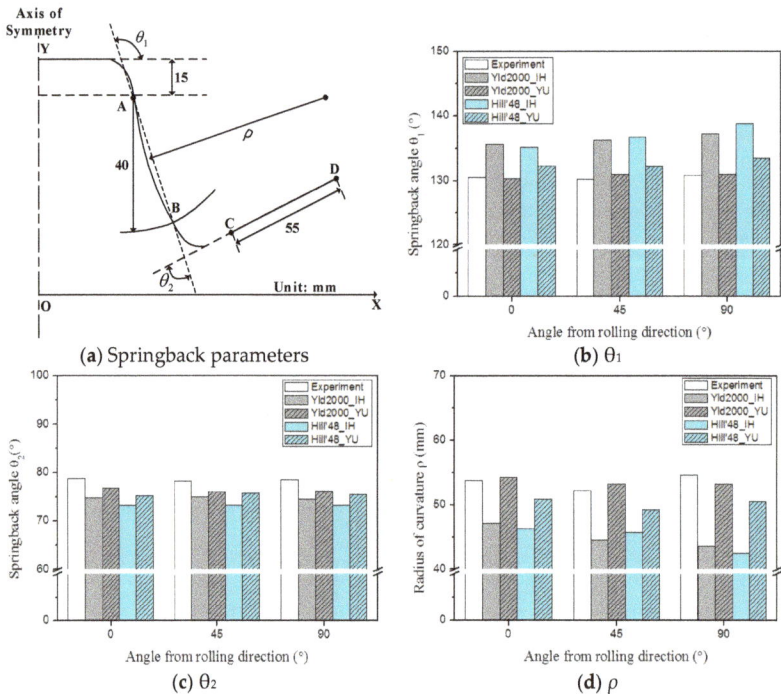

(a) Springback parameters

(b) θ_1

(c) θ_2

(d) ρ

Figure 11. Comparison of springback with springback parameters for U-bending test.

Comparing the predicted and experimental results from the U-bending test, the hardening model was the predominant influence on springback prediction accuracy. When an isotropic hardening model was used, the predicted shape differed significantly from the experimental results. However, when the Yoshida-Uemori model was used, the predicted shape was similar to the experimental results. This demonstrates that the Yoshida–Uemori model resulted in better springback predictions relative to the isotropic hardening model, which is consistent with improved approximations of the reverse-loading curves. For the U-bending test, the deformation mode of the sheet was a uniaxial tension mode, and the sheet was deformed with nonlinear loading conditions. The anisotropic behavior in the uniaxial tension mode could be described well by both the Hill'48 and Yld2000-2d yield functions. However, the hardening behavior from the nonlinear loading conditions is only described by the Yoshida-Uemori model because this model effectively considers changes in the elastic modulus due to pre-strain, the Bauschinger effect, and transient behavior. Furthermore, since the inflow amount of the test specimen was large during the U-bending test, the tension-compression behavior is repeated at the wall of the test specimen as the experiment progresses, as shown in Figure 12. This increases the importance of considering nonlinear loading conditions.

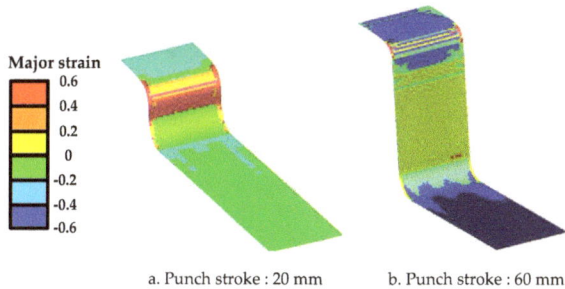

a. Punch stroke : 20 mm   b. Punch stroke : 60 mm

Figure 12. Tension-compression behavior of the U-bending test.

4.2. Springback Prediction for T-Shape Drawing Test

In order to investigate the effects of the constitutive equations on springback prediction accuracy, the predicted and experimental results of the T-shape drawing test were compared, as shown in Figure 13. The springback prediction accuracies in T-shape drawing displayed the same tendencies observed in the U-bending test. The results of the Yld2000-2d yield function and Yoshida-Uemori model combination demonstrated an agreement rate of 82.21%, whereas the combined result of the Hill'48 yield function and isotropic hardening model showed low prediction accuracy with an agreement rate of 73.54%. In this study, the agreement rate was defined as follows:

$$\text{Agreement rate} = \frac{A_{\pm 0.5}}{A_{\text{total}}} \times 100(\%) \tag{13}$$

where $A_{\pm 0.5}$ represents the area within an allowable tolerance of ± 0.5 mm and A_{total} represents the total area of the manufactured part. The allowable tolerance was that acceptable variation from the specified dimensions. In this study, the differences between the analytical and experimental results defined using the Geomagic Design X program should be within ± 0.5 mm.

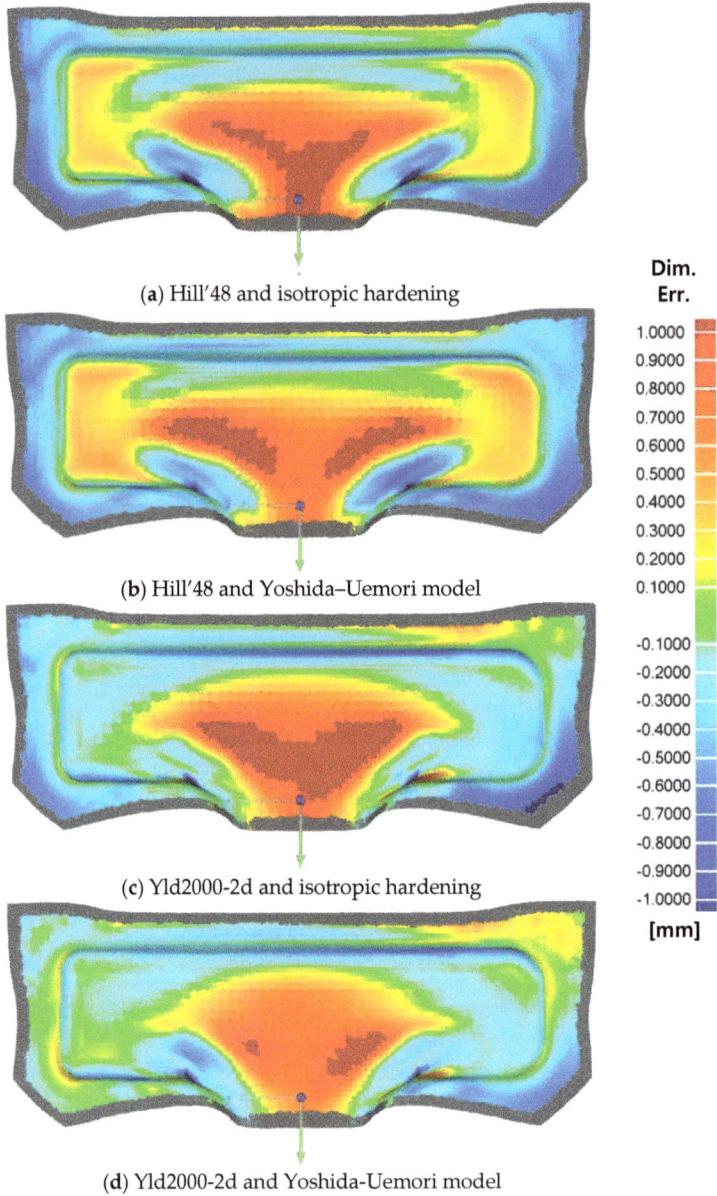

(a) Hill'48 and isotropic hardening

(b) Hill'48 and Yoshida–Uemori model

(c) Yld2000-2d and isotropic hardening

(d) Yld2000-2d and Yoshida-Uemori model

Dim. Err.

1.0000
0.9000
0.8000
0.7000
0.6000
0.5000
0.4000
0.3000
0.2000
0.1000

-0.1000
-0.2000
-0.3000
-0.4000
-0.5000
-0.6000
-0.7000
-0.8000
-0.9000
-1.0000

[mm]

Figure 13. Comparison between predictions and experiments for T-shape drawing test.

When the predicted and experimental results for the T-shape drawing test were compared, it was observed that the yield function was the predominant influence on springback prediction accuracy. Although the uniaxial deformation mode is dominant in U-bending, T-shape drawing has various deformation modes. Additionally, since the inflow amount of the test specimen is small during the T-shape drawing, it is more important to consider the biaxial Lankford value and yield stress than to consider the non-linear condition. When the Hill'48 yield function was used, the predicted shapes differed greatly from those observed experimentally. However, when Yld2000-2d was used, the predicted shape was similar to the experimental results. In the T-shape drawing test, various deformation modes such as the biaxial tension, plane strain, and deep drawing modes were represented, as shown in Figure 14, and the sheet was deformed with an approximately linear loading condition. The hardening behavior in the linear loading condition could be described well by both the isotropic and Yoshida-Uemori hardening models. However, the yield behaviors for various deformation modes are only described by the Yld2000-2d yield function because it considers Lankford values and yield stresses according to the rolling direction and biaxial deformation mode.

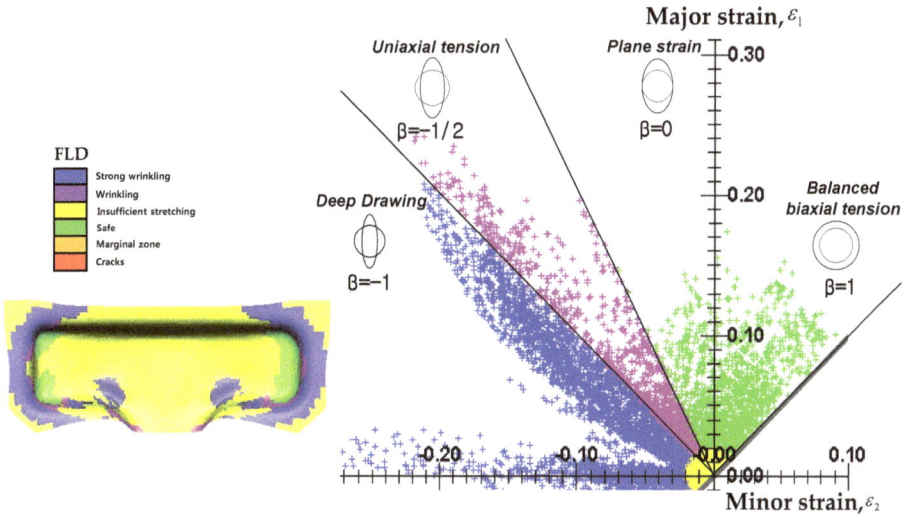

Figure 14. T-shape drawing with various deformation modes.

5. Conclusions

In this study, FEA and experiments were conducted to evaluate the effect of constitutive equations on springback prediction accuracy for the cold stamping of a TRIP1180 sheet. Based on the experimental and analytical results, the following conclusions can be drawn:

1. Uniaxial tension, bulge, tension-compression, and loading-unloading tests were conducted to investigate anisotropy, nonlinear hardening behavior, and changes in the elastic modulus of a TRIP1180 sheet. The material constants of various constitutive equations were determined based on the experimental results, and were implemented in FE simulations for modeling and analyzing springback.
2. FE simulations and experiments were performed to evaluate springback behavior in U-bending and T-shape drawing tests. In both cases, the Yld2000-2d yield function and Yoshida-Uemori model showed excellent prediction accuracy, whereas the Hill'48 yield function and isotropic hardening model showed low prediction accuracy.
3. In the U-bending test, the hardening model had a more dominant influence on the prediction accuracy of springback than the yield function due to the nonlinear loading conditions. The hardening behavior observed under nonlinear loading conditions was only described by the Yoshida-Uemori model, because it effectively considered changes in the elastic modulus due to the pre-strain, the Bauschinger effect, and transient behavior.
4. In the T-shape drawing test, the yield function had a more dominant influence on the prediction accuracy of springback than the hardening model because various deformation modes were

present. The yield behavior for various deformation modes was only described by the Yld2000-2d yield function because this function considered the Lankford values and yield stresses according to the rolling direction and biaxial deformation mode.

5. To predict the springback present in AHSS cold stamping, it is necessary to use appropriate constitutive equations according to the forming process. Furthermore, these constitutive equations need to accurately describe the yield behavior, elastic modulus changes, and hardening behavior for a variety of deformation modes.

Acknowledgments: This work was supported by the Small and Medium Business Administration of Korea (SMBA) grant funded by the Korea government (MOTIE) (No. S2315965).

Author Contributions: Ki-Young Seo and Byung-Min Kim conceived and designed the experiments; Ji Hoon Kim interpreted results; Ki-Young Seo and Jae-Hong Kim wrote and revised the manuscript; Hyunk-Seok Lee measured the experiment specimens and supplied the materials.

Conflicts of Interest: The authors declare no conflict of interest.

References

1. Hisashi, H.; Nakagawa, T. Recent trends in sheet metals and their formability in manufacturing automotive panels. *J. Mater. Process. Technol.* **1994**, *115*, 2–8.
2. Chen, P.; Koc, M. Simulation of springback variation in forming of advanced high strength steels. *J. Mater. Process. Technol.* **2007**, *190*, 189–198. [CrossRef]
3. Chaboche, J.L. Constitutive equations for cyclic plasticity and cyclic viscoplasticity. *Int. J. Plast.* **1989**, *5*, 247–302. [CrossRef]
4. Lee, M.G.; Chung, K.; Kim, D.; Kim, C.; Wenner, M.L.; Barlat, F. Spring-back evaluation of automotive sheets based on isotropic–kinematic hardening laws and non-quadratic anisotropic yield functions, Part I: Theory and formulation. *Int. J. Plast.* **2005**, *21*, 861–882.
5. Zang, S.L.; Guo, C.; Thuillier, S.; Lee, M.G. A model of one-surface cyclic plasticity and its application to springback prediction. *Int. J. Mech. Sci.* **2011**, *53*, 425–435. [CrossRef]
6. Larsson, R.; Bjoerklund, O.; Nilsson, L.; Simonsson, K. A study of high strength steels undergoing non-linear strain paths—Experiments and modelling. *J. Mater. Process. Technol.* **2011**, *211*, 122–132. [CrossRef]
7. Eggertsen, P.A.; Mattiasson, K. On the modeling of the bending-unbending behaviour for accurate springback predictions. *Int. J. Mech. Sci.* **2009**, *51*, 547–563. [CrossRef]
8. Kim, B.M.; Lee, H.S.; Kim, J.H.; Kang, G.S.; Ko, D.C. Development of Seat Side Frame by Sheet Forming of DP980 with Die Compensation. *Int. J. Prec. Eng. Manufacturing.* **2017**, *18*, 115–120.
9. Yoshida, F. Material models for accurate simulation of sheet metal forming and springback. In Proceedings of the AIP Conference Proceedings 1252, Pohang, Korea, 13–17 June 2010.
10. Zhu, Y.X.; Liu, Y.L.; Yang, H.; Li, H.P. Development and application of the material constitutive model in springback prediction of cold-bending. *Mater. Des.* **2012**, *42*, 245–258. [CrossRef]
11. Lee, M.G.; Kim, D.; Kim, C.; Wenner, M.L.; Wagoner, R.H.; Chung, K. Spring-back evaluation of automotive sheets based on isotropic-kinematic hardening laws and non-quadratic anisotropic yield functions Part II: Characterization of material properties. *Int. J. Plast.* **2005**, *21*, 883–914.
12. Hill, R. A theory of the yielding and plastic flow of anisotropic metals. *Proc. R. Soc. Lond.* **1948**, *193*, 281–297. [CrossRef]
13. Barlat, F.; Brem, J.C.; Yoon, J.W.; Chung, K.; Dick, R.E.; Lege, D.J.; Pourboghrat, F.; Choi, S.H.; Chu, E. Plane stress yield function for aluminum alloy sheets—Part I: Theory. *Int. J. Plast.* **2003**, *19*, 1297–1319. [CrossRef]
14. In *SEP1240: Testing and Documentation Guideline for the Experimental Determination of Mechanical Properties of Steel Sheets for CAE-Calculations*, 1st ed.; Verlag Stahleisen GmbH: Düsseldorf, Germany, 2006.
15. Swift, H.W. Plastic instability under plane stresses. *J. Mech. Phys. Solids.* **1952**, *1*, 1–18. [CrossRef]
16. Yoshida, F.; Uemori, T. A model of large-strain cyclic plasticity describing the bauschinger effect and workhardening stagnation. *Int. J. Plast.* **2002**, *18*, 661–686. [CrossRef]
17. Chongthairungruang, B.; Uthaisangsuk, V.; Suranuntchai, S.; Jiratheranat, S. Springback prediction in sheet metal forming of high strength steels. *Mater. Des.* **2013**, *50*, 253–266. [CrossRef]
18. Nelder, J.A.; Mead, R. A simplex method for function minimization. *Comput. J.* **1965**, *7*, 308–313. [CrossRef]

19. Chatti, S.; Fathallah, R.A. A study of the variations in elastic modulus and its effect on springback prediction. *Int. J. Mater. Form.* **2014**, *7*, 19–29. [CrossRef]
20. Ghaei, A.; Green, D.E.; Aryanpour, A. Springback simulation of advanced high strength steels considering nonlinear elastic unloading–reloading behavior. *Mater. Des.* **2015**, *88*, 461–470. [CrossRef]
21. Lee, M.G.; Kim, D.; Kim, C.; Wenner, M.L.; Wagoner, R.H.; Chung, K. Spring-back evaluation of automotive sheets based on isotropic-kinematic hardening laws and non-quadratic anisotropic yield functions, Part III: Applications. *Int. J. Plast.* **2005**, *21*, 915–953. [CrossRef]
22. Ouakdi, E.H.; Louahdi, R.; Khirani, D.; Tabourot, L. Evaluation of springback under the effect of holding force and die radius in a stretch bending test. *Mater. Des.* **2012**, *35*, 106–112. [CrossRef]
23. Jung, J.B.; Jun, S.W.; Lee, H.S.; Kim, B.M.; Lee, M.G.; Kim, J.H. Anisotropic hardening behaviour and and springback of advanced high-strength steels. *Metals* **2017**, *7*, 480. [CrossRef]
24. Gomes, C.; Onipede, O.; Lovell, M. Investigation of springback in high strength anisotropic steels. *J. Mater. Process. Technol.* **2005**, *159*, 91–98. [CrossRef]
25. Makinouchi, A.; Nakamachi, E.; Onate, E.; Wagoner, R.H. NUMISHEET'93 Benchmark Problem. In Proceedings of the 2nd International Conference on Numerical Simulation of 3D Sheet Metal Forming Processes-Verification of Simulation with Experiment, Isehara, Japan, 31 August–2 September 1993.

![metals logo] *metals*

MDPI

Article

Generation of a Layer of Severe Plastic Deformation near Friction Surfaces in Upsetting of Steel Specimens

Sergei Alexandrov [1,2,*], Leposava Šidjanin [3], Dragiša Vilotić [3], Dejan Movrin [3] and Lihui Lang [1]

[1] School of Mechanical Engineering and Automation, Beihang University, No. 37 Xueyuan Road, Beijing 100191, China; lang@buaa.edu.cn

[2] Institute for Problems in Mechanics, Russian Academy of Sciences, 101-1 Prospect Vernadskogo, Moscow 119526, Russia

[3] University of Novi Sad Faculty of Technical Sciences Trg Dositeja Obradovica 6, 21000 Novi Sad, Serbia; lepas@uns.ac.rs (L.Š.); vilotic@uns.ac.rs (D.V.); movrin@uns.ac.rs (D.M.)

* Correspondence: sergei@buaa.edu.cn; Tel.: +7-495-4343665

Received: 30 November 2017; Accepted: 11 January 2018; Published: 19 January 2018

Abstract: Narrow layers of severe plastic deformation are often generated near frictional interfaces in deformation processes as a result of shear deformation caused by friction. This results in material behavior that is very different from that encountered in conventional tests. To develop models capable of predicting the behavior of material near frictional surfaces, it is necessary to design and carry out tests that account for typical features of deformation processes in a narrow sub-surface layer. In the present paper, upsetting of steel specimens between conical and flat dies is used as such a test. The objective of the paper is to correlate the thickness of the layer of severe plastic deformation generated near the friction surface and the die angle using a new criterion for determining the boundary between the layer of severe plastic deformation and the bulk.

Keywords: friction; sliding; upsetting; fine grain layer

1. Introduction

The interface between tool and workpiece in metal forming processes is crucial to both friction and heat transfer [1]. As a result, this interface controls the evolution of microstructure during the process of deformation. In particular, narrow layers of severe plastic deformation are often generated in the vicinity of frictional interfaces in metal forming processes. A complete review of results related to the generation of such layers and published before 1987 has been presented in [2]. In recent years there has been considerable interest in studying material behavior in the vicinity of frictional interfaces in deformation processes [3–9]. One reason for that is that the narrow sub-surface layers affect the performance of structures and machine parts under service conditions [10–13]. Physical properties of these layers can be improved by appropriate heat treatment [14]. To this end, however, a method of predicting these properties after metal forming processes is required. The conditions under which the material is being deformed within the sub-surface layer are completely different from that encountered in conventional material tests. Therefore, the latter cannot be used to determine the flow stress and other constitutive equations within the layer [15]. As a consequence, numerical methods cannot be used for studying metal forming processes in which a layer of severe plastic deformation is generated near frictional interfaces. For example, attempts to model actual high strain gradients in the vicinity of friction surfaces with traditional finite elements have had difficulty representing such gradients [16]. A possible way to overcome this difficulty is to develop a theory that takes into account that the thickness of the layer of severe plastic deformation is very small as compared with other dimensions that classify the workpiece. The conceptual approach here might be somehow similar to that used in the mechanics of cracks (see, for example, [17]). In the latter, linear elastic solutions are supposed to be valid everywhere including the vicinity of crack tips where stresses found from these solutions approach infinity. Then, the stress intensity factor is used instead of stresses to describe physical processes in a small region near crack tips. If a conceptually similar approach were developed for predicting the evolution of material properties near frictional interfaces in metal forming processes then a rigid perfectly plastic solid would play the role of linear elasticity in the mechanics of cracks [18]. In particular, in the case of rigid perfectly plastic solids the equivalent strain rate approaches infinity in the vicinity of maximum friction surfaces and its magnitude in a narrow region near the surface is controlled by the strain rate intensity factor [19]. It is therefore reasonable to assume that the strain rate intensity factor controls the evolution of material properties in the sub-surface layer. Such theories have been proposed in [8,20,21]. For a further development of these theories it is necessary to collect more data from independent experiments to correlate the strain rate intensity factor and properties of the sub-surface layer. Such experiments should be designed rather than chosen according to common practice. In the present paper, upsetting of hollow cylinders between conical and flat dies is used to generate a layer of severe plastic deformation in the vicinity of the frictional interface.

2. Upsetting Test

Upsetting between flat and conical dies is used to generate a layer of severe plastic deformation in the vicinity of the frictional interface between the conical die and workpiece. No lubricant is used on the surface of the conical die to increase friction and, as a result, to get a more pronounced layer of severe plastic deformation. On the other hand, the surface of the flat die is treated to minimize friction. In particular, this surface is lubricated by mineral oil for cold forging. A schematic diagram of the experimental setup is shown in Figure 1.

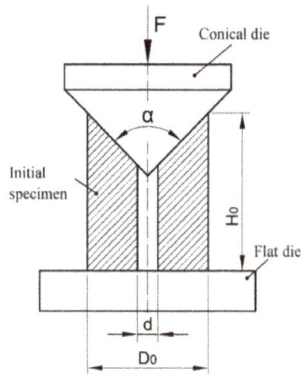

Figure 1. Experimental setup.

It is seen from this figure that one end of the specimen exactly fits the conical die at the initial instant. Because of this design of specimens, the layer of severe plastic deformation starts to generate over the entire friction surface at the initial instant. The only design parameter adopted in the present study is the die angle α (Figure 1). In particular, three dies with $\alpha = 60$ deg, $\alpha = 90$ deg, and $\alpha = 120$ deg are used. Parameters H_0, D_0 and d are fixed. In particular, $H_0 = 35$ mm, $D_0 = 32$ mm and $d = 5$ mm. The specimens are made of normalized C45E steel. Its nominal chemical composition is shown in Table 1. The initial microstructure is uniform over the volume of specimens. The initial microstructure is illustrated in Figure 2. Three nominally identical specimens are tested using each type of the conical die. Upsetting is conducted on a hydraulic press. As an illustration, the $\alpha = 60$ deg die, an initial specimen and the specimen after upsetting are shown in Figure 3.

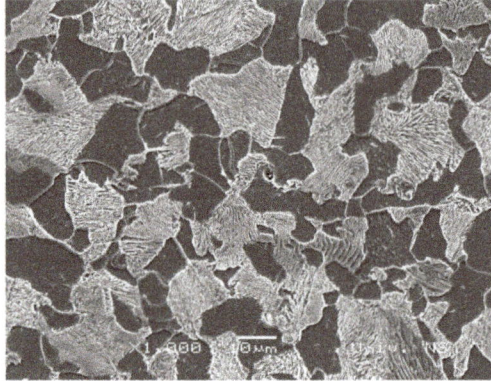

Figure 2. Illustration of the initial microstructure of samples (the average equivalent diameters of pearlite colonies and ferrite grains are 20 μm and 13.5 μm respectively).

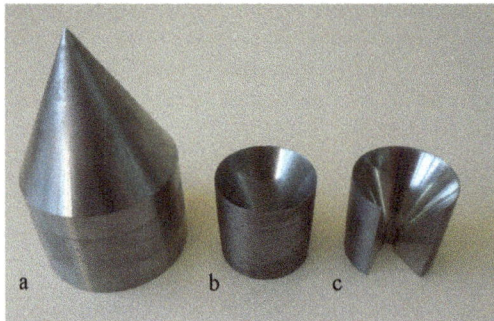

Figure 3. Illustration of (**a**) α = 60 deg die, (**b**) initial specimen and (**c**) specimen after upsetting.

Table 1. Nominal chemical composition of C45E steel (mass fraction, %).

C	Mn	Si	S	P	Cr	Ni	Cu	Mo	V	Al
0.44	0.42	0.23	0.010	0.018	0.006	0.042	0.066	0.008	0.001	0.022

3. Metallographic Observations near the Friction Surface

A standard technique (mechanical grinding and polishing, followed by etching with 3% nital) is used to prepare samples for metallographic studies. The surface of samples is examined with a scanning electron microscope (SEM) JSM 6440LV, produced by JEOL (Tokyo, Japan), operated at 25 KV. The initial microstructure (Figure 2) is classified by the average equivalent circular diameters of pearlite colonies and ferrite grains. Those are 20 μm and 13.5 μm, respectively. Each of these average values is found based on 100 measurements with the use of an ImageJ image analyzer. The microstructure of specimens after upsetting is studied in the vicinity of the friction surface between the conical die and workpiece. It will be seen later that the distribution of microstructure is highly non-uniform in the direction normal to the friction surface. In particular, a narrow layer of severe plastic deformation is generated near the friction surface. The present study focuses on the variation of the thickness of this layer along the intersection of the friction surface and a generic meridian plane and on the dependence of this variation on the angle α. In order to determine the thickness of the layer of severe plastic deformation, it is necessary to have a criterion that identifies severe plastic deformation. It is evident that it is impossible to introduce such a criterion in an unambiguous way. This situation is similar to that in the mechanics of fluids where a criterion for the boundary layer thickness is required [22]. In the case under consideration, an appropriate criterion can be based on metallographic observations. As the deformation proceeds the shape of each ferrite grain in the vicinity of the friction surface changes such that lateral dimensions (the dimensions that are approximately parallel to the friction surface) quickly become large compared to the third dimension (the thickness of ferrite grains). The shape of ferrite grains at the end of upsetting with the $\alpha = 60$ deg die is illustrated in Figure 4. The layer of severe plastic deformation is clearly seen in this figure. The suggested criterion for the thickness of this layer is that the thickness of ferrite grains in the layer is less or equal to 2 µm. The thickness of the layer determined according to this criterion is shown at 5 points. This method is used to determine the variation of the thickness of the layer of severe plastic deformation along the friction surface in all specimens after upsetting. In what follows, δ denotes the thickness of the layer of severe plastic deformation, S denotes the distance along the friction surface in a generic meridian plane and $S = 0$ corresponds to the outer surface of specimens (Figure 5).

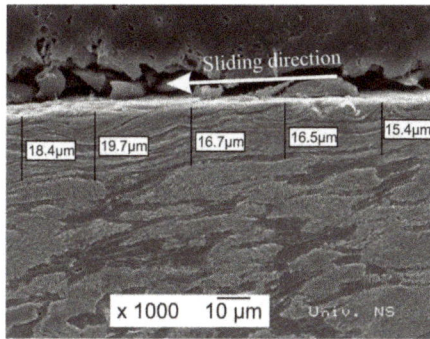

Figure 4. Illustration of the method used to determine the thickness of the layer of severe plastic deformation near friction surfaces.

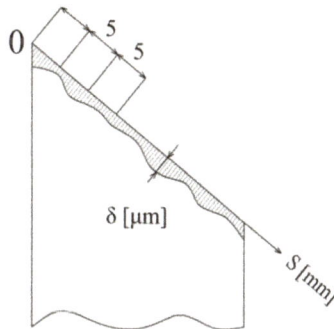

Figure 5. Definition for "S".

It has been found that fracture occurs at $S = S_f$. The value of S_f depends on α. In particular, $S_f = 40$ mm (or about 75% of the total length) for $\alpha = 60$ deg, $S_f = 20$ mm (or about 53% of the total length) for $\alpha = 90$ deg and $S_f = 25$ mm (or about 80% of the total length) for $\alpha = 120$ deg. The microstructure of the material in the vicinity of the point $S = S_f$ is shown in Figure 6. The microstructure of the material within the layer of severe plastic deformation is illustrated at several values of S in the range 5 mm $\leq S$ < S_f in Figure 7 after upsetting with the $\alpha = 60$ deg die, in Figure 8 after upsetting with the $\alpha = 90$ deg die and in Figure 9 after upsetting with the $\alpha = 120$ deg die. The thickness of the layer of severe plastic deformation at these values of S has been determined as illustrated in Figure 4 for the $\alpha = 60$ deg die. The measured values of δ and its average value for one specimen of each series are summarized in Table 2. In this table, R^2 is defined as [23]

$$R^2 = \frac{SS_r}{SS_{to}} = 1 - \frac{SS_e}{SS_{to}} \tag{1}$$

where R^2 is coefficient of determination, SS_r regression sum of squares, SS_{to} total sum of squares and SS_e error sum of square.

The variation of δ with S for all three series of specimens is depicted in Figure 10. The values of δ shown in this figure have been averaged over three specimens of each series.

Table 2. Measured values of δ for one specimen of each series.

	$\alpha = 60$ Deg		$\alpha = 90$ Deg		$\alpha = 120$ Deg	
S	δ and its value averaged over 5 measurements	R^2	δ and its value averaged over 5 measurements	R^2	δ and its value averaged over 5 measurements	R^2
0	-		-		-	
5	(6.6; 6.2; 6.6; 4.7; 5) 5.8		(11.1; 25; 13.5; 12.7; 12.9) 12.5		(10.1; 17.7; 26; 18.8; 10.9) 16.7	
10	(8.5; 9; 9.5; 9.5; 9) 9.1		(14.5; 15; 15; 17.5; 15.5) 15.5		(28; 25; 24; 25; 30) 26.4	
15	(11; 10.5; 10.5; 12; 13.5) 11.5	0.932	(18.4; 19.7; 16.5; 16.2; 16.7) 17.5	1	(38; 30; 30; 31; 32) 32.2	0.677
20	(12.5; 10.5; 12; 11.5; 10) 11.3		Crack appears		(20.6; 21.4; 29.8; 25.4; 25) 24.44	
25	(15.5; 17.3; 13; 15.5; 15) 15.4		-		Crack appears	
30	(28; 29; 33; 35; 35.5) 32.1		-		-	
40	Crack appears		-		-	

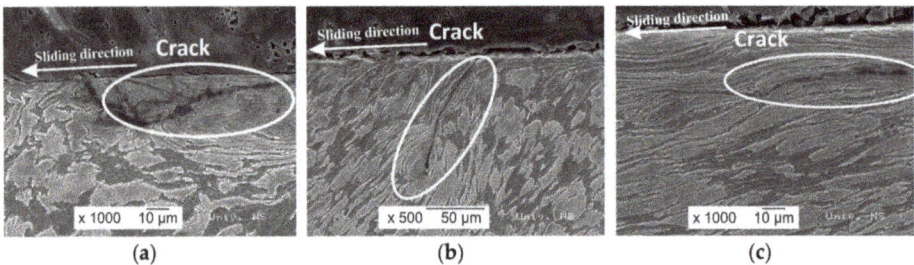

Figure 6. Appearance of cracks in upsetting with (a) the $\alpha = 60$ deg die at $S = 40$ mm, (b) the $\alpha = 90$ deg die at $S = 20$ mm and (c) the $\alpha = 120$ deg die at $S = 25$ mm.

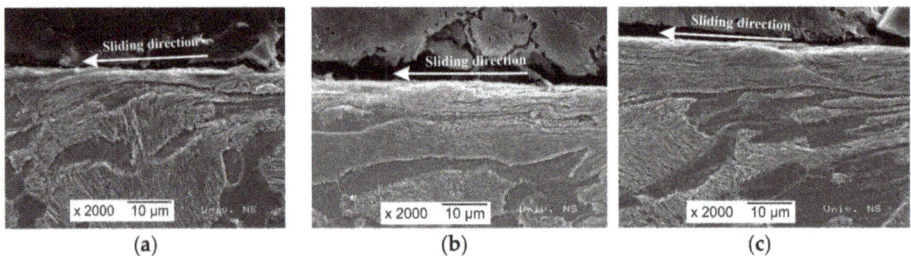

Figure 7. *Cont.*

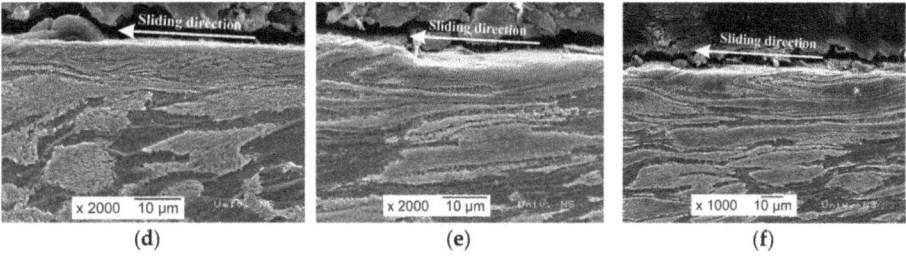

Figure 7. The thickness of the layer of severe plastic deformation after upsetting with the $\alpha = 60$ deg die at (**a**) $S = 5$ mm, (**b**) $S = 10$ mm, (**c**) $S = 15$ mm, (**d**) $S = 20$ mm, (**e**) $S = 25$ mm, and (**f**) $S = 30$ mm.

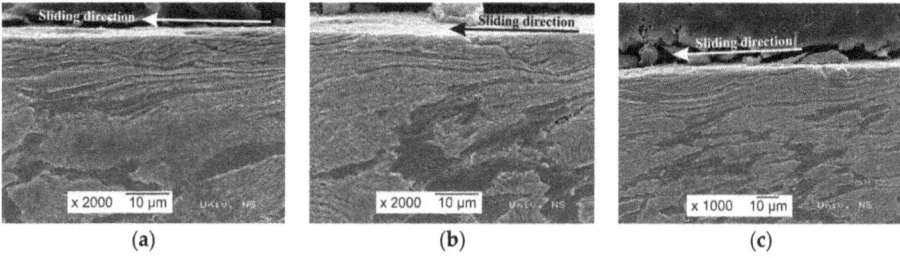

Figure 8. The thickness of the layer of severe plastic deformation after upsetting with the $\alpha = 90$ deg die at (**a**) $S = 5$ mm, (**b**) $S = 10$ mm and (**c**) $S = 15$ mm.

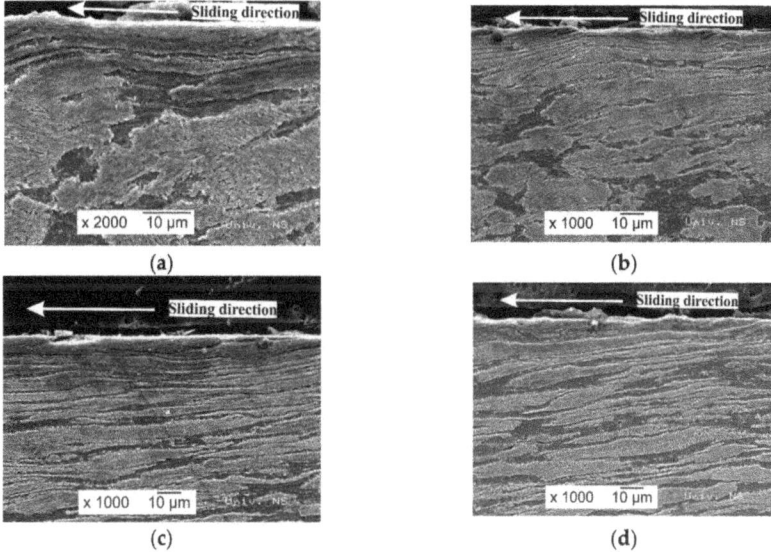

Figure 9. The thickness of the layer of severe plastic deformation after upsetting with the $\alpha = 120$ deg die at (**a**) $S = 5$ mm, (**b**) $S = 10$ mm, (**c**) $S = 15$ mm and (**d**) $S = 20$ mm.

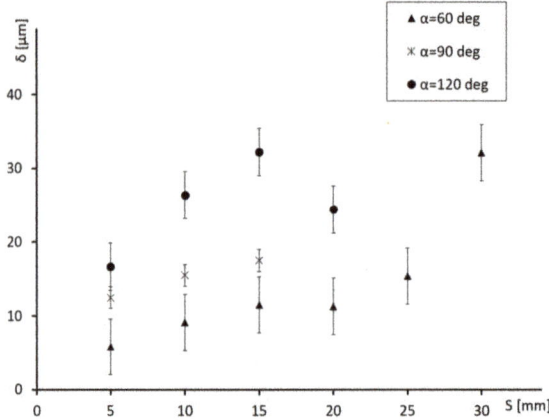

Figure 10. Variation of the thickness of the layer of severe plastic deformation with S and α.

4. Conclusions and Discussion

Upsetting of hollow cylinders between conical and flat dies has been conducted to generate a layer of severe plastic deformation in the vicinity of the friction interface between the conical die and workpiece. It is seen from Figures 7–9 that the thickness of this layer depends on both α and S. Therefore, the test proposed can be used to reveal a possible correlation between this thickness and the strain rate intensity factor. This correlation is required for the development of the theories proposed in [8,20,21].

The tendencies in the behavior of the curves shown in Figure 10 are that the thickness of the layer of severe plastic deformation increases as both α and S increase. An exception is the thickness of this layer measured at S = 20 mm after upsetting with the α = 120 deg die. This deviation from the general trend can be explained by the existence of a dead region of the workpiece that sticks to the die. Upsetting between α = 120 deg and flat dies is rather similar to upsetting between two flat (i.e., α = 180 deg) dies. If friction is high enough, the radial velocity at the friction surfaces changes its sense between the outer and inner traction free surfaces in upsetting of disks between two flat dies [24]. Therefore, dead regions inevitably appear in upsetting of disks between two flat dies.

In addition to experimental data, the theories [8,20,21] require the theoretical value of the strain intensity factor. This factor is defined as [19]

$$\xi_{eq} = \frac{D}{\sqrt{z}} + o\left(\frac{1}{\sqrt{z}}\right) \tag{2}$$

as $z \to 0$. Here ξ_{eq} is the equivalent strain rate (quadratic invariant of the strain ate tensor), D is the strain rate intensity factor, z is the normal distance to the maximum friction surface. It is seen from (1) that the gradient of the equivalent strain rate is very high in the vicinity of frictional interfaces and that the strain rate intensity factor controls the magnitude of the equivalent strain rate in a narrow layer near frictional interfaces. Since the equivalent strain rate is responsible for the evolution of material properties, this theoretical result is in qualitative agreement with the experimental results shown in Figures 4 and 7–9. However, it is also seen from (2) that the strain rate intensity factor is the coefficient of the singular term. Therefore, conventional finite element methods are not capable of determining the strain rate intensity factor [25]. In particular, using the commercial package ABAQUS an upsetting process has been analyzed in [26]. All the finite element analyses presented in this paper failed to converge in the case of the maximum friction law. Probably, the extended finite element method [27] can be used for this purpose. However, to the best of authors' knowledge, no attempts have been made to determine the strain rate intensity factor by means of this method. To date the only accurate method for calculating the strain rate intensity factor is based on the method of characteristics [28]. However, its validity is restricted to plane strain problems. Therefore, there is an urgent need for the development of a numerical method for calculating the strain rate intensity factor in axisymmetric flow. Solutions found by this method might be used in conjunction with the experimental results shown in Figure 10 to provide necessary input for the theories proposed in [8,20,21].

Author Contributions: Sergei Alexandrov is responsible for the development of the general concept, Leposava Šidjanin is responsible for metallographic observation, Dragiša Vilotić is responsible for upsetting test, Dejan Movrin is responsible for conducting experimental research, and Lihui Lang is responsible for the interpretation of experimental results.

Conflicts of Interest: The authors declare no conflict of interests.

References

1. Beynon, J.H. Tribology of hot metal forming. *Tribol. Int.* **1998**, *31*, 73–77. [CrossRef]
2. Griffiths, B.J. Mechanisms of White Layer Generation with Reference to Machining and Deformation Processes. *J. Tribol.* **1987**, *109*, 525–530. [CrossRef]
3. Kim, Y.-T.; Ikeda, K. Flow behavior of the billet surface layer in porthole die extrusion of aluminum. *Metal. Mater. Trans. A* **2000**, *31*, 1635–1643. [CrossRef]
4. Hosoda, K.; Asakawa, M.; Kajino, S.; Maeda, Y. Effect of die semi-angle and multi-pass drawing on additional shear strain layer. *Wire J. Int.* **2008**, *41*, 68–73.
5. Wideroe, F.; Welo, T. Conditions for Sticking Friction between Aluminium Alloy AA6060 and Tool Steel in Hot Forming. *Key Eng. Mater.* **2011**, *491*, 121–128. [CrossRef]
6. Sanabria, V.; Müller, S.; Gall, S.; Reimers, W. Investigation of Friction Boundary Conditions during Extrusion of Aluminium and Magnesium Alloys. *Key Eng. Mater.* **2014**, *611–612*, 997–1004. [CrossRef]
7. Sanabria, V.; Mueller, S.; Reimers, W. Microstructure Evolution of Friction Boundary Layer during Extrusion of AA 6060. *Procedia Eng.* **2014**, *81*, 586–591. [CrossRef]
8. Alexandrov, S.; Jeng, Y.-R.; Hwang, Y.-M. Generation of a Fine Grain Layer in the Vicinity of Frictional Interfaces in Direct Extrusion of AZ31 Alloy. *J. Manuf. Sci. Eng.* **2015**, *137*. [CrossRef]
9. Hwang, Y.-M.; Huang, T.-H.; Alexandrov, S. Manufacture of Gradient Microstructures of Magnesium Alloys Using Two-Stage Extrusion Dies. *Steel Res. Int.* **2015**, *86*, 956–961. [CrossRef]
10. Griffiths, B.J.; Furze, D.C. Tribological Advantages of White Layers Produced by Machining. *J. Tribol.* **1987**, *109*, 338–342. [CrossRef]
11. Warren, A.W.; Guo, Y.B. Numerical Investigation on the Effects of Machining-Induced White Layer during Rolling Contact. *Tribol. Trans.* **2005**, *48*, 436–441. [CrossRef]
12. Kajino, S.; Asakawa, M. Effect of "additional shear strain layer" on tensile strength and microstructure of fine drawn wire. *J. Mater. Process. Technol.* **2006**, *177*, 704–708. [CrossRef]
13. Choi, Y. Influence of a white layer on the performance of hard machined surfaces in rolling contact. *J. Eng. Manuf.* **2010**, *224*, 1207–1215. [CrossRef]
14. Wu, X.; Yang, M.; Yuan, F.; Wu, G.; Wei, Y.; Huang, X.; Zhu, Y. Heterogeneous lamella structure unites ultrafine-grain strength with coarse-grain ductility. *Proc. Natl. Acad. Sci. USA* **2015**, *112*, 14501–14505. [CrossRef] [PubMed]
15. Jaspers, S.P.F.C.; Dautzenberg, J.H. Material behaviour in metal cutting: Strains, strain rates and temperatures in chip formation. *J. Mater. Process. Technol.* **2002**, *121*, 123–135. [CrossRef]
16. Appleby, E.J.; Lu, C.Y.; Rao, R.S.; Devenpeck, M.L.; Wright, P.K.; Richmond, O. Strip drawing: A theoretical-experimental comparison. *Int. J. Mech. Sci.* **1984**, *26*, 351–362. [CrossRef]
17. Kanninen, M.F.; Popelar, C.H. *Advanced Fracture Mechanics*, 1st ed.; Oxford University Press: New York, NY, USA, 1985; ISBN 978-0195035322.
18. Alexandrov, S. Interrelation between Constitutive Laws and Fracture in the Vicinity of Friction Surfaces. In *Physical Aspects of Fracture*; Bouchaud, E., Jeulin, D., Prioul, C., Roux, S., Eds.; Springer: Dordrecht, The Netherlands, 2001; pp. 179–190. ISBN 978-0-7923-7147-2.
19. Alexandrov, S.; Richmond, O. Singular plastic flow fields near surfaces of maximum friction stress. *Int. J. Non-Linear Mech.* **2001**, *36*, 1–11. [CrossRef]

20. Alexandrov, S.E.; Goldstein, R.V. On Constructing Constitutive Equations in Material Thin Layer Near Friction Surfaces in Material Forming Processes. *Dokl. Phys.* **2015**, *60*, 39–41. [CrossRef]
21. Goldstein, R.V.; Alexandrov, S.E. An approach to prediction of microstructure formation near friction surfaces at large plastic strains. *Phys. Mesomech.* **2015**, *18*, 223–227. [CrossRef]
22. Batchelor, G.K. *An Introduction to Fluid Dynamics*; Cambridge University Press: New York, NY, USA, 1999; ISBN 0-521-66396-2.
23. Vining, G.G.; Kowalski, S. *Statistical Method for Engineers*; Cengage Learning: Boston, MA, USA, 2011; ISBN 978-0-S38-73S18-6.
24. Male, A.T.; Cockcroft, M.G. A method for the determination of the coefficient of friction of metals under conditions of bulk plastic deformation. *J. Inst. Met.* **1964**, *93*, 38–46. [CrossRef]
25. Facchinetti, M.; Miszuris, W. Analysis of the maximum friction condition for green body forming in an ANSYS environment. *J. Eur. Ceram. Soc.* **2016**, *36*, 2295–2302. [CrossRef]
26. Chen, J.-S.; Pan, C.; Roque, C.M.O.L.; Wang, H.-P. A Lagrangian reproducing kernel particle method for metal forming analysis. *Comput. Mech.* **1998**, *22*, 289–307. [CrossRef]
27. Fries, T.-P.; Belytschko, T. The extended/generalized finite element method: An overview of the method and its applications. *Int. J. Numer. Methods Eng.* **2010**, *84*, 253–304. [CrossRef]
28. Alexandrov, S.; Kuo, C.-Y.; Jeng, Y.-R. A numerical method for determining the strain rate intensity factor under plane strain conditions. *Contin. Mech. Thermodyn.* **2016**, *28*, 977–992. [CrossRef]

metals

MDPI

Article

Sensitivity Analysis of Oxide Scale Influence on General Carbon Steels during Hot Forging

Bernd-Arno Behrens [1], Alexander Chugreev [1], Birgit Awiszus [2], Marcel Graf [2], Rudolf Kawalla [3], Madlen Ullmann [3], Grzegorz Korpala [3] and Hendrik Wester [1,*]

[1]   Institute of Forming Technology and Machines, Leibniz Universität Hannover, 30823 Garbsen, Germany; behrens@ifum.uni-hannover.de (B.-A.B.); chugreev@ifum.uni-hannover.de (A.C.)
[2]   Professorship Virtual Production Engineering, Technische Universität Chemnitz, 09126 Chemnitz, Germany; birgit.awiszus@mb.tu-chemnitz.de (B.A.); marcel.graf@mb.tu-chemnitz.de (M.G.)
[3]   Institute of Metal Forming, Technische Universität Freiberg, 09599 Freiberg, Germany; Rudolf.Kawalla@imf.tu-freiberg.de (R.K.); Madlen.Ullmann@imf.tu-freiberg.de (M.U.); Grzegorz.Korpala@imf.tu-freiberg.de (G.K.)
*   Correspondence: wester@ifum.uni-hannover.de; Tel.: +49-511-762-3405

Received: 8 December 2017; Accepted: 12 February 2018; Published: 14 February 2018

Abstract: Increasing product requirements have made numerical simulation into a vital tool for the time- and cost-efficient process design. In order to accurately model hot forging processes with finite, element-based numerical methods, reliable models are required, which take the material behaviour, surface phenomena of die and workpiece, and machine kinematics into account. In hot forging processes, the surface properties are strongly affected by the growth of oxide scale, which influences the material flow, friction, and product quality of the finished component. The influence of different carbon contents on material behaviour is investigated by considering three different steel grades (C15, C45, and C60). For a general description of the material behaviour, an empirical approach is used to implement mathematical functions for expressing the relationship between flow stress and dominant influence variables like alloying elements, initial microstructure, and reheating mode. The deformation behaviour of oxide scale is separately modelled for each component with parameterized flow curves. The main focus of this work lies in the consideration of different materials as well as the calculation and assignment of their material properties in dependence on current process parameters by application of subroutines. The validated model is used to carry out the influence of various oxide scale parameters, like the scale thickness and the composition, on the hot forging process. Therefore, selected parameters have been varied within a numerical sensitivity analysis. The results show a strong influence of oxide scale on the friction behaviour as well as on the material flow during hot forging.

Keywords: hot forging; finite-element; oxide scale

1. Introduction

In the field of bulk metal forming, hot forging is a widely-used process for the production of high-performance parts with complex shapes. The preheating of semi-finished parts to temperatures above 800 °C leads to a significant reduction in required forming forces as well as an increase in the material formability. Furthermore, the process efficiency can be increased by the use of process heat energy for a direct thermomechanical treatment to produce parts with locally-adapted properties [1]. However, a major disadvantage is the appearance of oxidation effects on the preheated steel surfaces. The scale layer itself has an inhomogeneous structure. On a steel surface, it typically consists of three different iron oxides, namely wuestite, magnetite, and haematite. The development and growth of these oxides depend on the steel matrix properties like the used alloying concept and surface quality as well as the oxidation time, atmosphere, and temperature. The formation of oxide scale leads to material losses of up to 3% by weight and requires further process steps for removal and rework. During the forging process, the presence of oxide scale influences friction as well as material flow and can lead to an increase in die wear [2]. The tooling and setup cost up to 15% of total production costs in bulk metal forming and extensive investigations on tool failure have shown that more than 70% is caused by die wear [3]. Therefore, die wear has a decisive influence on the entire efficiency of a hot forging process [4].

The oxidation process on heated steel surfaces is significantly influenced by alloying elements. The influence of different alloying elements on the formation of an oxide scale layer is described in [5]. Comparison of Si-steel with IF-steel and S355-steel have shown that the alloying elements Si and P lead to a delayed oxidation at lower temperatures and to a significantly increased oxidation rate for temperatures above 1100 °C. Previous studies investigated the influence of carbon content on the formation of oxide scale and have shown a relationship between temperature, carbon content, and growth of oxide scale. Temperatures higher than 700 °C in combination with a high carbon content lead to a decrease in oxide scale layer thickness [6,7]. The composition of oxide scale is variable. The oxidation behaviour of pure iron in air and oxygen atmosphere has been examined in numerous studies [8,9]. They pointed out that the classic oxide scale structure at temperatures above 700 °C consists of an extremely thin cover layer of haematite, a thin intermediate layer of magnetite, and a thick inner layer of wuestite directly on the steel surface. With a decrease in temperature below 650 °C, the layer thickness of magnetite and haematite increases whereas that of wuestite decreases. Until about 580 °C, wuestite is still the major phase. A further decrease in temperature below 570 °C leads to an unstable wuestite layer. Thus the oxide scale layer consists of two iron oxides, a thick layer of magnetite with about 80 wt %, and a thin haematite cover layer [10]. In the temperature range between 700 °C and 1250 °C, the oxide scale composition is nearly constant with a ratio of 1:4:95 for the layers of haematite, magnetite, and wuestite, respectively [11].

Experimental studies in the temperature range between 900 °C and 1200 °C have shown strong varying forming behaviour of the different iron oxides wuestite, magnetite, and haematite, which build up the typical oxide scale layer on general steel. The yield stresses are observed to be strongly temperature-dependent as known for metals. The yield stress of wuestite has been found to be low as compared to magnetite, whereas the highest bearable strain was found for wuestite. Maximum yield stress has been observed for haematite. Moreover, it has been found to be the hardest oxide scale component at room temperature with a Vickers hardness of 1000 HV10 as compared to magnetite (600 HV10) and wuestite (400 HV10). Furthermore, the investigations have shown a significant and partly contrary influence of the strain rate on the forming behaviour. Contrary to haematite, magnetite and wuestite show an increase in yield stress with increasing strain rate [12,13]. Further investigations on synthesized iron oxides at high temperatures have shown a strong dependency of hardness on temperature. The hardness of all the oxides decreases with an increase in temperature, whereby the strongest reduction was observed for magnetite and haematite [14].

The development of an oxide scale layer on the steel surface in a metal forming process results in changes in surface properties. In particular, friction conditions in the contact zone between workpiece and tool are affected during the forming process, which has a significant influence on the material flow [15]. Nevertheless, only analytical and phenomenological approaches for mono materials have been published so far. Various research groups have investigated friction with regard to different oxide scale conditions which were induced by a defined furnace temperature and holding times in variable atmospheres. They showed that a thin oxide scale layer has a positive influence on the friction properties in contrast to thicker ones, which are harder and brittle [16–18]. Furthermore, there have been experimental investigations indicating a decrease of friction coefficient with increasing oxide scale layer thickness [19–21].

In addition, the forming behaviour of the steel matrix has a significant influence on the forging process. Parameters with a significant effect on the yield stress are process-specific parameters (e.g., strain rate and temperature) as well as material-specific parameters (e.g., alloying concept, microstructure, and forming history) [22]. Experimental investigations have shown a significant influence of total carbon content on the material flow behaviour. An increase in carbon content has led to a reduction in the yield stress, particularly at high temperatures [23]. A Hensel–Spittel-based flow curve model which takes initial grain size, temperature, strain rate, and carbon content into account has been presented by Korpala et al. [24].

The main focus of the presented work is on the investigation of the oxide scale influence on the hot forging process, in particular on the plastic deformation and the material flow. Therefore, a numerical sensitivity analysis has been performed regarding the effect of oxide scale thickness, composition, and different friction conditions. This analysis is based on a numerical model for description of thin surface layers, which has been validated by means of an experimental ring compression test.

2. Materials and Methods

2.1. Experimental Procedure

The use of finite element FE-simulation requires detailed mathematical models for a realistic description of the material behaviour by consideration of process parameters like strain, strain rate, and temperature, as well as material-specific parameters like carbon content and initial grain size. Therefore, various experimental tests like the cylindrical compression test and the ring compression test have been performed. At least three repetitions have been performed for each experimental test.

The ring compression test is a standard procedure to investigate friction properties in forging processes. In the scope of this research project, ring compression tests have been performed in order to validate the numerical model as well as to determine the friction properties between the oxide scale layer and dies. The ring samples used had an outer diameter of 9 mm, an inner diameter of 4.5 mm, and a height of 3 mm. Figure 1a shows a deformed test sample. The experimental setup is provided in Figure 1b. After the ring compression test, the inner diameter and height were measured at different positions in order to calculate averaged values (Figure 1c).

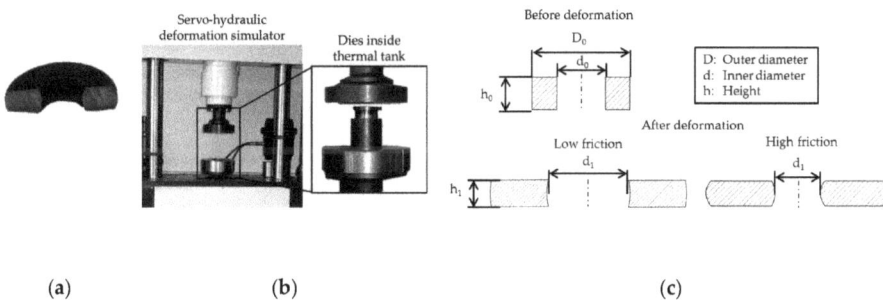

(a) (b) (c)

Figure 1. Deformed ring compression sample (**a**); Experimental ring compression setup on servo-hydraulic deformation simulator WUMSI with thermal tank (**b**); Schematic representation of ring compression test for different friction conditions (**c**).

The mechanical behaviour of different general steel grades with varying carbon content (C15, C45, C60) were examined by uniaxial compression tests. These unalloyed carbon steel grades were chosen in order to measure the influence of carbon content on forming behaviour as well as the growth of oxide scale layer without interference by other alloy elements. An overview of the chemical composition is shown in Table 1.

Table 1. Chemical composition of the steel samples in wt %.

Steel Grade	C	Si	Mn	P	S	Cr	Mo	Ni	Cu	Al
C15	0.160	0.210	0.420	0.001	<0.001	0.100	0.045	0.087	0.108	0.017
C45	0.490	0.200	0.420	0.001	0.001	0.100	0.034	0.081	0.096	0.021
C60	0.63	0.200	0.420	<0.001	<0.001	0.100	0.035	0.076	0.100	0.017

The used samples (diameter 10 mm, length 18 mm) showed a preformed and annealed initial microstructure. Within the experimental procedure, the samples were austenized at various temperatures and subsequently brought on different deformation temperatures. The experimental compression tests were performed on a servo-hydraulic deformation simulator (WUMSI, 400 kN, an in-house development of TU Freiberg, Germany and WPM Leipzig, Markkleeberg, Germany) with varying strain rates between 1 s^{-1} and 20 s^{-1}. The strain was measured by means of a linear variable differential transformer (LVDT) sensor which had been integrated into the cylinder. The tools made out of Al_2O_3 had been lubricated with graphite to reduce the influence of friction between the tools and the workpiece. An overview of the considered parameters is shown in Figure 2.

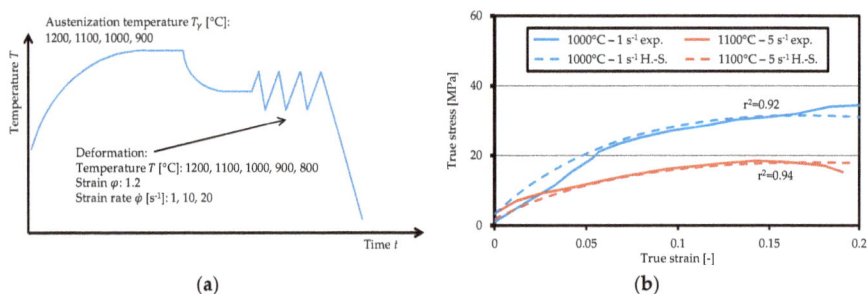

(a)   (b)

Figure 2. Schematic representation of the experimental procedure for the determination of flow curves for the steel matrix (**a**); Exemplary results of experimental cylindrical compression test and calculated true stress (H.-S.) for the pure iron oxide wuestite (based on [13]) (**b**).

Due to the fact that the mechanical properties of the iron oxides wuestite, magnetite, and haematite show a strong variation, separate flow curve models for each iron oxide were derived and parametrized based on the results of the experimental compression tests. In order to separately derive mechanical properties for each iron oxide, the required samples were manufactured with powder metallurgy. Therefore, pure oxide powders of haematite, magnetite, and wuestite were compressed and sintered into cylindrical samples. Subsequently, the samples were deformed in the temperature range of hot forming (900–1150 °C) and different strain rates up to 10 s^{-1}. Experimental results for the iron oxide wuestite at a temperature of 1000 °C and a strain rate of 1 s^{-1}, as well as a temperature of 1100 °C and a strain rate of 5 s^{-1} are shown in Figure 2b. Detailed information regarding the powder metallurgy process route as well as experimental data for various temperatures and strain rates are presented in [13].

2.2. Aspects of the Numerical Model and Its Implementation

This paper focuses on the numerical investigation of thin oxide scale layers in a hot forging by the use of the finite element method. The developed numerical model is based on a multi-material approach in order to take the strongly varying mechanical properties of the steel matrix and the iron oxides in the oxide scale layer into account. Due to the fact that an experimental characterisation of the interface between oxide scale and steel matrix is very challenging, the contact between steel matrix and oxide scale surfaces is modelled as a glued contact type, which is similar to a tying between node and surface. In order to reduce the computational time the 2D axial symmetry has been considered, the matrix as well as the oxide scale layer have been discretised with four-node, isoparametric, quadrilateral elements. The element stiffness is described by using four Gaussian points and a full integration scheme. The number of elements is about 20,000 but strongly depends on the oxide scale thickness. To avoid extensive element distortion, a remeshing criterion has been implemented. A schematic representation of the numerical model which had been set up in the commercial FE-software Simufact Forming v14.0.1 (Simufact Engineering Gmbh, Hamburg, Germany, 2017), which is based on the implicit MSC.Marc solver, is provided in Figure 3.

For a description of varying mechanical properties, the model is based on a multi-material approach. The calculation of actual yield stress with regard to the local temperature, strain, and strain rate, as well as material allocation, is carried out with a user subroutine which is scripted in FORTRAN with the FE solver. Each calculation process is linked with a specific element ID to ensure the correct assignment of calculated data to the correct numerical element. The user subroutine is called for each element in each iteration of every solver increment. The required data like local temperature, strain, and strain rate, as well as a material ID and element ID, are provided by the solver. Subsequently, the calculated yield stress is transferred back to the solver, linked with the specific element ID.

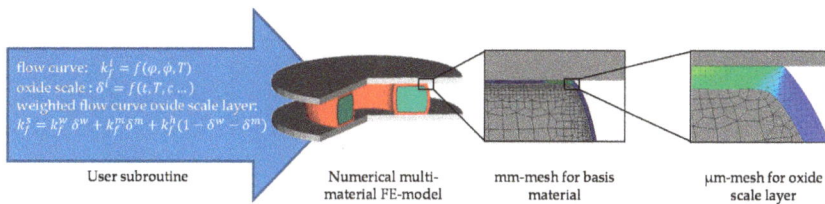

Figure 3. Schematic representation of the numerical model for thin surface layers and user subroutine.

In order to describe the thin surface layer consisting of three different materials, the oxide scale is considered to be a smeared continuum. Based on this approach, materials with non-uniform properties can be described as a homogenous continuum [25]. Therefore, single yield stresses are calculated for each of the iron oxides, wuestite, magnetite, and haematite, separately. For this purpose, three different material models parameterized with the findings of oxide scale characterisation are implemented as functions into the user subroutine. Each function calculates the flow stress for one of the three oxide scale parts depending on the current temperature, true strain, and strain rate, as well as material-specific parameters. With regard to the assumption of a smeared continuum, a weighting of the individual flow stresses is required in order to calculate a homogenous flow stress for the oxide scale layer. The weighted oxide scale flow stress is calculated depending on the oxide scale composition and is given by:

$$k_f^s = k_f^w \, \delta^w + k_f^m \delta^m + k_f^h (1 - \delta^w - \delta^m) \tag{1}$$

whereby k_f^s represents the global oxide scale flow stress, and k_f^w, k_f^m and k_f^h are the flow stress of the constituent oxide scale components. The terms δ^w and δ^h are the mass fractions of wuestite and magnetite, respectively. All numerical results of the oxide scale layer presented in this paper have been calculated based on the smeared approach described above and under assumption of various initial oxide scale compositions.

In general, the mass fractions depend on the temperature, time, and carbon content of the matrix material. Furthermore, within this common research project, an Arrhenius-based approach has been developed to describe the oxide scale growing process on carbon steel under consideration of carbon content [24]. This approach will be implemented into the user subroutine. A representation of the dataflow inside the user subroutine, as well as between the user subroutine and Simufact Forming solver, is given in Figure 4.

By integration of the user subroutine into the Simufact Forming GUI, key parameters like carbon content of used steel and oxide scale composition can be easily adjusted by the user. Due to the modular structure of the user subroutine, it can be easily extended to take other phenomena like separation of oxide scale parts into consideration.

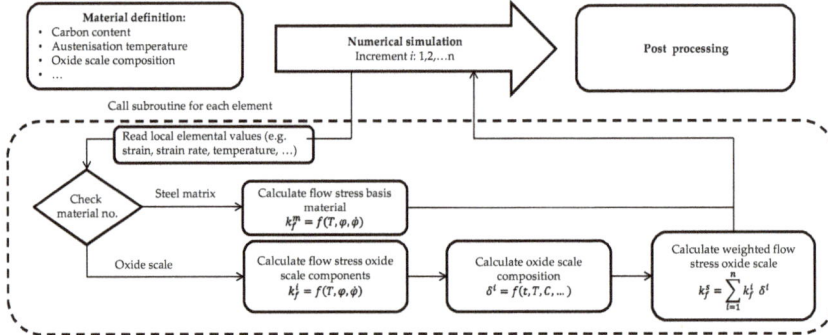

Figure 4. Dataflow inside the user subroutine as well as between the user subroutine and the numerical solver.

Based on experimental data, the models have been parameterized via regression analysis using the least square method. The onset of yielding is described by means of von Mises criterion. The plastic behaviour of the carbon steel (C15, C45, C60) is based on the following two Hensel–Spittel approaches according to [26]:

$$HS_a = A_a C\%^{m_{C_1}} \varphi^{m_a} e^{-m_{a1} T} \varphi^{m_{a2}} \vartheta^{m_{a\varphi}} c\%^{m_{C_2}} \dot{\varphi}^{m_{a3}} \tag{2}$$

$$HS_b = A_b C\%^{m_{C3}} e^{-m_{b1} T} \varphi^{m_{b2}} \dot{\varphi}^{m_{b3}} \tag{3}$$

The material hardening depending on temperature T, strain φ and strain rate $\dot{\varphi}$ at the beginning of the deformation process is described by the first term (HS_a) whereas the second term (HS_b) takes the material softening at higher strains caused by recrystallization into consideration. Both terms are weighted by a transition function δ:

$$\delta_{HS} = 0.5 + \pi^{-1} \tan^{-1} \left[w_1 T^{-w_2} \left(\varphi - \varphi_k T_{\gamma}^{w_{\gamma}} \varphi^{\dot{w}_4} \right) \right] \tag{4}$$

In addition to temperature, strain, and strain rate, this model takes into account the carbon content C as well as the initial grain size. Thus it is possible to describe various steel grades in a more general way regarding the concentration of the alloying element carbon. The initial grain size is considered with the help of austenization temperature T_γ, therefore the effect of preheating is also taken into account. An overview of the derived material-specific parameters A, m, w is presented in Tables 2 and 3.

Table 2. Material-specific parameters for the flow curve terms HS_a and HS_b.

A_a	m_{C_1}	m_{a1}	m_{a2}	$m_{a\varphi}$	m_{C_2}	m_{a3}	A_b	m_{C_3}	m_{b1}	m_{b2}	m_{b3}
3275	0.03662	0.0027	0.41618	−0.0421	−0.08004	0.07959	894	−0.01367	0.00276	0.00236	0.17777

Table 3. Material-specific parameters for the transition function δ_{HS}.

w_1	w_2	w_4	w_γ	φ_k
1.83501	0.00015	0.13325	0.048	0.15951

Exemplary calculated flow curves for the steels C15, C45, and C60 at various temperatures and a strain rate of 10 s^{-1}, using the approach for general steel, are presented in Figure 5a. Hereby, an austenization temperature of 1100 °C had been chosen. A comparison between experimentally-measured and calculated flow curves is shown for the steel grade C15 at different temperatures and a strain rate of 10 s^{-1} in Figure 5b. The comparison provides a good qualitative agreement. The calculated correlation coefficient (r^2) for the flow curve models is between 0.994 and 0.997. A validation for all steel grades has been presented in [24].

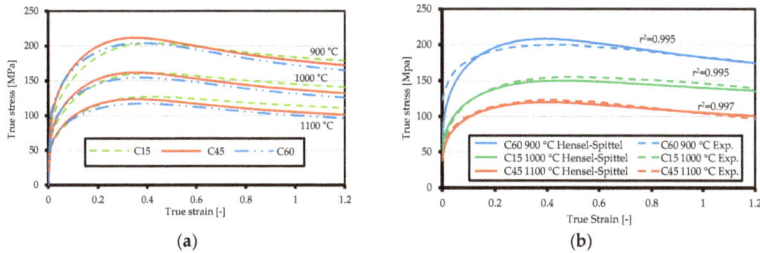

Figure 5. Calculated flow curves for the steel grades C15, C45, and C60 at different temperatures (900 °C, 1000 °C, 1100 °C) and a strain rate of 10 s^{-1} (a) Comparison between experimentally-measured and calculated flow curves for the steel grades C15, C45, and C60 at different temperatures and a strain rate of 10 s^{-1} (b).

The description of plastic behaviour of the iron oxides is based on Hensel–Spittel flow curve equations and thereby the current yield stress is expressed as a function of temperature, strain, and strain rate [13]. The flow curves have been parametrized based on the findings of cylindrical compression tests for the pure iron oxide samples. In the temperature range relevant for hot forging, the flow curves for the iron oxides, and in particular for wuestite, are lower in comparison with the calculated yield stresses of carbon steel. Exemplary calculated flow curves at a temperature of 1000 °C and various strain rates for the iron oxides wuestite, magnetite, and haematite are presented in Figure 6. An exemplary comparison between experimental data and calculated true stress based on the Hensel–Spittel (H.–S.) approach for the iron oxide wuestite is presented in Figure 2b. The calculated correlation coefficient (r^2) for all strain rates and temperatures is between 0.72 and 0.76 [13]. The variation of r^2 is due to the challenging experimental procedure for the pure iron oxides.

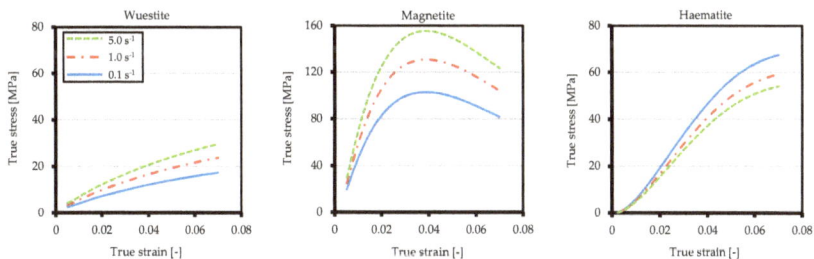

Figure 6. Exemplary calculated flow curves for wuestite, magnetite, and haematite at a temperature of 1000 °C.

3. Results and Discussions

3.1. Ring Compression Test and Model Validation

The presented numerical model has been validated by comparing numerical results with experimentally-measured force displacement curves of ring compression tests. Therefore, the samples were heated to a temperature of 1000 °C and 900 °C with an oxidation time of 30 s. The preheating led to a specific oxide scale layer with an initial thickness of 50 μm, which had been used as an initial condition for numerical simulation. The initial thicknesses were measured on the basis of metallography recordings. Based on the findings of Tominaga [27] and Sun [28], an initial oxide scale composition of 64 wt % wuestite, 30 wt % magnetite, and 6 wt % haematite had been calculated for the oxidation temperature of 1000 °C. The dependency of oxide scale composition on oxidation temperature is shown in Figure 7, where the blue lines indicate the calculated composition at 1000 °C.

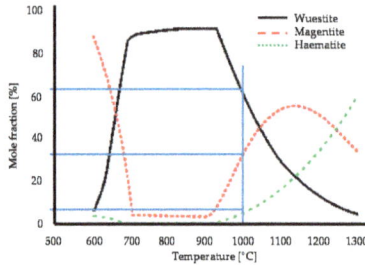

Figure 7. Oxide scale composition depending on oxidation temperature (based on [28]). The mole fractions for an temperature of 1000 °C are indicated by blue lines.

The heated sample was subsequently compressed with a stroke of up to 60% of its initial height. Due to the used isothermal containment, the tool temperature had been kept equivalent to workpiece temperature and the tools had been modelled as heat-conducting rigid bodies. The time–stroke relationship was calculated based on the experimental data with an average punch speed of 3.2 mm·s^{-1}. By evaluation of experimental ring compression tests, the friction factor (m) was found to be 0.65 (C15 and C60) and 0.7 (C45). For this purpose, the deformed samples had been geometrically measured and the ratios of initial inner diameter to deformed inner diameter as well as initial height to deformed height had been calculated and filled in a friction nomogram according to the work of Male and Cockcroft [29]. Experimental data for heat transfer coefficients were not measured in the current research project. Previous research studies predicted a heat transfer coefficient for non-fractured oxide scale which is 10–15 times lower than that of steel [30,31]. Based on these data, a heat transfer coefficient of 1400 W/(m^2·K) has been assumed for numerical simulation. A comparison of numerically-calculated and experimentally-measured force-displacement curves is presented in Figure 8a and shows a good qualitative agreement, thus it can be concluded that the used numerical boundary conditions are applicable.

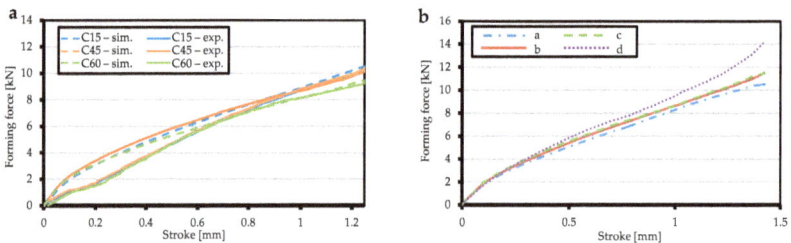

Figure 8. Comparison of experimentally-measured and numerically-predicted force-displacement curves for the steel grades C15, C45, C60 at a temperature of 1000 °C, an average forming speed of 3.2 mm s^{-1}, and the calculated ratio of 64:30:6 for the layers wuestite, magnetite, and haematite (**a**); Comparison of numerically-calculated force-displacement curves for the steel C15 at a temperature of 1000 °C, an average forming speed of 3.2 mm s^{-1}, and various ratios of haematite, magnetite, and wuestite a: 1:4:95; b: 6:30:64; c: 5:40:55 and d: without oxide scale. (**b**).

Particularly at the beginning of the forging process, there is an offset between measured and calculated force. This can be caused either by the machine stiffness in the experimental tests or due to the used oxide scale material model. On the basis of complex oxide scale material behaviour, the material characterisation had been examined to specific temperatures, strains, and strain rates. Therefore, the implemented flow curve model needs to be extrapolated for higher strains as well as a wider temperature range. Furthermore, in a complex hot forging process, the oxide scale is brittle and porous, thus it can be compressed and undergo rupture [13]. Subsequently, in order to examine the influence of a varying oxide scale composition, force displacement curves have also been calculated for different compositions of oxide scale as well as oxide scale (Figure 8b). The further boundary conditions have been kept unchanged. Although the mechanical properties of iron oxides differ widely, the global influence of oxide scale composition on force-displacement curve characteristics is small compared to the one without an oxide scale layer.

In addition to the comparison of force-displacement curves, the experimentally-measured and numerically-calculated inner diameters and heights after ring compression tests as well as the percentage deviation are presented in Table 4. The results show a good agreement.

Table 4. Comparison between experimentally-measured and numerically-calculated height and inner diameter after ring compression tests at 1000 °C and 900 °C.

Oxidation Temperature Oxide Scale Thickness	Steel Grade	Experimental (mm)		Simulation (mm)		Deviation (%)	
		Inner Diameter	Height	Inner Diameter	Height	Inner Diameter	Height
Oxidation temperature: 1000 °C Oxide scale thickness 50 μm	C15	3.7	1.61	3.91	1.63	5.1	1.2
	C45	3.75	1.67	3.86	1.7	2.9	1.8
	C60	3.8	1.68	3.77	1.66	−0.79	−1.2
Oxidation temperature: 900 °C Oxide scale thickness: 30 μm	C15	3.65	1.59	3.74	1.59	2.47	1.89
	C45	3.71	1.82	4.00	1.82	7.82	−1.1
	C60	3.8	1.67	3.81	1.68	0.26	0.6

3.2. Sensitivity Analysis

The numerical model presented in this paper was used to perform a sensitivity analysis regarding the influence of an oxide scale layer on the hot forging process. Therefore, influential parameters like layer thickness, friction conditions, and oxide scale composition have been studied. The carbon of the steel grade mainly influences the growth rate of the oxide scale layer. The mechanical behaviour of the oxide scale layer is regardless of the considered steel grade. Therefore, exemplary results for the steel grade C15 at 1000 °C are presented.

3.2.1. Layer Thickness

The growth of an oxide scale layer directly before or during the forming process is particularly influenced by oxidation time and temperature as well as the steel matrix. As the mechanical properties of oxide scale strongly deviate from the steel matrix, the forming process is influenced depending on oxide scale volume. The numerical simulations were carried out with ring compression samples and three different layer thicknesses (30 μm, 50 μm, 100 μm) as well as an unscaled variant. The further boundary conditions have been kept constant. Initially, a friction factor $m = 0.65$ had been assumed, derived from experimental ring compression tests for the steel grade C15.

The final ring profiles as well as the material flow in x-direction as contour plot are presented in Figure 9a. The impact of different oxide scale thicknesses on material flow and component shape can be seen between the oxide scaled variants and the variant without an oxide scale layer as well as between the scaled variants themselves. The variant without an oxide scale layer shows a conventional material flow, known from the ring compression test with high friction. The material flow is divided and the change of inner diameter is related to the friction conditions, whereby the sides form a convex shape. With an increasing friction factor, the decrease of the inner diameter is intensified [32]. However, the scaled variants show a light concave up to a straight shape at the inner diameter, which increases with increasing layer thickness. The shape at the outer diameter turns from slight convex to a flat surface with increasing layer thickness. The numerical results indicate an increased sliding behaviour in the contact zone of workpiece and die as a result of the oxide scale, which leads to changes in friction conditions and material flow. The oxide scale layer seems to act like an additional lubricant, whereby the effect is intensified with increasing layer thickness. It can be assumed that the yield stresses of the oxide scale and the steel matrix decisively differ and thus provoke the changes in material flow as well as an increased sliding. Furthermore, the different yield stresses result in an outflow of the oxide scale in the contact zone between the die and the steel matrix.

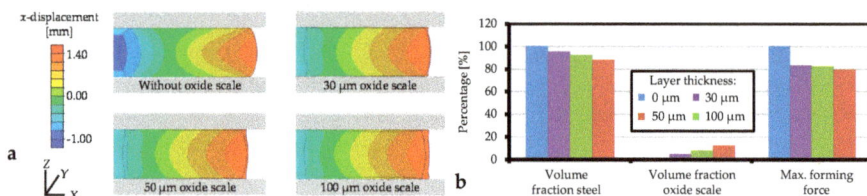

Figure 9. Final component shape and x-displacement depending on oxide scale thickness (**a**); influence of oxide scale on maximum forming force (**b**).

The volume fractions of steel matrix, oxide scale, and the maximum forming forces are shown as percentages in Figure 9b. The bar chart shows the influence of thin oxide scale layers as well. The volume fraction of a 50-μm oxide scale layer is just about 10% but results in a forming force reduction of nearly 20%. This strong reduction at even small amounts of oxide scale is not just due to lower yield stresses of the oxide scale. It indicates an improved sliding as well. Nonetheless, it must be taken into account that the results are calculated based on the approach of non-fractured oxide scale. In complex forging processes, the oxide scale can rupture due to its brittleness, which also has an influence on sliding behaviour. Therefore, it is necessary to implement a specific damage criteria into the user subroutine in order to take into account rupture of oxide scale as well as oxide scale detachment. Furthermore, new material characterisation methods are required in order to measure rupture as well as the detachment of the thin and brittle oxide scale layer at elevated temperatures, which might be more challenging.

3.2.2. Friction Conditions

In order to carry out further examinations on the influence of oxide scale on the sliding behaviour in the contact zone between workpiece and die, additional simulations for a variant without oxide scale and a variant with an oxide scale layer of 50 μm at 1000 °C have been performed while varying the friction factor m. Calculated results for the final component shape and material flow in x-direction for three variants with and without oxide scale and a varying friction factor are shown in Figure 10.

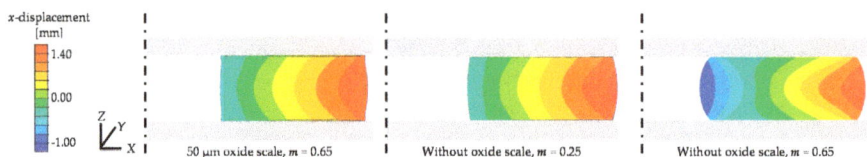

Figure 10. Final component shape and x-displacement depending on the friction factor *m* and the oxide scale.

The results clarify the assumptions of an improved sliding behaviour due to oxide scale. The variant without an oxide scale layer and a very low friction factor of $m = 0.25$ shows an almost identical final shape to the variant with oxide scale and a friction factor of $m = 0.65$. The material flow also shows a relatively good agreement, whereby a higher flow is observed for the variant without oxide scale. The calculated force–displacement curves of the variants mentioned above are presented in Figure 11. A comparison between the variant without oxide scale considering $m = 0.25$ and the variant with oxide scale considering $m = 0.65$ shows a good qualitative agreement compared to the variant without oxide scale and $m = 0.65$. Thus it can be assumed that the oxide scale reduces the friction and acts like a lubricant.

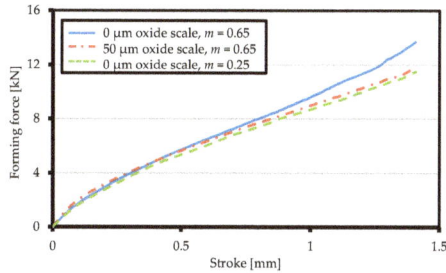

Figure 11. Numerically-calculated force–displacement curves for ring compression tests of variants with and without oxide scale for different friction factors.

In addition, further numerically-calculated variants with and without oxide scale and varying friction factors (low, medium, high friction) are compared in Figure 12 to point out the effect of the oxide scale presence on the sliding behaviour. The variants without oxide scale show, as expected, a decreasing inner diameter and increasing material flow inwards with increasing friction factor. Only the variant with low friction ($m = 0.1$) shows a concave shape at the inner diameter. However, only the scaled variant with high friction ($m = 0.85$) exhibits a decreased inner diameter as well as a concave shape. All scaled variants show an intensified material flow outwards, as compared with the variants without oxide scale. Thus, the oxide scale seems to dampen the friction influence.

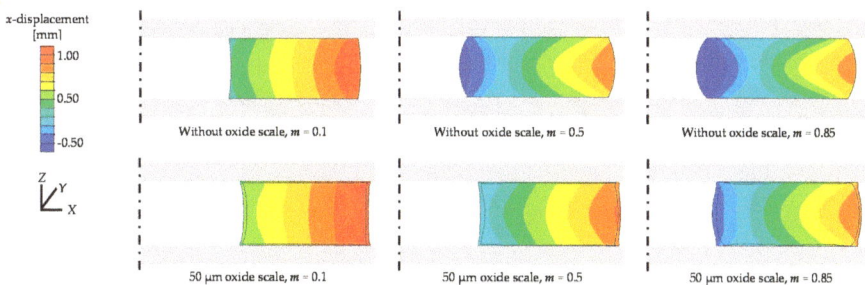

Figure 12. Comparison between numerically-calculated variants with and without oxide scale and varying friction.

3.2.3. Oxide Scale Composition

The composition of an oxide scale layer changes depending on the process and material-specific parameters like oxidation time and temperature. Due to the fact that mechanical properties of iron oxides strongly deviate from each other, the forming process is influenced by their composition.

The final component shape as well as calculated v. Mises stresses are presented in Figure 13 with regard to varying oxide scale compositions.

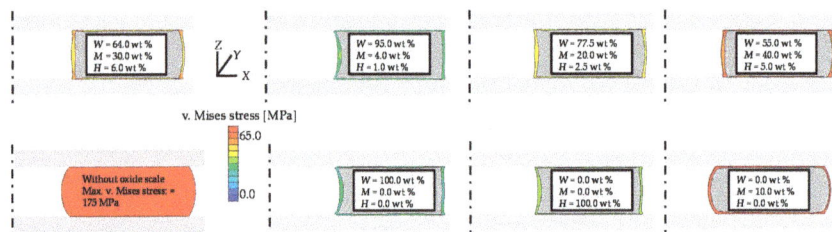

Figure 13. Final component shape and v. Mises stress depending on mass fractions of wuestite (*W*), magnetite (*M*), and haematite (*H*).

Four variants with different combinations of the iron oxides and a thickness of 50 μm as well as one variant without oxide scale were calculated. Additionally, three variants with pure oxides (wuestite, magnetite, haematite) were calculated. The further boundary conditions have been kept constant. The calculated v. Mises stress inside the steel matrix was higher than inside the oxide scale layer regardless of the oxide scale composition. The variant, with a layer consisting of 100 wt % magnetite, exhibited the highest v. Mises stress as compared with the other pure oxide variants. Furthermore, a strong influence of oxide scale composition on resulting stresses can be seen. A reduction of wuestite mass fraction or increase of the mass fractions of magnetite and haematite led to an increase of stresses inside the oxide scale layer. This clarifies the above-mentioned strong variations of oxide scale mechanical properties. The influence on material flow as well as sliding behaviour was reduced by an increase of yield stresses of oxide scale and the accompanying alignment with the steel matrix. The variant with the lowest mass fraction of wuestite and highest v. Mises stresses showed the lowest deviations of final shape as compared to the variant without oxide scale.

Furthermore, experimental and numerically-calculated height and inner diameter after compression have been determined under consideration of varying oxide scale composition. An overview is shown in Table 5. The variants with different oxide scale compositions exhibited a strong variation of resulting inner diameter which is consistent with the assumption that the oxide scale itself and the composition influence the material flow. The variant with the highest mass fraction of haematite showed the closest agreement with experimentally-measured values. The differences between the experimentally-measured and numerically-calculated values could have been caused by sliding of the oxide scale on the steel matrix as well as fracture. In this regard, further experimental investigations on the oxide scale composition have to be carried out.

Table 5. Experimental and numerically-calculated height and inner diameter after ring compression tests for the variants with steel grade C15 at a temperature of 1000 °C.

Variant	Height (mm)	Inner Diameter (mm)
Experimental ring compression test (averaged)	1.62	3.7
Simulation; W = 64.0 wt %; M = 30.0 wt %; H = 6.0 wt % (calculated composition for 1000 °C based on [28])	1.63	3.91
Simulation; W = 55.0 wt %; M = 40.0 wt %; H = 5.0 wt %	1.63	3.89
Simulation; W = 77.50 wt %; M = 20.0 wt %; H = 2.5 wt %	1.63	4.25
Simulation; W = 95.0 wt %; M = 4.0 wt %; H = 1.0 wt %	1.63	4.41
Simulation; without oxide scale	1.62	2.67

4. Conclusions

Based on a multi-material approach, an FE-model describing the oxide scale material behaviour in hot forging has been developed. Four different material models for both oxide scale and steel have been implemented in Simufact Forming by means of user subroutines. This enables an accurate description of the oxide scale material flow behaviour depending on temperature, strain, and strain rate. Furthermore, the implemented model for general steel grades takes into account the influence of varying carbon content as well as the initial microstructure. The influence of carbon content on the yield stress can be seen in the Hensel–Spittel coefficients. However, no significant influence of carbon content on the forming behaviour has been identified. The developed numerical model has been validated by comparing the results of the performed ring compression tests and the numerical simulation. Within a numerical sensitivity study, influential parameters of oxide scale layer like layer thickness, friction behavior, and oxide scale composition have been varied. The numerical results show a decisive influence of the oxide scale layer on a hot forging process. The comparison of variants with and without an oxide scale shows that the presence of an oxide scale layer in the contact zone between die and steel matrix acts like an additional lubricant. The findings indicate that, beside others, the differences between yield stress in steel matrix and oxide scale layer lead to an improved sliding behaviour. Nevertheless, it must be taken into account that the results are calculated based on the approach of non-fractured oxide scale. Therefore, further investigations on the adhesion interface between oxide scale and metal matrix as well as oxide scale rupture will be carried out in order to examine its behaviour during the hot forging process. For this purpose, the presented user subroutine will be extended with a damage criteria. This will enable more detailed description of the oxide scale influence on the hot forging process.

Acknowledgments: This work was part of the cooperation project "General modelling of material behaviour and surface modifications for FEM analysis of die forging of carbon steels" (GR4872/1-1, UL471/1-1, BO3616/10-1) funded by the German Research Foundation (DFG). The authors would like to thank the DFG for its financial support.

Author Contributions: Rudolf Kawalla, Grzegorz Korpala and Madlen Ullmann designed and performed the experiments for general steel; Birgit Awiszus and Marcel Graf designed and performed the experiments for iron oxides; Bernd-Arno Behrens, Alexander Chugreev and Hendrik Wester designed the numerical model and implemented subroutines as well as performed the sensitivity analysis; Hendrik Wester wrote the paper.

Conflicts of Interest: The authors declare no conflict of interest.

References

1. Fischer, M.U.A.; Dickert, H.H.; Bleck, W.; Huskic, A.; Kazhai, M.; Hadifi, T.; Bouguecha, A.; Behrens, B.-A.; Labanove, N.; Felde, N.; et al. EcoForge: Energieeffiziente Prozesskette zur Herstellung von Hochleistungs-Schmiedebauteilen. *HTM J. Heat Treat. Mater.* **2014**, *69*, 209–219. (In German) [CrossRef]

2. Luong, L.; Heijkoop, T. The influence of scale on friction in hot metal working. *Wear* **1981**, *71*, 93–102. [CrossRef]

3. Behrens, B.-A.; Bouguecha, A.; Vucetic, M.; Chugreev, A. Advanced wear simulation for bulk metal forming processes. In Proceedings of the Numiform 2016: The 12th International Conference on Numerical Methods in Industrial Forming Processes, Troyes, France, 4–7 July 2016; Volume 80. [CrossRef]

4. Behrens, B.-A. Finite element analysis of die wear in hot forging processes. *CIRP Ann. Manuf. Technol.* **2008**, *57*, 305–308. [CrossRef]

5. Kawalla, R.; Steinert, F. Untersuchung des Einflusses von Prozessparametern in der Fertigstraße auf die Tertiärzunderausbildung. *Mat.-wiss. u. Werkstofftech.* **2007**, *38*, 36–42. [CrossRef]

6. Krzyzanowski, M.; Beynon, J.; Farrugia, D. *Oxide Scale Behavior in High Temperature Metal Processing*; Wiley-VCH Verlag GmbH & Co. KGaA: Weinheim, Germany, 2010; ISBN 978-35-2-732518-4.

7. Malik, A.U.; Whittle, D.P. Oxidation of Fe-C alloys in the temperature range 600–850 °C. *Oxid. Met.* **1981**, *16*, 339–353. [CrossRef]

8. Birks, N.; Frederik, S.; Meier, G.H. *Introduction to High Temperature Oxidation of Metals*, 2nd ed.; Cambridge University Press: Cambridge, UK, 2006; ISBN 978-05-2-148517-3.

9. Kubaschewski, O.; Hopkins, B.E. Oxidation of metals and alloys. *Mater. Corros.* **1954**, *11*, 108–114. [CrossRef]
10. Brauns, E.; Rahmel, A.; Christmann, H. Die Verschiebung des Nonvarianzpunktes zwischen Eisen, Wüstit, Magnetit und Sauerstoff im System Eisen—Sauerstoff durch Legierungselemente oder fremde Oxyde—Auswirkungen auf das Verhalten von Eisenlegierungen beim Verzundern. *Arch. Eisenhttenwes.* **1959**, *30*, 553–564. [CrossRef]
11. Garnaud, G.; Rapp, R.A. Thickness of the oxide scale layers formed during the oxidation of iron. *Oxid. Met.* **1977**, *11*, 193–198. [CrossRef]
12. Graf, M.; Kawalla, R. Scale behaviour and deformation properties of oxide scale during hot rolling of steel. *Key Eng. Mater.* **2012**, *504–506*, 546–551. [CrossRef]
13. Graf, M. *Modellierung des Zunderverhaltens Entlang der Prozesskette Warmband, TU Bergakademie Freiberg*; Freiberger Forschungsheft B353: Freiberg, Germany, 2013; ISBN 978-38-6-012480-2.
14. Takeda, M.; Onishi, T.; Nakakubo, S.; Fujimoto, S. Physical properties of iron-oxide scales on Si-containing steels at high temperature. *Mater. Trans.* **2009**, *50*, 2242–2246. [CrossRef]
15. Behrens, B.-A.; Bouguecha, A.; Hadifi, T.; Mielke, J. Advanced friction modeling for bulk metal forming processes. *Prod. Eng.* **2011**, *5*, 621–627. [CrossRef]
16. Barnes, D.J.; Wilson, J.E.; Stott, F.H. The influence of oxide films on the friction and wear of Fe-5% Cr alloy in controlled environments. *Wear* **1977**, *45*, 161–176. [CrossRef]
17. Vergne, C.; Boher, C.; Gras, R.; Levailant, C. Influence of oxides on friction in hot rolling: Experimental investigations and tribological modelling. *Wear* **2000**, *260*, 957–975. [CrossRef]
18. Hinsley, C.F.; Male, A.T.; Rowe, G.W. Frictional properties of metal oxides at high temperatures. *Wear* **1968**, *11*, 233–238. [CrossRef]
19. Munther, P.A.; Lenard, J.G. The effect of scaling on interfacial friction in hot rolling of steels. *J. Mater. Process. Technol.* **1993**, *37*, 3–36. [CrossRef]
20. Tingle, E.D. The importance of surface oxide films in the friction and lubrication of metals. *Trans. Faraday Soc.* **1950**, *46*, 93–102. [CrossRef]
21. Matsumoto, R.; Osumi, Y.; Utsunomiya, H. Reduction of friction of steel covered with oxide scale in hot forging. *J. Mater. Process. Technol.* **2014**, *214*, 651–659. [CrossRef]
22. Graf, M.; Ullmann, M.; Korpalla, G.; Kawalla, R. Materialkennwerte als Basis für die nummerische simulation von Warmumformprozessen. In Proceedings of the 22. Verformungskundliches Kolloquium, Planneralm, Germany, February 2013; pp. 49–55.
23. Wray, P.J. Effect of carbon content on the plastic flow of plain carbon steels at elevated temperatures. *Metall. Trans. A* **1982**, *13*, 125–134. [CrossRef]
24. Korpała, G.; Ullmann, M.; Graf, M.; Wester, H.; Bouguecha, A.; Awiszus, B.; Behrens, B.-A.; Kawalla, R. Modelling the influence of carbon content on material behavior during forging. *AIP Conf. Proc.* **2017**, *1896*, 190013. [CrossRef]
25. Behrens, B.-A.; Kawalla, R.; Awiszus, B.; Bouguecha, A.; Ullmann, M.; Graf, M.; Bonk, C.; Chugreev, A.; Wester, H. Numerical investigation of the oxide scale deformation behaviour with consideration of carbon content during hot forging. *Procedia Eng.* **2017**, *207*, 526–531. [CrossRef]
26. Korpala, G. *Einfluss der Chemischen Zusammensetzung auf die Mechanischen Eigenschaften von Unlegiertem Bainitischen Stahl mit Restaustenit*; Freiberger Forschungshefte: Freiberg, Germany, 2016.
27. Tominaga, J.; Wakimoto, K.; Mori, T.; Murakami, M.; Yoshimura, T. Manufacture of wire rods with good descaling property. *Trans. Iron Steel Inst. Jpn.* **1982**, *22*, 646–656. [CrossRef]
28. Sun, W. A Study on the Characteristics of Oxide Scale in Hot Rolling of Steel. University of Wollongong Thesis Collection. 2005. Available online: http://ro.uow.edu.au/theses/440 (accessed on 13 February 2018).
29. Male, A.T.; Cockcroft, M.G. A method for the determination of the coefficient of friction of metals under conditions of bulk plastic deformation. *J. Inst. Met.* **1964**, *93*, 38–46.
30. Krzyzanowski, M.; Beyon, J.H. Oxide Behaviour in hot rolling. In *Metal Forming Science and Practice*; Lenard, J., Ed.; Elsevier: Amsterdam, the Netherlands, 2002; pp. 259–295. ISBN 978-00-8-053631-6.

Metals **2018**, *8*, 140

31. Frolish, M.F.; Krzyzanowski, M.; Beyon, J.H. Oxide scale behaviour on aluminium and steel under hot working conditions. *J. Mater. Process. Technol.* **2006**, *177*, 36–40. [CrossRef]

32. Koch, S.; Vucetic, M.; Hübner, S.; Bouguecha, A.; Behrens, B.-A. Superimposed oscillating and non-oscillating ring compression tests for sheet-bulk metal forming technology. *Appl. Mech. Mater.* **2015**, *794*, 89–96. [CrossRef]

metals

MDPI

Article

Hot Deformation Behavior of As-Cast 30Cr2Ni4MoV Steel Using Processing Maps

Peng Zhou [1,2], Qingxian Ma [1,2,*] and Jianbin Luo [1,2]

[1]   Department of Mechanical Engineering, Tsinghua University, Beijing 100084, China;
     zpsarm@foxmail.com (P.Z.); luojblqw@mail.tsinghua.edu.cn (J.L.)
[2]   Key Laboratory for Advanced Materials Processing Technology of Ministry of Education,
     Tsinghua University, Beijing 100084, China
*   Correspondence: maqxdme@mail.tsinghua.edu.cn; Tel./Fax: +86-10-6277-1476

Academic Editors: Myoung-Gyu Lee and Yannis P. Korkolis
Received: 5 December 2016; Accepted: 19 January 2017; Published: 9 February 2017

Abstract: The hot deformation behavior of as-cast 30Cr2Ni4MoV steel was characterized using processing maps in the temperature range 850 to 1200 °C and strain rate range 0.01 to 10 s^{-1}. Based on the obtained flow curves, the power dissipation maps at different strains were developed and the effect of the strain on the efficiency of power dissipation was discussed in detail. The processing maps at different strains were obtained by superimposing the instability maps on the power dissipation maps. According to the processing map and the metallographic observation, the optimum domain of hot deformation was in the temperature range of 950–1200 °C and strain rate range of 0.03–0.5 s^{-1}, with a peak efficiency of 0.41 at 1100 °C and 0.25 s^{-1} which were the optimum hot working parameters.

Keywords: 30Cr2Ni4MoV steel; hot deformation; processing map; microstructure

1. Introduction

In order to evaluate the explicit microstructural response of the material to the processing parameters, which include strain rate, deformation temperature and true strain, and to solve the problems related to workability and microstructural control in materials during hot deformation, the processing map was developed in 1984 based on the dynamic materials modeling (DMM) by Prasad [1]. Based on the processing maps generated using data of flow stress as a function of temperature and strain rate over a wide range, several domains safe for processing and regimes of flow instabilities and cracking can be identified [2]. With the information obtained from the processing map, the guideline for optimizing hot processing parameters can be determined, and the damage processes and instability processed can be avoided. The processing maps have been widely investigated in the production of titanium alloys [3], magnesium alloys [4,5], aluminium alloys [6], nickel alloys [7] and steels [8,9].

30Cr2Ni4MoV steel has attracted extensive attention for its good properties in terms of strength, toughness and wear resistance, and has been widely used in the production of an ultra-super-critical power cycle generator. In recent years, much research has been carried out to characterize the hot working behavior of 30Cr2Ni4MoV steel. Chen et al. evaluated the effects of the strain rate, temperature and initial grain size on the behavior of dynamic recrystallization (DRX) and meta-dynamic recrystallization (mDRX) [10,11]. Liu et al. investigated the microstructure evolution of 30Cr2Ni4MoV steel during multi-pass hot deformation under different deformation conditions [12].

In this study, the hot compression tests of 30Cr2Ni4MoV steel were carried out at the temperatures from 850 to 1200 °C and strain rates from 0.01 to 10 s^{-1} on a Gleeble-1500 thermo-simulation machine. Based on the experimental flow stress, the processing maps were developed at different strains. Finally, the optimum hot formation processing parameters for 30Cr2Ni4MoV steel were obtained.

2. Experimental Procedure

2.1. Hot Deformation Tests

The composition of the 30Cr2Ni4MoV steel used in this research, which was directly sampled from 600 t ingot, with a composition of 0.28C-0.02Mn-0.01Si-0.003P-0.003S-1.72Cr-0.41Mo-3.63Ni-0.11V-(bal.)Fe, and all values given in wt %. The steel was machined into cylindrical specimens which were 12 mm in height and 8 mm in diameter. One-hit isothermal compression tests were performed on a Gleeble-1500 thermal mechanical simulation tester (Dynamic Systems Inc., Poestenkill, NY, USA) in Tsinghua University. In order to reduce frictional effects during compression and avoid the sticking problem in quenching, the Ta pieces with a thickness of 0.5 mm were positioned between the anvils and the specimens. The specimens were firstly preheated at 1200 °C for 5 min to obtain the same initial grain size and homogeneous microstructure before compression [13]. After the structure uniformity, they were then cooled to the test deformation temperature at 10 °C/s and held for 1 min prior to deformation for the purpose of temperature gradient elimination. Deformation temperature ranging from 850 °C to 1200 °C in increments of 50 °C were chosen for these compression tests. A deformation of strain $\varepsilon = 0.7$ was applied at strain rates ranging from 0.01 s^{-1} to 10 s^{-1}, which was followed by water quenching to preserve the deformed austenite microstructure for metallographic observation. The polished surfaces were etched using a saturation picric acid for 7 min water bath heating at 70 °C. All optical micrographs were obtained from the center of the longitudinal sections of the specimens and the original grain size after soaking at 1200 °C for 5 min was 266.1 μm.

2.2. Processing Map Establishment

The processing map is generated using data of flow stress as a function of temperature and strain rate over a wide range obtained from the hot compression test based on the theory of DMM [1,2,14,15]. According to the DMM, the workpiece essentially dissipates power during hot deformation, which may be represented as a sum of two complementary parts: G and J [1,2]. The G represents the power dissipated by plastic deformation, most of which is converted into viscoplastic heat and the rest is stored as defect power. By contrast, the J is the energy related to microstructure changing such as dynamic recovery (DRV) and DRX. The power partitioning of the two parts above is decided by the strain rate sensitivity (m) and the power dissipation through microstructure changes can be represented by a non-dimensional parameter, the efficiency of power dissipation (η), which is defined as:

$$\eta = \frac{2m}{m+1} \tag{1}$$

where the strain rate sensitivity (m) can be calculated by the following equation:

$$m = \frac{\partial J}{\partial G} = \frac{\partial(\ln \sigma)}{\partial(\ln \dot{\varepsilon})} \tag{2}$$

where σ is the flow stress and $\dot{\varepsilon}$ is the strain rate. The power dissipation map, where the various domains may be correlated with specific microstructural mechanisms, can be obtained based on the variation of η with temperature and strain rate. The value of the efficiency of dissipation in low stacking fault energy metals is about 0.3–0.5 for the DRX and 0.15–0.25 for the DRV [2].

The continuum instability criterion based on the extremum principle of irreversible thermodynamic is used as follows [16]:

$$\xi(\dot{\varepsilon}) = \frac{\partial \ln[m/(m+1)]}{\partial(\ln \dot{\varepsilon})} + m < 0 \tag{3}$$

The parameter $\xi(\dot{\varepsilon})$ can be plotted as a contour map in a frame of temperature and strain rate, and flow instabilities are predicted to occur when $\xi(\dot{\varepsilon})$ is negative. Such a plot is called an instability map that can be superimposed on the power dissipation map to obtain a processing map.

3. Results and Discussion

3.1. Flow Curve Behavior

Figure 1 shows the typical flow curves obtained at different deformation temperatures and strain rates. All the flow curves are fitted with a 7th order or higher polynomial and smoothed from the yield strain, which is identified on the flow curve in terms of a 2% offset in the total strain [17]. It is found that the temperature and the strain rate have significant effects on the flow behavior and the stress increases with increasing strain rate (Figure 1a), while it is decreased with the increasing temperature (Figure 1b). According to the shape of the flow curves, they can be divided into two categories: (1) a progressive stress increasing with the increasing strain that reveals a DRV mechanism for the deformation process, such as 950 °C and 0.1 s^{-1}; (2) the stress increases with the increasing strain to a peak and then decreases with further increasing strain until a steady stress is attained, which results from the occurrence of the DRX, such as 1200 °C and 0.1 s^{-1}.

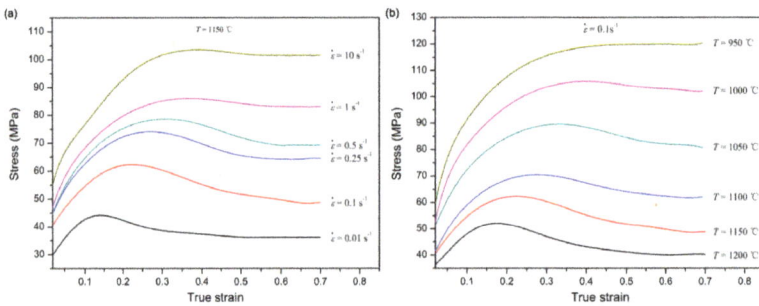

Figure 1. Typical fitted and smoothed experiment flow curves under different deformation conditions: (a) $T = 1150$ °C; (b) $\dot{\varepsilon} = 0.1$ s^{-1}

When the metals are subjected to plastic deformation at elevated temperature, the flow behavior is determined by the competition between dynamic softening and work hardening. At the initial stage of the deformation, the work hardening that resulted from the dislocation density increasing dominates the process which leads to a gradual increase of the stress. When the sample is deformed at a high temperature and low strain rate, as the deformation proceeded, the stress increases to a peak stress and a low work hardening rate, which resulted from the occurrence of DRX. Then, the softening due to DRX dominates the process, which results in a decrease of the stress until a steady stress is reached. By contrast, when the temperature is lower than 950 °C and the strain rate is higher than 0.1 s^{-1}, the softening caused by DRV cannot completely counteract the work hardening. Consequently, the stress progressively increases with the increasing strain.

The flow stress data obtained at different temperatures, strain rates, and strain are shown in Table 1, which is the input to the processing map establishment.

Table 1. Flow stress values (in MPa) of 25CrMo4 at different strain rates and temperatures for various strains.

Strain	Strain Rate (s^{-1})	Temperature (°C)							
		850	900	950	1000	1050	1100	1150	1200
	0.01	119.2	98.5	77.4	63.4	55.1	47.7	43.8	37.4
	0.1	145.6	119.8	96.2	86.0	72.5	62.1	54.6	47.8
0.1	0.25	149.4	129.7	108.2	95.1	78.1	68.9	62.9	54.5
	0.5	157.4	135.4	119.9	108.8	89.9	75.6	64.8	58.3
	1	166.9	141.9	127.9	113.0	93.2	81.3	68.7	58.0
	10	188.4	159.2	140.4	124.7	103.8	92.0	76.8	66.7

Table 1. *Cont.*

Strain	Strain Rate (s⁻¹)	Temperature (°C)							
		850	900	950	1000	1050	1100	1150	1200
0.2	0.01	137.4	114.9	86.8	73.6	61.5	51.3	42.2	34.6
	0.1	169.4	142.8	117.7	100.2	84.6	71.7	62.1	51.8
	0.25	175.0	156.0	129.6	113.3	93.8	79.3	70.2	61.4
	0.5	186.4	162.5	139.4	122.9	103.8	85.3	75.3	67.3
	1	199.4	168.5	151.1	131.1	109.8	94.8	79.5	66.9
	10	221.5	187.4	164.4	148.1	124.9	110.3	93.0	80.4
0.3	0.01	146.3	121.5	94.6	76.2	59.4	47.2	38.8	32.3
	0.1	181.9	153.9	125.3	104.5	87.6	71.5	60.1	46.7
	0.25	188.4	170.9	139.3	122.2	97.5	83.0	73.6	59.1
	0.5	201.7	178.8	149.4	134.5	111.2	91.6	78.6	68.3
	1	213.6	181.3	161.3	140.6	118.0	102.4	85.1	70.4
	10	236.3	202.3	178.0	160.8	136.6	120.9	101.5	88.1
0.4	0.01	148.9	117.4	87.0	71.6	55.5	44.7	37.2	30.8
	0.1	187.9	154.4	126.3	105.7	86.4	67.2	54.9	43.1
	0.25	194.3	176.6	142.4	125.7	98.7	80.8	69.5	54.1
	0.5	209.1	188.7	155.2	140.0	113.9	93.4	76.3	64.4
	1	221.2	188.1	167.4	146.1	122.0	105.5	85.7	68.8
	10	238.9	207.0	181.6	164.8	139.7	123.6	103.4	89.1
0.5	0.01	151.8	121.8	87.5	67.1	53.7	42.1	34.9	29.2
	0.1	194.0	160.0	123.6	104.2	84.4	65.2	51.6	41.0
	0.25	197.1	178.2	142.3	127.0	96.3	77.5	65.5	51.5
	0.5	217.0	190.8	159.1	143.6	114.2	92.5	72.0	60.7
	1	227.9	194.0	172.7	149.1	124.5	106.6	84.0	66.4
	10	239.8	209.1	183.0	164.4	139.5	123.2	102.0	87.0
0.6	0.01	155.2	116.1	82.4	65.5	52.3	41.8	34.5	28.8
	0.1	201.7	158.0	125.5	102.7	81.9	63.4	49.3	40.0
	0.25	202.9	180.8	144.1	125.5	94.7	77.6	64.1	51.1
	0.5	226.1	199.2	161.6	142.3	114.2	92.1	68.8	59.2
	1	235.3	199.7	176.5	153.6	127.4	108.4	83.0	65.4
	10	244.0	213.5	187.9	167.8	142.1	125.3	101.4	85.8
0.7	0.01	159.7	120.8	82.1	65.5	52.3	42.6	35.3	29.1
	0.1	210.8	165.4	125.3	101.8	81.3	63.2	48.3	40.0
	0.25	214.6	187.2	148.2	129.3	94.4	77.6	64.4	51.7
	0.5	237.3	199.9	167.0	153.0	114.2	92.2	67.0	59.2
	1	246.5	207.8	181.0	156.0	129.0	108.5	83.0	65.4
	10	248.5	217.5	190.2	167.3	140.5	122.4	97.4	80.9

3.2. Processing Map Establishment

The flow stress data as a function of temperature, strain rate and strain can be obtained from the fitted and smoothed flow curves at strains of 0.2, 0.4, 0.6 and 0.7. According to Equation (2), the strain rate sensitivity parameter m can be obtained by plotting the ln σ versus ln $\dot{\varepsilon}$ and then differentiating the third-order polynomial fitted curve (Figure 2). Once the values of m at different deformation conditions are determined, the values of η under strain of 0.2, 0.4, 0.6 and 0.7 can be calculated with the aid of Equation (1) and then the parameter $\xi(\dot{\varepsilon})$ can be obtained by Equation (3). Finally, the power dissipation map at different strains can be constructed based on the values of η as a function of temperature and strain rate, as well as the processing map by means of superimposition of the instability map on the power dissipation map.

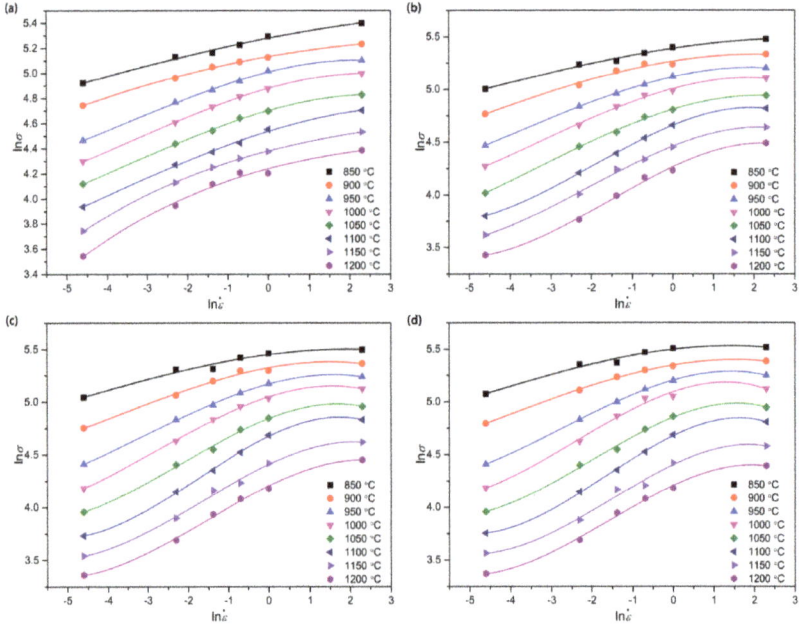

Figure 2. Polynomial fitted curves of lnσ versus ln ε̇ at strains of (**a**) 0.2; (**b**) 0.4; (**c**) 0.6; and (**d**) 0.7.

3.3. Variation of the Values of m

As described in Equations (1) and (3), the values of η and ξ(ε̇) are related with the values of *m* under different deformation conditions. In order to describe the variation of *m* with deformation temperature and strain rate, the 3D surfaces at strains of 0.2, 0.4, 0.6 and 0.7 are plotted as shown in Figure 3. It is found that the values of *m* vary irregularly with deformation temperature and strain rate, which is in agreement with the previous report [18]. However, the negative values of *m* can be observed at some regions, such as 1000 °C and 10 s^{-1} shown in Figure 3d. As pointed out previously [18], the negative *m*-values are usually a result from the occurrence of deformation twinning, shear band formation, dynamic strain aging or initiation and growth of micro-cracks that can lead to instabilities.

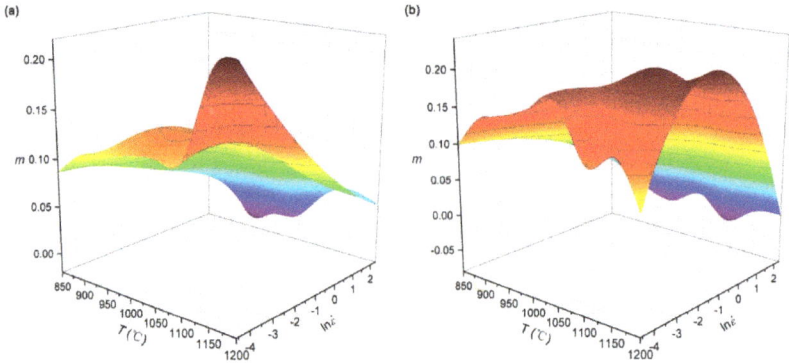

Figure 3. *Cont.*

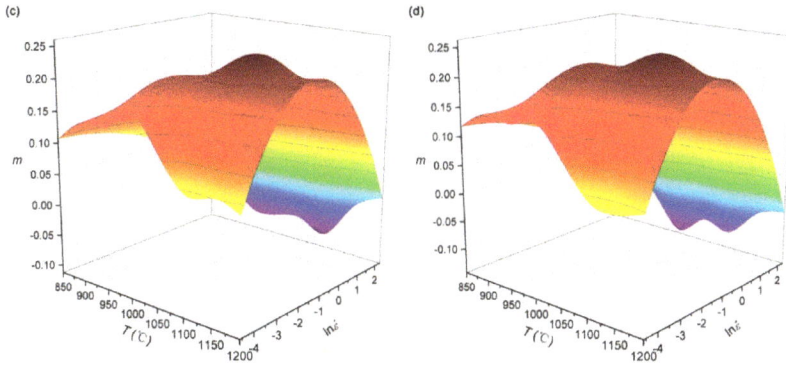

Figure 3. The 3D surfaces of *m*-value versus strain rate and temperature at true strains of (**a**) 0.2; (**b**) 0.4; (**c**) 0.6; and (**d**) 0.7.

3.4. Power Dissipation Map

3.4.1. Effect of the Strain on the Efficiency of Power Dissipation (η)

Figure 4 shows the dependence of the efficiency of power dissipation on the strain of the 30Cr2Ni4MoV steel under different temperatures and strain rates. It is found that the efficiency of power dissipation is sensitive to the deformation conditions. When the deformation temperatures are 1200 °C (Figure 4a), 1100 °C (Figure 4b) and 1000 °C (Figure 4c), the variation of η with deformation conditions can be divided into three categories:

(1) The values of η at the strain rate of 0.01 s^{-1} increase with the increasing strain to a peak and then decrease with further increasing strain. The phenomenon is related to the microstructure evolution during the deformation. At the beginning, the values of η increase to a peak, which means the completion of the DRX. Then, the growth of the recrystallized grains leads to the decrease of the η values. However, it should be pointed out that the values of η increase with the increasing strain, and no peak appears when the deformation temperature is 950 °C, which is a result of the absence of the DRX.

(2) The values of η at the strain rates range of 0.1–1 s^{-1} increase with increasing strain. This is attributed to the increase of the volume of the DRX. However, when the steel is deformed at 1000 °C with strain rates of 1 and 10 s^{-1}, the maximum values of η is less than 0.3, which means that DRV takes place during deformation.

(3) The values of η at the strain rate of 10 s^{-1} decrease with the increasing strain. The occurrence of the unstable flow bands is responsible for the decrease of the power dissipation efficiency.

However, when the deformation temperature is 900 °C (Figure 4d), the values of η are less than 0.3 and change very little at the strain rates range of 0.01–1 s^{-1}. This is because it is difficult for DRX to take place at 900 °C, and the DRV is the main softening mechanism. When the strain rate is 10 s^{-1}, the values of η decrease with the increasing stain, which results from the occurrence of the flow localization [19].

3.4.2. Processing Map Analysis and Microstructure Evolution

Figure 5 shows the processing maps of the 30Cr2Ni4MoV steel, which are obtained by superimposing the instability maps on the power dissipation efficiency maps, at temperatures in the range of 850–1200 °C and strain rates in the range of 0.01–10 s^{-1} when the strains are reached 0.2, 0.4, 0.6 and 0.7, respectively. The contours represent constant efficiency and the shade areas denote the unsafe domains obtained according to Equation (3). A comparison between Figure 5a,b reveals that the unsafe domain increases with the increase of strain. However, the map is not significantly affected by strain when the strain is larger than 0.4.

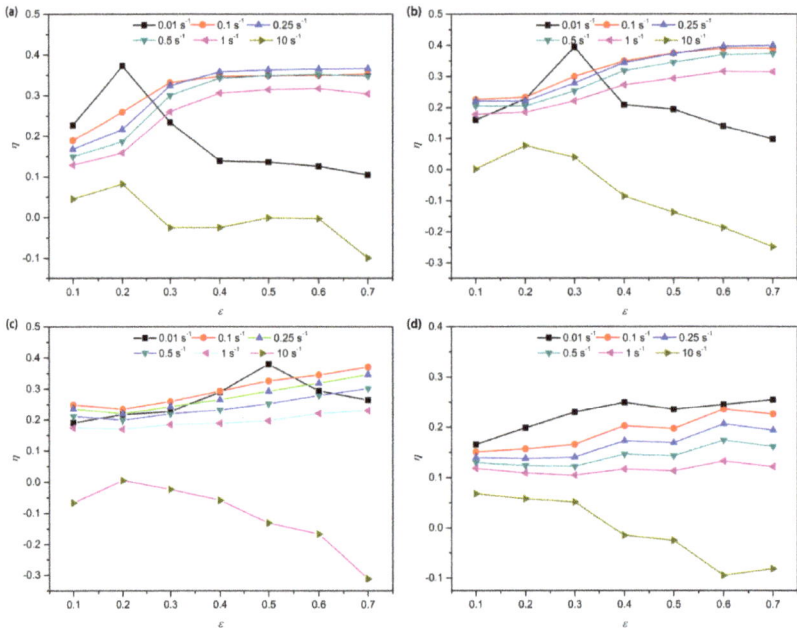

Figure 4. Effects of strain on the efficiency of power dissipation of the 30Cr2Ni4MoV steel at temperatures of (**a**) 1200 °C; (**b**) 1100 °C; (**c**) 1000 °C; and (**d**) 900 °C.

In the instability domain defined by Equation (3), the occurrence of flow localization, adiabatic shear bands and dynamic ageing affect the formability of the material during shaping and degrade the mechanical properties of the product [1,2]. By contrast, in the "safe" regime, the DRX and DRV occur and result in the microstructure evolution during deformation. The processing map of the 30Cr2Ni4MoV steel developed at the strain of 0.7 (Figure 5d) is divided into four domains marked as A, B, C and D, respectively. These domains are interpreted based on the above considerations and validated with the help of the microstructural examination.

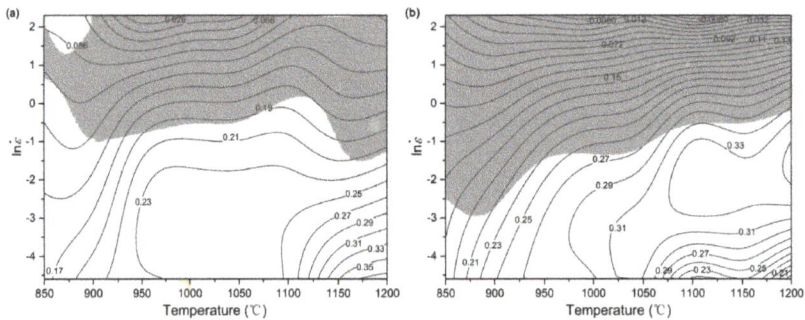

Figure 5. *Cont.*

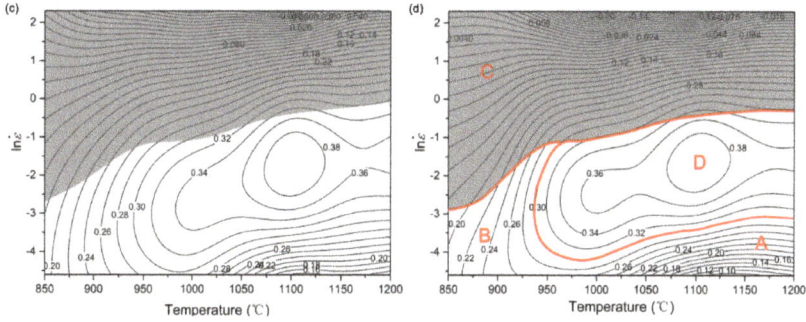

Figure 5. Processing maps of 30Cr2Ni4MoV steel at strains of (**a**) 0.2; (**b**) 0.4; (**c**) 0.6; and (**d**) 0.7.

Domain A occurs at the temperature range of 1050–1200 °C and the strain rate range of 0.01–0.03 s^{-1}. As shown in Figure 4a,b and Figure 5a–d, the values of η in this domain increase firstly and then decrease with increasing strain, such as 1200 °C and 0.01 s^{-1}. Figure 6a–c shows the optical microstructure of the samples compressed at 1200 °C and 0.01 s^{-1} with strains of 0.3, 0.5 and 0.7, respectively. The average grain sizes are 73, 82 and 101 μm, respectively. Therefore, the decrease of the values of η results from the growth of the recrystallized grains, which can be attributed to the enhanced mobility of grain boundaries under high temperature and low strain rate. In order to prevent the coarse grain structure, domain A is not recommended as a feasible deformation zone.

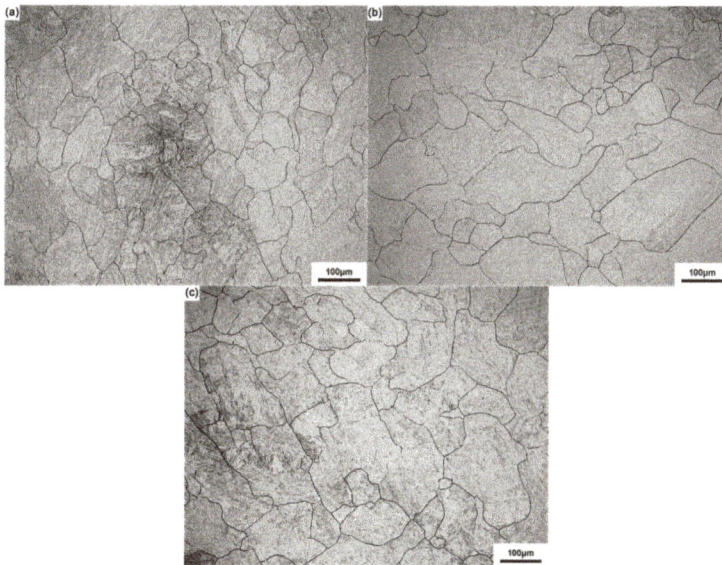

Figure 6. The optical microstructure of the 30Cr2Ni4MoV steel deformed at 1200 °C and 0.01 s^{-1} to strains of (**a**) 0.3; (**b**) 0.5; and (**c**) 0.7.

Domain B occurs at the temperature range of 900–1000 °C and the strain rate range of 0.01–0.03 s^{-1}. The values of η in this domain are less than 0.3, and this domain is a stability area. The optical micrographs of the samples deformed at the conditions of 850 °C/0.01 s^{-1} and 950 °C/0.01 s^{-1} are given in Figure 7a,b, respectively. The microstructure obtained at the conditions of 850 °C/0.01 s^{-1} represents the dynamic recovery followed by static recrystallization [20]. Similar microstructures are also interpreted to represent the DRX process [21]. When the deformation condition is 950 °C/0.01 s^{-1}, the typical "necklace structure" suggested that the DRX could be the primary mechanism during compression. It should be pointed out that the annealing twins play an important role during the nucleation and subsequent growth of recrystallized growth as shown in Figure 7a. For the 30Cr2Ni4MoV steel under consideration in this study, the DRV hardly takes place due to its low stacking fault energy. Therefore, the domains with relative low values of η less than 0.3 do not necessarily indicate the absence of the occurrence of DRX. The dependence of the work hardening rate on the stress under the above deformation conditions are shown in Figure 8. It is found that the work hardening rate is linearly related to the stress when the deformation condition is 850 °C/0.01 s^{-1}. However, when the deformation condition is 950 °C/0.01 s^{-1}, the plot is a typical curve that indicates the occurrence of the DRX. Such plots confirm that the DRV takes place when the deformation condition is 850 °C/0.01 s^{-1} and the DRX dominates the deformation process when the deformation condition is 950 °C/0.01 s^{-1}. On account of the partial DRX and necklace structure, domain B is not the recommended zone for deformation.

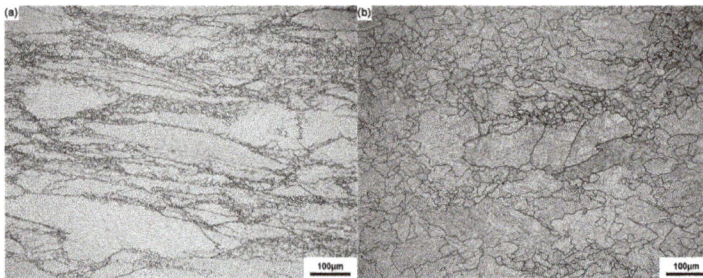

Figure 7. The microstructure of the deformed 30Cr2Ni4MoV steel to a strain of 0.7 at different conditions: (**a**) 850 °C/0.01 s^{-1} and (**b**) 950 °C/0.01 s^{-1}.

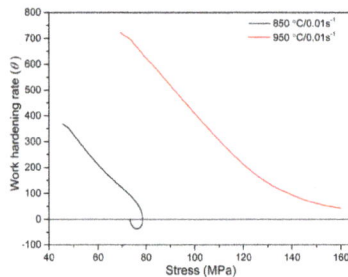

Figure 8. The dependence of work hardening rate on the stress at 0.01 s^{-1} and at different temperatures (850 °C and 950 °C).

Domain C is the instability region with the values of η lower than 0.3. The microstructures of the samples deformed at 850 °C with strain rates of 0.1, 1 and 10 s^{-1} are given in Figure 9a–c, respectively. All of the microstructures exhibit flow bands that manifest the occurrence of flow instability. This is attributed to the uneven temperature distribution, which results from the limited time for the heat transfer under high strain rate in the deformed specimen. The deformation preferably takes place in the part with high temperature, and the flow localization band and inhomogeneous microstructure appear. Furthermore, the intensity of the bands increases with an increase in strain rate. Additionally, it is observed that these bands are preferential sites for the nucleation of DRX. The flow bands and the partial DRX structure degrade the properties of the product. Hence, this domain where the flow instability occurs is to be avoided during the forming of 30Cr2Ni4MoV steel.

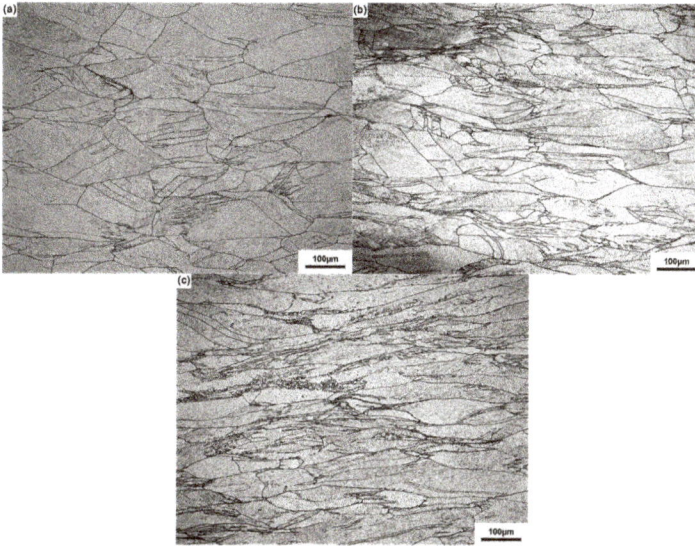

Figure 9. The microstructure of the deformed 30Cr2Ni4MoV steel to a strain of 0.7 at 850 °C and at different strain rates: (**a**) 0. 1 s^{-1}; (**b**) 1 s^{-1} and (**c**) 10 s^{-1}.

Domain D takes place at the temperature range of 950–1200 °C and the strain rate range of 0.03–0.5 s^{-1}. As the strain increases, the values of η in this domain increase and the peak value of η is 0.41. The microstructures of specimens deformed at strain rates of 0.1 and 0.25 s^{-1} are shown in Figure 10. It is found that all of these microstructures exhibit the typical recrystallized grain microstructure with a more refined grain size than that in the initial sample. The effect of the temperature on the grain size is shown in Figure 10a–c, and the influence of the strain rate is shown in Figure 10b,d. It is clear that as the temperature increases, the recrystallized grain size increases. This is attributed to the fact that higher temperature can provide more energy for the grain growth. When the strain rate is increased, the recrystallized grain size decreases. This is associated with the increasing rate of nucleation of recrystallization and decreasing time for the recrystallized grain growth, which both result from the increasing strain rate. In this domain, the variation of the recrystallized grain size with strain rate and temperature is shown in Figure 11. Based on the variation of η and the microstructure examination in this domain, it can be confirmed that the optimized process parameter for hot deformation of 30Cr2Ni4MoV steel is 1100 °C and 0.25 s^{-1}, for which the recrystallized grain size is 35 μm.

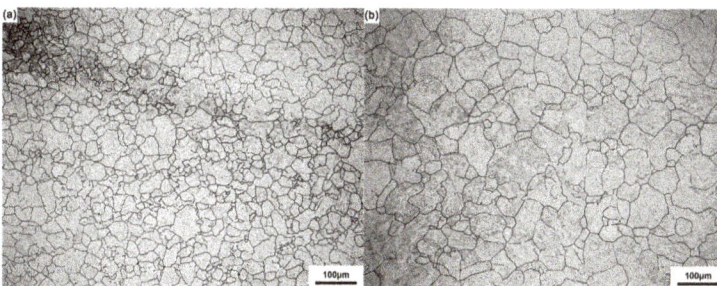

Figure 10. *Cont.*

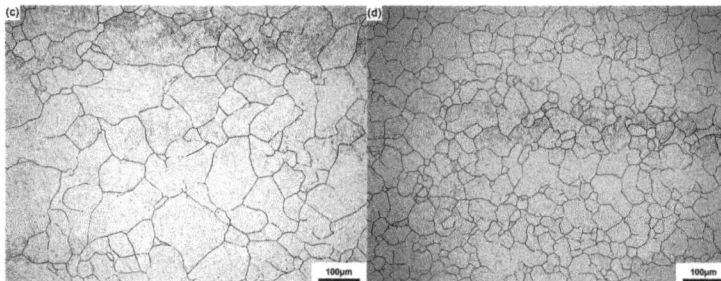

Figure 10. The microstructure of the deformed 30Cr2Ni4MoV steel to a strain of 0.7 at: (**a**) 1000 °C/0.1 s^{-1}; (**b**) 1100 °C/0.1 s^{-1}; (**c**) 1200 °C/0.1 s^{-1}; and (**d**) 1100 °C/0.25 s^{-1}.

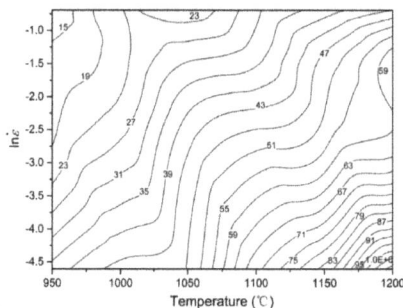

Figure 11. Contour map representing isograin size contours (marked as μm) of the deformed samples of 30Cr2Ni4MoV steel.

4. Conclusions

The hot deformation behavior of as-cast 30Cr2Ni4MoV steel has been studied in the temperature range 850 to 1200 °C and strain rate range 0.01 to 10 s^{-1} with a view toward optimizing the workability and controlling the microstructure. Processing maps under different strains have been developed for this purpose. The following conclusions are drawn:

(1) The strain rate sensitivity varies irregularly with deformation temperature and strain rate, and negative values of strain rate sensitivity can be observed.
(2) When the DRX and DRV take place, the value of the efficiency of power dissipation increases with the increasing strain. By contrast, the value of the efficiency of power dissipation decreases with the increasing strain when the flow localization occurs.
(3) The optimum domain for hot deformation is in the temperature range of 950–1200 °C and strain rate range of 0.03–0.5 s^{-1} with a peak efficiency of 0.41 at 1100 °C and 0.25 s^{-1}.

Acknowledgments: The authors gratefully acknowledge financial support from National Basic Research Program of China (2011CB012903).

Author Contributions: Peng Zhou conceived, designed and performed the experiments. Peng Zhou analyzed the data and wrote the paper. Qingxian Ma and Jianbin Luo contributed with advice on method of analysis.

Conflicts of Interest: The authors declare no conflict of interest.

References

1. Prasad, Y.V.R.K.; Gegel, H.L.; Doraivelu, S.M.; Malas, J.C.; Morgan, J.T.; Lark, K.A.; Barker, D.R. Modeling of dynamic materials behavior in hot deformation: Forging of Ti-6242. *Metall. Trans. A* **1984**, *15*, 1883–1892. [CrossRef]
2. Prasad, Y.V.R.K. Processing maps: A status report. *J. Mater. Eng. Perform.* **2003**, *12*, 638–645. [CrossRef]
3. Sun, Y.; Hu, L.X.; Ren, J.S. Investigation on the hot deformation behavior of powder metallurgy TiAl-based alloy using 3D processing map. *Mater. Charact.* **2015**, *100*, 163–169. [CrossRef]
4. Roostaei, M.; Parsa, M.H.; Mahmudi, R.; Mirzadeh, H. Hot compression behavior of GZ31 magnesium alloy. *J. Alloy. Compd.* **2015**, *631*, 1–6. [CrossRef]
5. Wu, H.Y.; Yang, J.C.; Zhu, F.J.; Wu, C.T. Hot compressive flow stress modeling of homogenized AZ61 Mg alloy using strain-dependent constitutive equations. *Mater. Sci. Eng. A* **2013**, *574*, 17–24. [CrossRef]
6. Chen, L.; Zhao, G.; Yu, J. Hot deformation behavior and constitutive modeling of homogenized 6026 aluminum alloy. *Mater. Des.* **2015**, *74*, 25–35. [CrossRef]
7. Wu, K.; Liu, G.; Hu, B.; Li, F.; Zhang, Y.; Tao, Y. Characterization of hot deformation behavior of a new Ni–Cr–Co based P/M superalloy. *Mater. Charact.* **2010**, *61*, 330–340. [CrossRef]
8. Akbari, Z.; Mirzadeh, H.; Cabrera, J.M. A simple constitutive model for predicting flow stress of medium carbon microalloyed steel during hot deformation. *Mater. Des.* **2015**, *77*, 126–131. [CrossRef]
9. Badjena, S.K. Dynamic recrystallization behavior of vanadium micro-alloyed forging medium carbon steel. *ISIJ Int.* **2014**, *54*, 650–656. [CrossRef]
10. Chen, F.; Cui, Z.S.; Sui, D.S.; Fu, B. Recrystallization of 30Cr2Ni4MoV ultra-super-critical rotor steel during hot deformation. Part III: Metadynamic recrystallization. *Mater. Sci. Eng. A* **2012**, *540*, 46–54. [CrossRef]
11. Chen, F.; Cui, Z.S.; Chen, S.J. Recrystallization of 30Cr2Ni4MoV ultra-super-critical rotor steel during hot deformation. Part I: Dynamic recrystallization. *Mater. Sci. Eng. A* **2011**, *528*, 5073–5080. [CrossRef]
12. Liu, X. Research on the Plastic Forming and Quality Control of Low-Pressure Rotor of Nuclear Steam Turbine. Ph.D. Thesis, Tsinghua University, Beijing, China, March 2010.
13. Chen, R.K. Study on Heat Treatment for Low Pressure Rotors of 30Cr2Ni4MoV Steel. Ph.D. Thesis, Shanghai Jiao Tong University, Shanghai, China, March 2012.
14. Kutumarao, V.V.; Rajagopalachary, T. Recent developments in modeling the hot working behavior of metallic materials. *Bull. Mater. Sci.* **1996**, *19*, 677–698. [CrossRef]
15. Murty, S.V.S.N.; Nageswara Rao, B.; Kashyap, B.P. Instability criteria for hot deformation of materials. *Int. Mater. Rev.* **2000**, *45*, 15–26. [CrossRef]
16. Prasad, Y.V.R.K.; Rao, K.P.; Sasidhara, S. *Hot Working Guide: A Compendium of Processing Maps*, 2nd ed.; ASM International: Materials Park, OH, USA, 1997; pp. 25–157.
17. Jonas, J.J.; Quelennec, X.; Jiang, L.; Martin, E. The Avrami kinetics of dynamic recrystallization. *Acta Mater.* **2009**, *57*, 2748–2756. [CrossRef]
18. Cai, Z.W.; Chen, F.X.; Ma, F.J.; Guo, J.Q. Dynamic recrystallization behavior and hot workability of AZ41M magnesium alloy during hot deformation. *J. Alloy. Compd.* **2016**, *670*, 55–63. [CrossRef]
19. Lin, Y.C.; Liu, G. Effects of strain on the workability of a high strength low alloy steel in hot compression. *Mater. Sci. Eng. A* **2009**, *523*, 139–144. [CrossRef]
20. Venugopal, S.; Mannan, S.L.; Prasad, Y.V.R.K. Optimization of hot workability in stainless steel-type AISI 304L using processing maps. *Metall. Mater. Trans. A* **1992**, *23*, 3093–3103. [CrossRef]
21. Roberts, W.; Boden, H.; Ahlblom, B. Dynamic recrystallization kinetics. *Metal Sci.* **2013**, *13*, 195–205. [CrossRef]

metals

MDPI

Article

Comparison of Hydrostatic Extrusion between Pressure-Load and Displacement-Load Models

Shengqiang Du [1], Xiang Zan [1,2,]*, Ping Li [1], Laima Luo [1,2], Xiaoyong Zhu [2] and Yucheng Wu [1,2,]*

[1] School of Materials Science and Engineering, Hefei University of Technology, Hefei 230009, China; 15255150652@163.com (S.D.); li_ping@hfut.edu.cn (P.L.); luolaima@126.com (L.L.)
[2] National–Local Joint Engineering Research Centre of Nonferrous Metals and Processing Technology, Hefei 230009, China; zhuxiaoyong@hfut.edu.cn
* Correspondence: zanx@hfut.edu.cn (X.Z.); ycwu@hfut.edu.cn (Y.W.);
  Tel.: +86-551-6290-1367 (X.Z.); +86-551-6290-1012 (Y.W.)

Academic Editors: Myoung-Gyu Lee and Yannis P. Korkolis
Received: 25 October 2016; Accepted: 27 February 2017; Published: 1 March 2017

Abstract: Two finite element analysis (FEA) models simulating hydrostatic extrusion (HE) are designed, one for the case under pressure load and another for the case under displacement load. Comparison is made of the equivalent stress distribution, stress state ratio distribution and extrusion pressure between the two models, which work at the same extrusion ratio (R) and the same die angle (2α). A uniform Von-Mises equivalent stress gradient distribution and stress state ratio gradient distribution are observed in the pressure-load model. A linear relationship is found between the extrusion pressure (P) and the logarithm of the extrusion ratio ($\ln R$), and a parabolic relationship between P and 2α, in both models. The P-value under pressure load is smaller than that under displacement load, though at the same R and α, and the difference between the two pressures becomes larger as R and α grow.

Keywords: hydrostatic extrusion; FEA; pressure load; die angle; extrusion ratio

1. Introduction

Hydrostatic extrusion (HE) is a unique forming method that was presented by Robertson in 1893 [1]. During the process, the material is surrounded by a high-pressure medium, which forms hydrostatic pressure conditions that improve the material's formability; thereby, larger amounts of deformation can be achieved as compared to the conventional extrusion process. The medium also ensures good lubricant conditions, and even generates dynamic lubrication between the die and the billet [2], and hence great surface quality. HE as a special severe plastic deformation (SPD) method has great advantages for large deformation processes and the forming processes of difficult-to-form materials.

By using the HE process, Ozaltin et al. [3] improved the strength of Ti-45Nb by 45% and also attained good plasticity by refining the grain. Also, by HE, Yu et al. [4] realized the deformation of AZ31 at 200–300 °C, at a maximum R of 31.5. Xue et al. [5] improved the properties of Zr-based metallic glass/porous tungsten phase composite; the breaking strength reaching 2112 MPa and the fracture strain reaching 53%. Kaszuwara et al. [6] densified Nd-Fe-B powder to the theoretical maximum density by HE. Kováč et al. [7] prepared MgB_2 wires by internal magnesium diffusion and HE. Skiba et al. [8] deformed GJL250 grey cast iron and GJS500 nodular cast iron by improved HE equipment with back pressure. Hydrostatic extrusion is widely used in the preparation of materials which are hard to deform. Finite element analysis (FEA) has also been used for investigations of the HE processes. Zhang et al. [9] simulated the HE process with tungsten alloy; the displacement load on the upper surface and a rigid boundary on the lateral surface of the billet were used instead of the pressure load of the pressure medium. Li et al. [10] simulated the HE process of W-40 wt. % Cu at

650–800 °C with simplified boundary conditions and calculated the linear relationship between the extrusion pressure and temperature, and proved the simulation results with experiments. Replacing the pressure of a pressure medium with a displacement load, accompanied by near-zero friction between the billet surface and the virtual rigid container, it is easy to model the HE process and improve the convergence rate effectively. However, without hydrostatic pressure, the simulation results reduce the accuracy and differ from real hydrostatic extrusion. In Li's work [10], a large gradient of equivalent stress distribution at the un-deformed region surrounded by the pressure medium was found, which was different from the real HE process where that region was in a hydrostatic state and the equivalent stress should be almost near zero. Thus, using the simplified displacement-load mode may introduce inaccuracy into the simulation results. Manafi and Saeidi [11] simulated 93 tungsten alloy by HE with a pressure boundary condition and found the optimized die angle. Peng et al. [12], by calculating the stress distribution in Nb/Cu composited by HE, investigated its interface bonding status. Manifi et al. [13] improved conventional backward extrusion by employing HE principles to reduce the extrusion load; the maximum load was reduced by 80% compared to the conventional back extrusion process. Kopp and Barton [14] improved the model of HE and analyzed the differences between experimentation and simulation.

The comparisons between the simulation and experiment were discussed in [10–14], but the comparison between the different simulation models has not been discussed in detail yet. In the present study, the pressure-load mode and displacement-load mode are used to simulate the HE process. In addition, the main work of this paper is comparing the deviation of the calculated results of the two modes under the same conditions and finding out the influence of the pressure. The judgments of the comparison are made through the theories of HE.

2. FEM Methods and Materials

The biggest difference between hydrostatic extrusion and conventional extrusion lies in the way of transferring loads from the punch to the billet. In HE process, the billet is tapered to match the die geometry, the gap between the billet and the container is filled with pressure medium, which surrounds the billet and conveys the extrusion force of the moving punch onto it, and the pressure medium is forced by its inherent pressure into the gap between the die and the billet, generating excellent lubrication on the contact surface (Figure 1a). In the present study, castor oil is used as the liquid pressure medium. The billet is tapered to match the die geometry before the extrusion in order to ensure the pressure medium staying in the container. In the real experiment, the gaps between the punch, container and the die were sealed by rubber and pure copper seal rings to ensure the system is under good sealing state. While, in the conventional extrusion, the billet is pressed by the punch directly (Figure 1b) and so deformed.

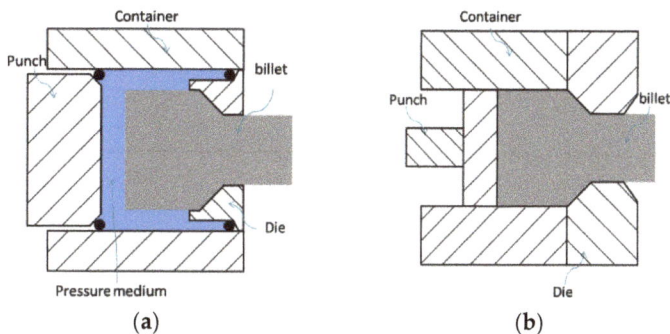

Figure 1. Principle of (**a**) hydrostatic extrusion and (**b**) conventional extrusion.

To simulate the HE process accurately, it has to take the fluid-structure interaction mode, which, however, is too complicated for large-scale calculation. So, the model developed in the present study is a partly simplified one, which improves the calculation efficiency and ensures the calculation precision. The numerical software ANSYS (V15.0, ANSYS Inc., Canonsburg, PA, USA) was used for the simulation. In the model, the pressure medium is replaced by uniform pressure loads, the die is partly replaced by rigid lines, and no friction is set between the billet and the fluid while friction between the billet and the die is in agreement with Coulomb's friction law. The model could be further simplified to 2-D because of the axial symmetry of both billet and die as columns. An eight-node plane element (PLANE 183) is used.

The pressure load mode is the mode replacing the pressure medium as a boundary condition of the billet, modeling only its pressure properties. The pressure is set to increase linearly with the time, replacing the effect of the punch pushing the pressure medium. So, in this model, the punch is not needed because the billet is deformed by the increased pressure. In the pressure load model, the central axis of the billet is the symmetrical axis of both billet and die, and the pressure load only exists over the un-deformed outer surface of the billet. The fillet at upper right of the billet is built to verify the uniform distribution of the pressure load. The coefficient of friction between the billet and the die is set as 0.05 (Figure 2a). A displacement load model which is commonly used is also established, for the sake of comparison (Figure 2b). In the displacement load mode, the displacement load with even speed is directly applied on the upper surface of the billet. The lateral surface of the billet is constrained by displacement constraint to ensure the materials cannot flow along the positive direction of the radius. Thus, the extrusion force can be calculated by the reaction force. Rigid lines are placed outside the un-deformed region instead of the displacement constraint and the coefficient of friction over this region is set as 0 which replaces the zero friction between billet and pressure medium. So, the displacement load mode is essentially a conventional extrusion mode without friction between the billet and the die. The material used to be deformed in both models is AA2024, which is a typical ideal elastoplastic material, whose specific parameters are given in Table 1. The two models are simulated at room temperature (298 K), and the parameters used for the simulations presented in this paper are given in Table 2.

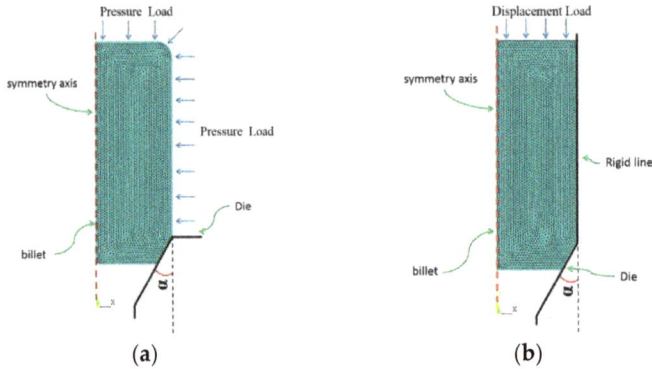

Figure 2. Models of (**a**) pressure load and (**b**) displacement load.

Table 1. Material properties of AA2024.

Material	Density	Poisson Ratio	Elastic Modulus	Yield Stress
AA2024	2.79 g/cm^3	0.3	71.7 GPa	340 MPa

Table 2. Simulation parameters of pressure load mode and displacement load mode.

Load Mode	Extrusion Ratio	Die Angle (2α)	Initial Height	Initial Diameter
Pressure load	2.25, 2.78, 4.00, 6.25	25°, 30°, 35°, 40°, 60°, 90°, 120°	80 mm	30 mm
Displacement load				

3. Results and Discussion

As shown in Figure 3, fluid pressure changed with time as the set pressure was distributed uniformly over the outer surface of the un-deformed region of the billet and decreases gradually and finally disappeared near the entrance of the die. The distributions of the pressure at different stages all proved that the hydrostatic pressure property was perfectly represented. The pressure changed linearly with time and reached a certain amount when the billet was deformed. The distribution of the fluid pressure, again as shown in Figure 3, proved that the pressure load model fits the real HE process well.

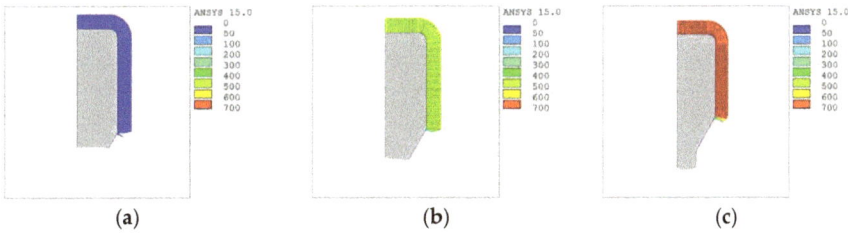

(a)   (b)   (c)

Figure 3. Fluid pressure distribution at (**a**) initial stage; (**b**) pressure-up stage and (**c**) stable stage.

3.1. Comparison of Distribution of Stress and Strain Field

The distribution information, including the equivalent strain, equivalent stress, etc., as calculated, was compared under the same scale bar, between the two models, thus making the difference much more obvious. The comparison was conducted at $R = 4.00$ and $2\alpha = 60°$.

The Von-Mises equivalent strain distributions of the two models (Figure 4) were found to be basically the same, proving that the comparison was conducted as the two billets were experiencing the same degree of deformation.

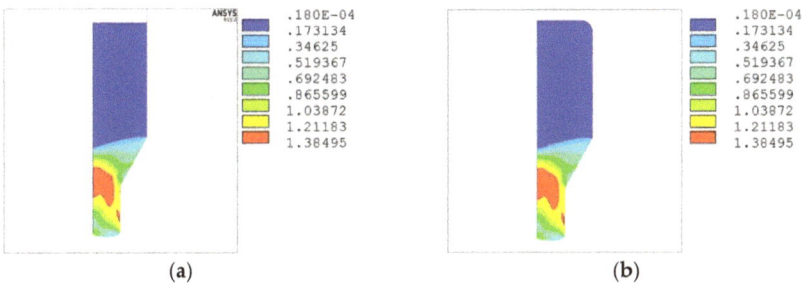

(a)   (b)

Figure 4. Von-Mises equivalent strain distribution of (**a**) displacement-load model and (**b**) pressure-load model.

The axial stress distributions (Figure 5) indicate that, on the surface of the billet under pressure load, the area of the compressed stress–concentrated region in the inlet region was larger and that of the tensile stress–concentrated region was smaller in the outlet region than those in the case of displacement load. These differences can affect the material's formability. So in the pressure-load model, there was less of a tendency to generate cracks on the surface of the billet when the material went through the inlet and outlet regions, and hence there was good surface quality, which is an important characteristic of HE.

(a)   (b)

Figure 5. Axial distribution of (**a**) displacement-load model and (**b**) pressure-load model.

The Von-Mises equivalent stress distributions of the two models were obviously different, just as shown in Figure 6. It was found that, under displacement load, the material was pressed by unequal σ_1, σ_2, σ_3, generating an exorbitant Von-Mises stress in the un-deformed region and a tiny Von-Mises stress in a small region only in the core of the billet at the inlet of the die, which is totally different from the situation of the real HE process (Figure 6a). The Von-Mises equivalent stress was extremely small in the un-deformed region under pressure load, because the billet was surrounded by hydrostatic pressure, which made the primary stress (σ_1, σ_2, σ_3) nearly equal. The value of the equivalent stress gradually reached the yield value as the deformation went on, and so there exists a gradient distribution of the equivalent stress in the inlet region, where the deformed and un-deformed regions are clearly demarcated (Figure 6b). In Reference [10], the stress distribution with a large value was found in the un-deformed region, proving the mode in that work was similar to the displacement load. In addition, this equivalent stress difference indicates that the pressure-load model is more suitable for HE analysis.

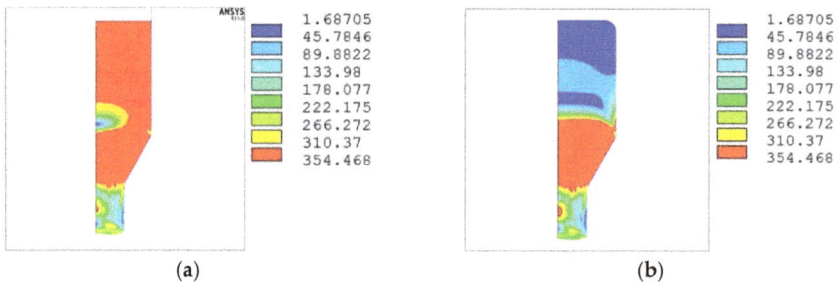

(a)   (b)

Figure 6. Von-Mises equivalent stress distribution of (**a**) displacement load and (**b**) pressure load.

The equivalent strain and stress distributions of the pressure-load model proved the un-deformed region was under hydrostatic pressure, ensuring no deformation was happening. Some experiments [15,16] were conducted to verify the materials' deformation behavior through HE. The billet was cut through the center along the extrusion axis and a grid was printed on the cut surface. Finally, the two parts were put together and extruded. After extrusion, no deformation was found at the grid in the part surrounded by the pressure medium, the un-deformed region. The experiments fit the simulation results well.

The deformed region can be distinguished by the stress state ratio distribution. The boundary between the deformed and un-deformed regions under displacement load was not stable, lower in the core and higher near the surface (Figure 7a). In the case of pressure load, the boundary was parallel to the top surface, shaped like the Von-Mises equivalent stress distribution, thus proving that the material flowed uniformly under the hydrostatic pressure (Figure 7b). The die limited the material's movement during the deformation, making the material flow more easily in the core but less near the surface. The hydrostatic pressure can effectively improve the material's flow, for the material flowed uniformly even where the die exerts its limitation. However, the displacement load cannot benefit the material's flow, hence the uneven flow and the deformation near the surface lagging behind that in the core.

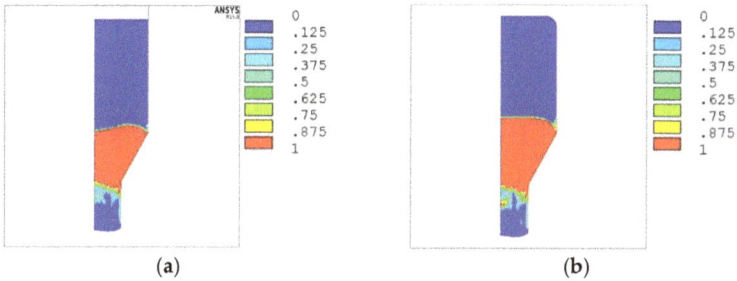

Figure 7. Stress state ratio distribution of (**a**) displacement load (**b**) and pressure load.

As can be seen in the contact pressure distribution, the material's flow near the surface differed between the two models. The contact pressure in the inlet region was higher under displacement load (Figure 8a) because uneven material flow results in more redundant work, so the load to achieve the same deformation is higher, hence the higher contact pressure. A lower contact pressure can be found under pressure load because the material deforms uniformly in this region and so lower redundant work is needed, hence the lower contact pressure (Figure 8b). Hydrostatic pressure can make the material flow uniformly, as was obviously shown in the pressure-load model.

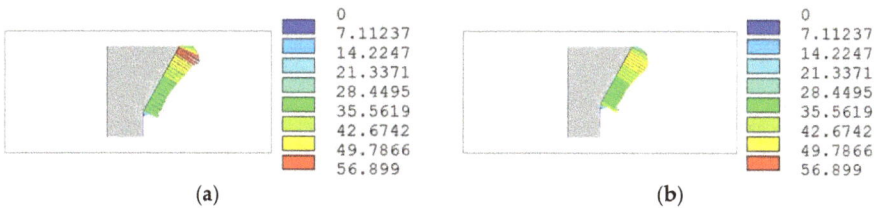

Figure 8. Contact pressure distribution of (**a**) displacement load and (**b**) pressure load.

3.2. Comparison of Extrusion Pressure

By analyzing the difference in HE, a further comparison was made between the two models. R is the deformation ratio, written as $R = D^2/d^2$. In the pressure-load model, the pressure increased linearly with the time; meanwhile, the billet was deformed. The pressure corresponding to the position when the bottom of the billet is pressed out of the die is the extrusion pressure (P), and this is the minimum pressure to complete the extrusion process. In the displacement-load model, the position with same deformation ratio can be found, and the extrusion pressure was calculated by the reaction force and the area of the contact surface. A linear relationship exists between P and $\ln R$ in both models; for instance, at $\alpha = 45°$, $P = 424\ln R + 197$ for displacement load and $P = 347\ln R + 160$ for pressure load, respectively (Figure 9a). The pressure gap between the two load models can be found under the same working conditions. The *P*-value is higher in the displacement-load model, because the material flows less uniformly, so that a higher *P* is required to overcome the redundant work. But under a pressure load, the *P*-value is lower because the hydrostatic pressure load can maintain uniform deformation during the process. The gap becomes larger as the *R* value increases, because the deformation uniformity decreases as the deformation ratio grows. However, the gap grows only a bit, indicating that the deformation ratio is not the main cause for the redundant work.

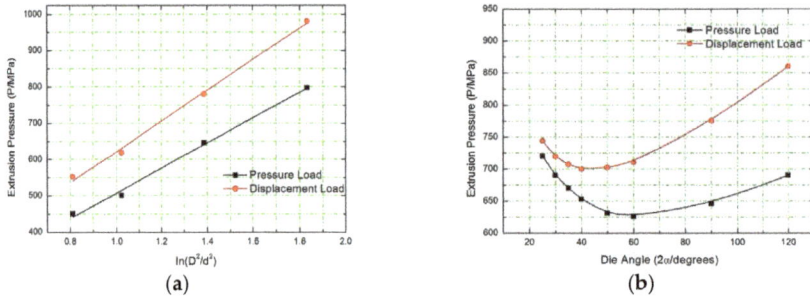

Figure 9. Relationship between (**a**) *P* and ln*R*; (**b**) *P* and 2α.

In both models, there exists a parabolic relationship between *P* and 2α (Figure 9b). Extrusion pressure first declined and then increase when 2α increased from 30° to 120°. The optimized die angle (2α), corresponding to the smallest *P*-value, was 40° and 60° in the displacement-load and the pressure-load model, respectively. The size of the optimized die angle depends on the redundant work and friction work during the deformation. The redundant work, resistant to non-uniform deformation, increased rapidly when the die angle grew. Meanwhile, the friction work also changed because both the contact pressure and contact area changed. As the die angle grew, the contact pressure increased whereas the contact area decreased, and the friction work first declined and then increased. The redundant work increase was lower than the friction work decrease at first, so the *P*-value declined, but as the die angle grew, the redundant work increase gradually grew faster than the friction work decrease, leading to the rise of the *P*-value. The pressure gap between the two models became larger rapidly as 2α grew, indicating that the redundant work increased faster in the displacement-load model. The gap grew rapidly with the die angle, indicating the angle was the main cause for the redundant work.

The relationship between *P* and 2α can be well explained by the stress state ratio distribution (Figure 10). As the die angle grew from 30° to 120°, the deformation region boundary in the displacement-load model increased more rapidly than in the pressure-load model and the shape of the boundary changed more drastically at the same time. The irregularity of the boundary indicated a non-uniform deformation, so more redundant work was generated, which in turn resulted in greater extrusion pressure to achieve the same deformation. So the extrusion pressure was lower in the pressure load model at the same die angle where the boundary was more stable. In addition, because of the hydrostatic pressure, little redundant work was generated in the pressure-load model, so the extrusion pressure value fluctuated in a small range at different die angles.

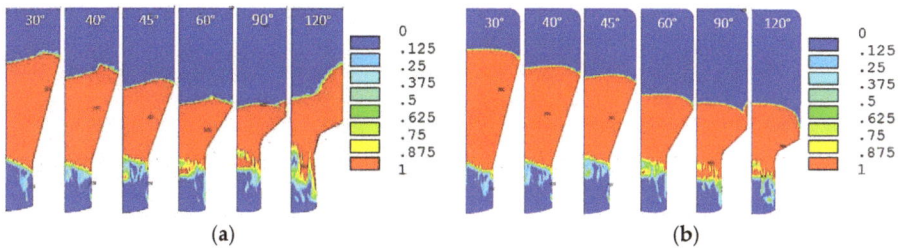

Figure 10. Stress state ratio distribution of (**a**) displacement-load model and (**b**) pressure-load model at $2\alpha = 30°, 40°, 45°, 60°, 90°$ and $120°$.

4. Conclusions

Through numerical simulation, two models, the displacement-load model and pressure-load model, are designed and compared. Both models can simulate the hydrostatic extrusion process to a certain extent, gaining similar results of the strain calculation. Under the pressure load, the value of the Von-Mises equivalent stress in the un-deformed region is very small, which proves that this region is under uniform hydrostatic pressure. The deformation boundary in the stress state ratio distribution is almost horizontal, which proves that the material is pressed by hydrostatic pressure. However, in the displacement-load model, the Von-Mises stress value is large, and the deformation boundary is irregular, proving that, with no hydrostatic pressure, the deformation is non-uniform. The relationship between P and R was found as $P = 347\ln R + 160$ or $P = 424\ln R + 197$ in the pressure-load and displacement-load model. In addition, the optimized die angle was $60°$ or $40°$, respectively. It can be proved that in the displacement-load model, the non-uniform material flow generates more redundant work, resulting in a higher extrusion pressure to achieve the same deformation. With the increase of the die angle, the abnormal growth of the extrusion pressure under displacement load indicates that the redundant work increases rapidly, which will lead to the deviation of the calculation results from the actual HE process. By comparing the data above, it is found that the numerical model with pressure load can simulate the hydrostatic extrusion process more accurately.

Acknowledgments: The authors would like to acknowledge the financial support from the National Magnetic Confinement Fusion Program with Grant No. 2014GB121001, the National Natural Science Foundation of China No. 51675154, and the research support from the Laboratory of Nonferrous Metal Material and Processing Engineering of Anhui Province.

Author Contributions: Shengqiang Du, Xiang Zan and Ping Li designed the simulation models; Shengqiang Du performed the simulation; Xiang Zan, Laima Luo and Xiaoyong Zhu contributed to analyze the simulation results; Yucheng Wu provided support and contributed to the discussions; Shengqiang Du wrote the paper.

References

1. Robertson, J. Method of and Apparatus for Forming Metal Articles. British Patent No. 19 356, 14 October 1894.
2. Wilson, W.R.D.; Walowit, J.A. An isothermal hydrodynamic lubrication theory for hydrostatic extrusion and drawing processes with conical dies. *J. Lubr. Technol.* **1971**, *93*, 69–74. [CrossRef]
3. Ozaltin, K.; Chrominski, W.; Kulczyk, M.; Panigrahi, A.; Horky, J.; Zehetbauer, M.; Lewandowska, M. Enhancement of mechanical properties of biocompatible Ti–45Nb alloy by hydrostatic extrusion. *J. Mater. Sci.* **2014**, *49*, 6930–6936. [CrossRef]
4. Yu, Y.; Zhang, W.C.; Duan, X.R. Study on microstructure and properties of thin tube of AZ31 magnesium alloy by extrusion technology. *Powder Metall. Technol.* **2013**, *31*, 201–206. [CrossRef]
5. Xue, Y.F.; Cai, H.N.; Wang, L.; Wang, F.C.; Zhang, H.F. Strength-improved Zr-based metallic glass/porous tungsten phase composite by hydrostatic extrusion. *Appl. Phys. Lett.* **2007**, *90*, 081901. [CrossRef]

6. Kaszuwara, W.; Kulczyk, M.; Leonowicz, M.K.; Gizynski, T.; Michalski, B. Densification of Nd-Fe-B powders by hydrostatic extrusion. *IEEE Trans. Magn.* **2014**, *50*, 1–5. [CrossRef]
7. Kováč, P.; Hušek, I.; Melišek, T.; Kopera, L.; Kováč, J. Critical currents, I_c-anisotropy and stress tolerance of MgB$_2$ wires made by internal magnesium diffusion. *Sci. Technol.* **2014**, *27*, 88–93. [CrossRef]
8. Skiba, J.; Pachla, W.; Mazur, A.; Przybysz, S.; Kulczyk, M.; Przybysz, M.; Wróblewska, M. Press for hydrostatic extrusion with back-pressure and the properties of thus extruded materials. *J. Mater. Process. Technol.* **2014**, *214*, 67–74. [CrossRef]
9. Zhang, Z.H.; Wang, F.C.; Sun, M.Y.; Yang, R.; Li, S.K. Finite element analysis and experimental investigation of the hydrostatic extrusion process of deforming two-layer Cu/Al composite. *J. Beijing Inst. Technol.* **2013**, *22*, 544–549. (In Chinese).
10. Li, D.R.; Liu, Z.Y.; Yu, Y.; Wang, E.D. Numerical simulation of hot hydrostatic extrusion of W-40 wt. % Cu. *Mater. Sci. Eng. A* **2009**, *499*, 118–122. [CrossRef]
11. Manafi, B.; Saeidi, M. Deformation behavior of 93 Tungsten alloy under hydrostatic extrusion. *Elixir Mech. Eng.* **2014**, *76*, 28487–28492.
12. Peng, X.; Sumption, M.D.; Collings, E.W. Finite element modeling of hydrostatic extrusion for mono-core superconductor billets. *IEEE Trans. Appl. Supercond.* **2003**, *13*, 3434–3437. [CrossRef]
13. Manafi, B.; Shatermashhadi, V.; Abrinia, K.; Faraji, G.; Sanei, M. Development of a novel bulk plastic deformation method: Hydrostatic backward extrusion. *Int. J. Adv. Manuf. Technol.* **2016**, *82*, 1823–1830. [CrossRef]
14. Kopp, R.; Barton, G. Finite element modeling of hydrostatic extrusion of magnesium. *J. Technol. Plast Technol.* **2003**, *28*, 1–12.
15. Barton, G. Finite-Elemente Modellierung des Hydrostatischen Strangpressens von Magnesiumlegierungen. Ph.D. Thesis, Rheinisch-Westfaelische Technische Hochschule Aachen, Aachen, Germany, January 2009. Available online: http://publications.rwth-aachen.de/record/51193/files/Barton_Gabriel.pdf (accessed on 1 March 2017).
16. Kulczyk, M.; Przybysz, S.; Skiba, J.; Pachla, W. Severe plastic deformation induced in Al, Al-Si, Ag and Cu by hydrostatic extrusion. *Arch. Metall. Mater.* **2014**, *59*, 59–64. [CrossRef]

metals

MDPI

Article

Modeling the Constitutive Relationship of Al–0.62Mg–0.73Si Alloy Based on Artificial Neural Network

Ying Han [1,*], Shun Yan [1], Yu Sun [2] and Hua Chen [1]

[1] Key Laboratory of Advanced Structural Materials, Ministry of Education, Changchun University of Technology, Changchun 130012, China; yanshunccgy@163.com (S.Y.); chenhua@ccut.edu.cn (H.C.)
[2] National Key Laboratory for Precision Hot Processing of Metals, Harbin Institute of Technology, Harbin 150001, China; yusun@hit.edu.cn
* Correspondence: hanying_118@sina.com; Tel.: +86-431-8571-6396

Academic Editors: Myoung-Gyu Lee and Yannis P. Korkolis
Received: 18 February 2017; Accepted: 23 March 2017; Published: 26 March 2017

Abstract: In this work, the hot deformation behavior of 6A02 aluminum alloy was investigated by isothermal compression tests conducted in the temperature range of 683–783 K and strain-rate range of 0.001–1 s^{-1}. According to the obtained true stress–true strain curves, the constitutive relationship of the alloy was revealed by establishing the Arrhenius-type constitutive model and back-propagation (BP) neural network model. It is found that the flow characteristic of 6A02 aluminum alloy is closely related to deformation temperature and strain rate, and the true stress decreases with increasing temperatures and decreasing strain rates. The hot deformation activation energy is calculated to be 168.916 kJ mol^{-1}. The BP neural network model with one hidden layer and 20 neurons in the hidden layer is developed. The accuracy in prediction of the Arrhenius-type constitutive model and BP neural network model is eveluated by using statistics analysis method. It is demonstrated that the BP neural network model has better performance in predicting the flow stress.

Keywords: deformation behavior; constitutive model; BP neural network; aluminum alloy

1. Introduction

Deformation behaviors of metal materials are considered as the comprehensive presentation of material properties and processing parameters, such as temperature, strain rate and strain [1–4]. Since the flow behaviors of materials provide valuable information and crucial instructions for thermoplastic processing, an increasing number of researchers have paid great attention to this field. In order to reveal the internal influence of processing parameters on the flow characteristics of materials, some constitutive models have been established [5]. At present, mainly three kinds of constitutive models have been broadly accepted, including analytical models, phenomenal models and empirical models [6–8]. Specifically, analytical models are developed based on the plastic deformation theories of kinetics and dynamics of dislocation. As a result, these models require a thorough understanding of the physical mechanisms controlling deformation behavior of materials. However, it is almost impossible for analytical models to be employed in practical application due to coupled influence of various processing parameters. As for phenomenal models, though less closely related to physical theories, they still depend on explicit understanding of deformation mechanisms. Such models usually consist of several separate equations, each of which is responsible for a fixed processing domain. Therefore, the generality and simplicity of phenomenal models are often limited. Empirical models are aimed at researching the quantitative relationship between true stresses and processing parameters. Regression analysis technique is constantly employed to establish equations and statistical methods are performed to evaluate the performance of the model. Therefore, empirical models do not need details about physical or chemical revolution involved during the deformation process. The Arrhenius-type constitutive model is recognized to be one of the most commonly used empirical models and has been successfully applied to various alloys, including magnesium alloys [9], titanium alloys [10,11], aluminum alloys [12] and steels [13]. In contrast, the artificial neural network (ANN) model does not refer to any mathematical model, and it only learns from examples and recognizes patterns from a series of inputs and outputs without any prior assumptions about their nature. Since it is not involved in physical interpretation of plastic deformation mechanisms, the ANN model acts as a robust and intelligent data information treatment system. It has now been recognized as a powerful tool in the field of material science and increasingly applied by a growing number of scholars. For instance, Li et al. [14] compared the ANN model with the Arrhenius-type constitutive model regarding the hot deformation behavior of an Al-Zn-Mg alloy. Ji et al. [15] used a back-propagation neural network model, which was trained with Lavenberg-Marquardt learning algorithm, to study the high-temperature flow behavior of Aermet100 steel. In the research, the performance of the ANN model was evaluated by using a wide variety of standard statistical indices, which turned out that the extrapolation ability of neural network model was very high in the proximity of training domain. Li et al. [16] conducted a comprehensive and comparative study on Zerilli–Armstrong, Arrhenius-type and ANN models in terms of their prediction ability of hot deformation behavior of T24 steel. Quan et al. [17] applied ANN to predict the flow stress of as-cast Ti-6Al-2Zr-1Mo-1V alloy, which suggested that the ANN model has a good capacity to model complicated flow behavior of titanium alloy. Haghdadi et al. [18] utilized ANN to predict the hot deformation behavior of an A356 aluminum alloy, indicating the fact that the trained ANN model was a robust tool to characterize the high-temperature flow behavior of the studied alloy. Despite considerable research work regarding the application of the ANN model in terms of deformation behavior of various materials, the essential differences between the Arrhenius-type constitutive model and the ANN model has not been clearly revealed and application limitation of both methods should be explained.

6A02 aluminum alloy is one of the significant machinable aluminum alloys, which has been extensively used in the fields of aerospace and auto-industry due to its good ductility, high specific strength and satisfied corrosion resistance. In this work, the Arrhenius-type constitutive model and the ANN model are developed and a comprehensive comparison between them is conducted to study the high-temperature flow behavior of 6A02 aluminum alloy.

2. Materials and Methods

In the present work, the raw material was 6A02 aluminum alloy. The chemical composition and starting microstructure of the alloy are presented in Table 1 and Figure 1, respectively. It can be seen that the grains of 6A02 aluminum alloy are equiaxed and the average grain size is about 200 μm.

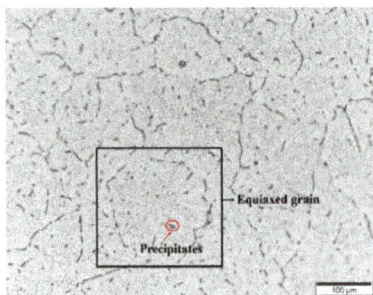

Figure 1. The starting microstructure of the 6A02 aluminum alloy.

Table 1. The chemical composition of 6A02 aluminum alloy.

Element	Mg	Si	Fe	Mn	Cu	Ni	Ti	Zn	Al
Wt %	0.62	0.73	<0.05	<0.05	<0.01	<0.01	<0.01	<0.01	Bal

The cylindrical compressive specimens were obtained from a bar and subsequently machined into 6 mm in diameter and 9 mm in height. Isothermal compression tests were carried out on the Gleeble-1500D simulator (Harbin Institute of Technology, Harbin, China) in the temperature range of 683–783 K with an interval of 20 K, and the strain rates were selected as 0.001, 0.01, 0.1 and $1\,\text{s}^{-1}$. The surfaces of specimens were grinded by sandpaper to remove the oxide layer and guarantee smoothness. Thermocouples were welded in the middle of specimens to measure the temperature during the experimental process. In addition, the graphite powder was applied on both surfaces of specimens to reduce the friction coefficient between the specimen ends and the anvils. The specimens were heated at a rate of 10 K/s up to deformation temperature, and held for 3 min to maintain a uniform temperature in the sample and reduce the material anisotropy. The testing specimens were subject to be compressed to a total true strain of approximately 0.6. The stress and strain variations were automatically recorded by a computer equipped with a data acquisition system. As the compression was completed, the deformed specimen was immediately quenched in water to preserve the hot deformation microstructure. The procedure of compression tests is shown in Figure 2.

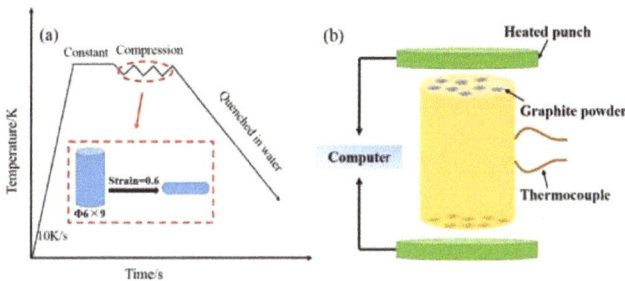

Figure 2. Schematic illustration of experimental procedure (**a**) and experimental device (**b**).

3. Results and Discussion

3.1. Flow Behavior

The true stress–true strain curves obtained from isothermal compression tests at the strain rate of $0.01\,\text{s}^{-1}$ and the temperature of 723 K are shown in Figure 3. It can be observed from this figure that the flow stress is strongly dependent on the strain rate and deformation temperature. At the strain hardening stage, dislocation multiplication plays a dominant role, the true stress increases rapidly with increasing strain up to a peak value in this stage. While at the dynamic softening stage, the true stress drops gradually due to dynamic recrystallization. Finally, the equilibrium between strain hardening and dynamic softening occurs and the true stress remains at a stable state. It is also noted from Figure 3b that the flow stress at low strain rates of 0.001 and $0.01\,\text{s}^{-1}$ are more likely to become stable, which illustrates that the completed dynamic recrystallization has been occurred. Generally, the relationship between true stress and processing parameters of the studied alloy are highly nonlinear. Moreover, similar deformation characteristics of the other alloys are also observed by many researchers [19–21].

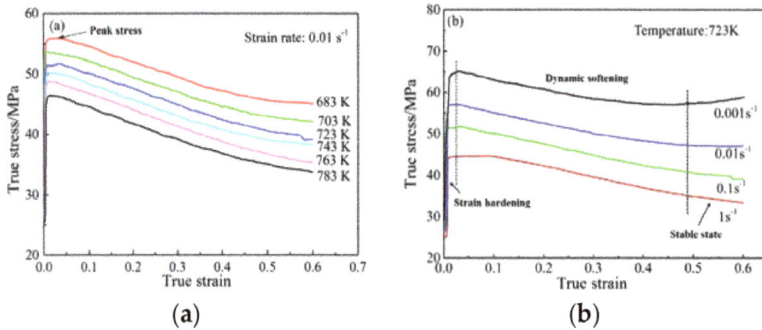

Figure 3. Typical true stress–true strain curves of 6A02 alloy at (a) $\dot{\varepsilon} = 0.01$ s^{-1} and (b) $T = 723$ K.

3.2. Arrhenius-Type Constitutive Model

During the hot deformation process of metals and alloys, the correlation between flow stress, strain rate and temperature can be presented by using the Arrhenius-type constitutive model presented as follows:

$$\dot{\varepsilon} = A[\sinh(\alpha\sigma)]^n \exp(-Q/RT) \tag{1}$$

where A, α and n are the material constants, $\dot{\varepsilon}$ is the strain rate (s^{-1}), σ represents the flow stress (MPa), Q refers to the activation energy for hot deformation (J mol^{-1}), T is the absolute temperature in Kelvin, and R is the gas constant (8.314 J mol^{-1} K^{-1}). According to the Taylor series deployment method, the Arrhenius constitutive equation can be interpreted as:

$$\dot{\varepsilon} = A_1\sigma^{n_1} \exp[-Q/RT] \ (\alpha\sigma < 1.2) \tag{2}$$

$$\dot{\varepsilon} = A_2 \exp(\beta\sigma) \exp[-Q/RT] \ (\alpha\sigma > 1.2) \tag{3}$$

$$\dot{\varepsilon} = A[\sinh(\alpha\sigma)^n] \exp[-Q/RT] \ (\alpha\sigma \text{ taking any value}) \tag{4}$$

where $A_1 = A\alpha^n$, $A_2 = A/2^n$, and $\alpha = \beta/n_1$.

The natural logarithm form of the equations above can be indicated as:

$$\ln\dot{\varepsilon} = \ln A_1 - Q/RT + n_1 \ln\sigma \tag{5}$$

$$\ln\dot{\varepsilon} = \ln A_2 - Q/RT + \beta\sigma \tag{6}$$

$$\ln\dot{\varepsilon} = \ln A - Q/RT + n\ln[\sinh(\alpha\sigma)] \tag{7}$$

The peak values of flow stress are presented in Table 2. With these experimental data fitted into Equations (5) and (6), the values of n_1 and β are derived from mean slope of the curves of $\ln\dot{\varepsilon}-\ln\sigma$ and $\ln\dot{\varepsilon}-\sigma$ in Figure 4a,b, and they are found to be 18.396 and 0.349, respectively. Therefore, α is determined as $\alpha = \beta/n_1 = 0.019$. Besides, the value of n is calculated by:

$$n = \left[\frac{\partial\ln\dot{\varepsilon}}{\partial\ln(\sinh(\alpha\sigma))}\right]_T \tag{8}$$

Table 2. Peak values of flow stress of 6A02 alloy during hot deformation.

Strain Rate/s^{-1}	Flow Stress/MPa					
	683 K	703 K	723 K	743 K	763 K	783 K
0.001	47.90	46.86	44.66	43.13	42.19	40.47
0.01	55.88	53.60	51.76	50.23	48.78	46.37
0.1	63.62	61.86	57.12	55.48	53.59	50.67
1	76.00	68.19	65.17	63.71	60.80	59.74

It is found that n turned out to be 13.962 from mean slope of the curves of $\ln\dot{\varepsilon}-\ln[\sinh(\alpha\sigma)]$ shown in Figure 4c. Q can be obtained by taking partial derivative of both sides of Equation (7) to $1/T$:

$$Q = nR\left[\frac{\partial\ln(\sinh(\alpha\sigma))}{\partial(1/T)}\right]_T = RnS \qquad (9)$$

The S value is found to be 1.455 from mean slope of $\ln[\sinh(\alpha\sigma)]-1000/T$ illustrated in Figure 4d. As a result, the value of Q can be determined as 168.916 kJ mol^{-1}. The Q value of the studied alloy is compared with that of other aluminum alloys calculated following the same procedure used for the present 6A02 alloy [22–26], which is presented in Figure 5. It can be observed that the Q value in this work is much smaller than most of the other kinds of aluminum alloys. In fact, 6A02 aluminum alloy is known to exhibit greater temperature dependence in the reduction of flow stress than other aluminum alloys, which is also proved by this work. While the hot deformation is a thermal activation course, activation energy is recognized as a significant indicator for expression of the energy required to overcome the barriers in the metal forming process at elevated temperatures, and it thereby directly reflects the difficulty degree for the hot deformation. As a result, the low value of Q means that the deformation for this alloy is easy at high temperatures and the dynamic recrystallization easily occurs. In addition, the large difference of Q values is visible in the same series of alloys such as 6A02 and 6A82, which demonstrates that the chemical composition and heat treatment state exert a decisive effect on the Q value of aluminum alloys. Yang et al. [23] suggested that the activation energy of Cu-rich alloys is higher than that of Cu-free alloys. The high Q value of 6A82 aluminum alloy may be related to the high Cu content.

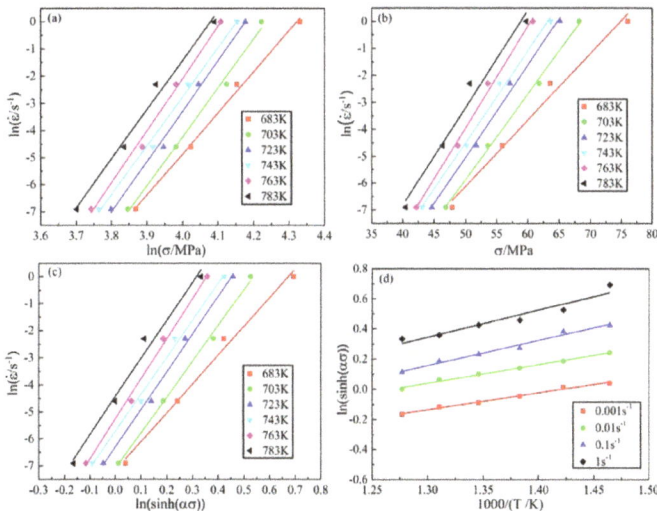

Figure 4. Relationship among stress, strain rate and temperature at a strain of 0.1: (a) $\ln\dot{\varepsilon}-\ln\sigma$; (b) $\ln\dot{\varepsilon}-\sigma$; (c) $\ln\dot{\varepsilon}-\ln[\sinh(\alpha\sigma)]$ and (d) $\ln[\sinh(\alpha\sigma)]-1000/T$.

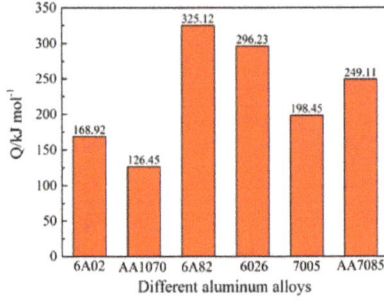

Figure 5. The comparison of Q values among different aluminum alloys.

It is acknowledged that the influence of deformation temperature and strain rate on the flow behavior can be evaluated by the Zener–Hollomon parameter [27], which can be defined as:

$$Z = \dot{\varepsilon} \exp(Q/RT) \tag{10}$$

Taking the natural logarithm form of Equation (10) leads to the following equation:

$$\ln Z = \ln \dot{\varepsilon} + Q/RT \tag{11}$$

According to Equation (7), Equation (11) can be adapted as:

$$\ln Z = \ln A + n \ln[\sinh(\alpha\sigma)] \tag{12}$$

The relationship between ln Z and ln[sinh($\alpha\sigma$)] is plotted in Figure 6 with experimental data at peak stress. The intercept corresponding to ln A is obtained as 21.701. It is noted that the correlation coefficient (R) for the linear regression of ln Z−ln[sinh($\alpha\sigma$)] reaches 0.981, which indicates that the hyperbolic-sine function is in good agreement with experimental results. Therefore, the Arrhenius-type constitutive equation of the studied aluminum alloy at peak stress can be developed as:

$$\dot{\varepsilon} = e^{21.701}[\sinh(0.019\sigma)]^{13.962} \exp(-168916/RT) \tag{13}$$

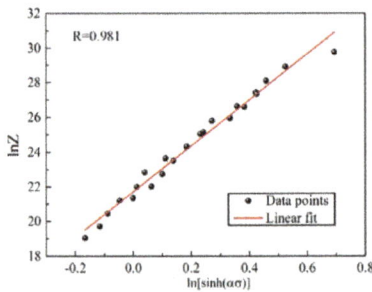

Figure 6. Relationship between flow stress and Zener–Hollomon parameter.

3.3. Artificial Neural Network Modeling

Back-propagation (BP) neural network is a forward multi-layer network of one-way transmission [28]. It can be known that a BP neural network is made up of input layer, hidden layer and output layer.

In the present work, the inputs were comprised of deformation temperature, strain rate and strain, while the true stress was identified as the output variable. Each data processing unit in the layer is called a neuron, whose property is usually controlled by an activation function. Tan-sig, log-sig and purelin are three derivable functions that can be employed as the activation. In order to obtain the best combination by two of them. The performance of different groups with 20 hidden layer neurons was revealed by mean square error (MSE), which is presented in Table 3. It was found that the MSE of the tansig-purelin group is the smallest, which was only 0.113. Therefore, tansig function is chosen as the input-hidden layer activation function and purelin is determined as the hidden-output layer activation function.

In the BP neural network structure, neurons of the same layer are not coupled with each other. When the input data passes through the network, the output is calculated and subsequently compared with the target. If the deviation exceeds the predetermined threshold, a reverse transmission process will initiate to correct the weight of each neutron. The above actions will not terminate until the deviation can be accepted.

Table 3. Performance of different activation function groups.

Input-Hidden Layer Activation Function	Hidden-Output Layer Activation Function	Mean Square Error (MSE)
tansig	tansig	0.222
tansig	purelin	0.113
tansig	logsig	0.289
purelin	tansig	0.139
purelin	purelin	1.862
purelin	logsig	0.167
logsig	tansig	0.732
logsig	purelin	0.491
logsig	logsig	1.094

In this work, a feed-forward neural network model trained with BP learning algorithm will be established; 192 data points as the input database was used. Before training the network, both input and output variables have to be normalized within the range from 0 to 1 in order to obtain a valid form for the neural network model to recognize, which can be treated as Equation (14):

$$X^* = \frac{X - 0.95X_{min}}{1.05X_{max} - 0.95X_{min}} \tag{14}$$

where X is the original data which refers to temperature, strain and flow stress; X^* is the normalized data of the corresponding X; X_{min} and X_{max} are the minimum and maximum values of X, respectively. Given that the strain rate changes sharply and the minimum of it is too small for the network to learn, the following equation has to be used for the normalization:

$$\dot{\varepsilon}^* = \frac{\ln \dot{\varepsilon} - 0.95 \ln \dot{\varepsilon}_{min}}{1.05 \ln \dot{\varepsilon}_{max} - 0.95 \ln \dot{\varepsilon}_{min}} \tag{15}$$

where $\dot{\varepsilon}$, $\dot{\varepsilon}_{min}$, $\dot{\varepsilon}_{max}$ and $\dot{\varepsilon}^*$ are the original, minimum, maximum and normalized strain rate, respectively. After pre-proceeding all the inputs, the architecture of the network needs to be elaborated. In order to develop a BP neural network model with desired generalization, 144 random data sets from true stress–true strain curves were adopted to train the model and the remaining 48 data sets were employed as the testing data sets.

The neutron number of the hidden layer is a crucial factor for the efficiency and accuracy of the BP neural network. Traditionally, the trial-and-error method is generally applied to determine the appropriate number of neutrons in the hidden layer. In the present investigation, the performance of the network with 8–25 hidden neurons were compared. The performance of different numbers of neurons was evaluated by MSE, as shown in Figure 7. It is revealed that 20 hidden-layer neurons presented the best performance and the MSE is only 0.113. Therefore, the number of hidden-layer neurons is determined to be 20.

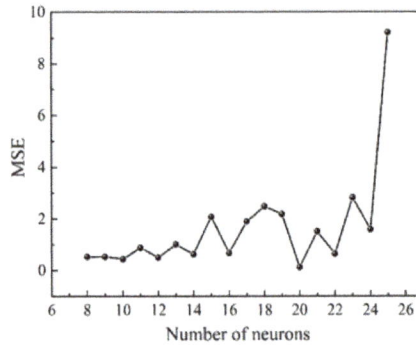

Figure 7. Performance of different numbers of neurons. MSE in the figure represents mean square error.

The accuracy of the established ANN model is further verified by a wide variety of standard statistical performance evaluation methods. The generalization property of the training and testing neural network is quantitatively verified in terms of correlation coefficient (R), relative error (δ), average absolute relative error (e_{AARE}), average root mean square error (e_{RMSE}), and scatter index (I_S). The above indexes are defined as listed, respectively.

$$R = \frac{\sum\limits_{i=1}^{N} \left(E_i - \overline{E}\right)\left(P_i - \overline{P}\right)}{\left(\sum\limits_{i=1}^{N} \left(E_i - \overline{E}\right)^2 \sum\limits_{i=1}^{N} \left(P - \overline{P_i}\right)^2\right)^{\frac{1}{2}}} \tag{16}$$

$$\delta = \frac{E_i - P_i}{E_i} \tag{17}$$

$$e_{AARE} = \frac{1}{N}\sum_{i=1}^{N}\left|\frac{E_i - P_i}{E_i}\right| \times 100\% \tag{18}$$

$$e_{RMSE} = \left[\frac{1}{N}\sum_{i=1}^{N}(E_i - P_i)^2\right]^{\frac{1}{2}} \tag{19}$$

$$I_s = \frac{e_{RMSE}}{\overline{E}} \tag{20}$$

where E_i is the experimental value and P_i is the predicted value from the ANN model. $\overline{E}$ and $\overline{P}$ are the mean values of E_i and P_i, respectively. N is the total number of data employed in the research.

The plots of experimental values versus predicted values obtained by the developed ANN model are shown in Figure 8 for both training and testing data sets. The correlation coefficient (R) is a powerful statistical tool to present the strength of the linear relationship between experimental and predicted values. It can be seen from this figure that the R values of training and testing are 0.999 and 0.952, respectively, which indicates satisfactory adaptation of the established ANN model. In addition, it is noted from Figure 9 that the average absolute relative error ($AARE$) and scatter index (I_S) of both training and testing network is small. Most of the relative errors for testing data sets are located in a quite narrow range from −10% to 10%. Therefore, it is illustrated that the artificial neural network model has been successfully developed to predict the flow behavior of the studied 6A02 aluminum alloy.

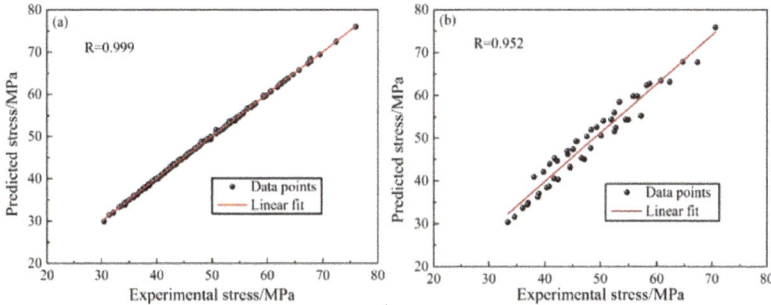

Figure 8. Comparison of flow stress values predicted by artificial neural network (ANN) model and experimental values for (**a**) training and (**b**) testing.

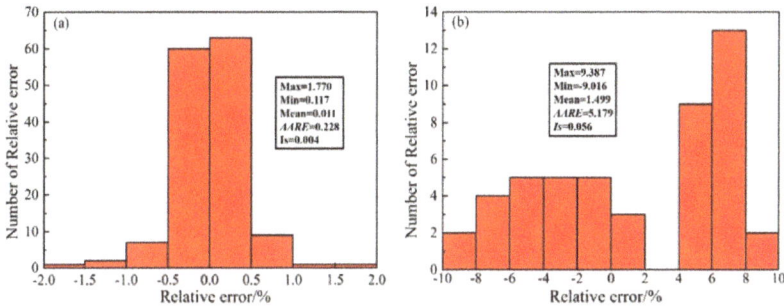

Figure 9. Distribution of relative errors between the developed ANN model and experimental results: (**a**) training sets and (**b**) testing sets.

3.4. Evaluation of Arrhenius-Type Constitutive Model and BP Neural Network

The relationship between experimental and predicted peak stress by the Arrhenius-type constitutive model is presented in Figure 10. It is noted that most of the data points settled randomly beside the fitting line and the error between experimental and predicted results is obvious. The reason for this consequence lies in accumulated error of all material constants such as n, Q and α, et al. Although the linear relationship is not so strong, the relative error can be quite small. The absolute relative errors (*ARE*) of peak stress at various deformation temperatures and strain rates are revealed in Figure 11. It can be observed that all of the *ARE* values are below 0.25% and they drop down as the strain rate increases. Generally, the Arrhenius-type constitutive model is capable of predicting flow stress with high accuracy. However, without the combination of strain in the equation, it can only predict peak stress or flow stress at a certain strain.

The relationship between predicted true stress by the developed ANN model and experimental results at 743 K/0.1 s^{-1} is illustrated in Figure 12. The predicted true stress is much closer to that of the experiment and the relative error is within the range of -0.004% to 0.016%, which indicates that the ANN model has satisfied prediction ability. Moreover, it is noticed that the value of relative error increases with increasing strain. Compared with the Arrhenius-type constitutive model, BP neural network is available to predict true stress in the entire strain range respectively. Additionally, the most significant advantage of BP neural network is to provide a full insight of the relationship between true stress and hot processing parameters such as strain rate, deformation temperature and strain.

In another words, despite the lack of physical theory support, BP neural network model possesses excellent predicted performance in terms of numerical simulation.

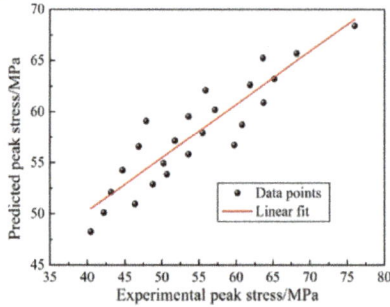

Figure 10. Relationship between predicted peak stress by Arrhenius-type constitutive model and experimental peak stress.

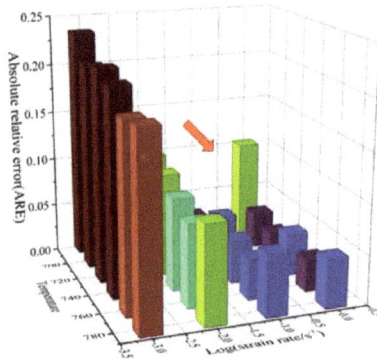

Figure 11. Distribution of absolute relative error (ARE) by Arrhenius-type constitutive model.

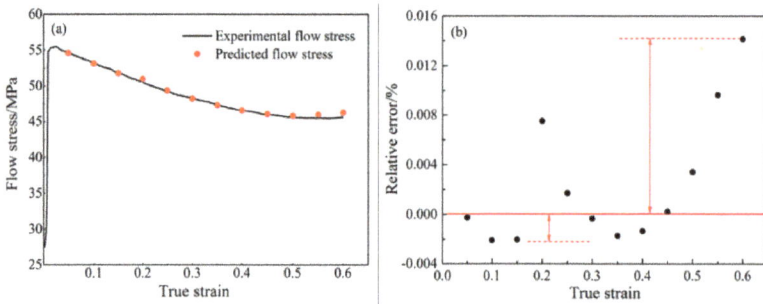

Figure 12. Relationship between predicted flow stress by the developed back-propagation (BP) neural network and experimental results at 743 K/0.1 s^{-1}.

4. Conclusions

The hot deformation behavior of 6A02 aluminum alloy was investigated by isothermal compression tests in the temperature range of 683–783 K at intervals of 20 K and strain-rate range of 0.001–1 s^{-1}. The Arrhenius-type constitutive model and BP neural network were established to

Metals **2017**, 7, 114

characterize the flow behavior of the experimental alloy, respectively. Furthermore, the generalization performance of both developed models were evaluated. The essential differences between the two methods were subsequently revealed.

(1) There is a nonlinear relationship between hot processing parameters and flow stress in the isothermal deformation of 6A02 aluminum alloy. The true stress decreases with increasing temperatures and decreasing strain rates. Besides, typical work hardening and dynamic softening features can be observed from the true stress–true strain curves of 6A02 aluminum alloy.

(2) The Arrhenius-type constitutive model is established to present the deformation behavior of the studied alloy. The activation energy (Q) is calculated to be 168.916 kJ mol^{-1}, which is much smaller than other kinds of aluminum alloy. Moreover, the absolute relative error by this model is no more than 0.25%, which demonstrates high accuracy of the established model.

(3) A back-propagation neural network with one hidden layer and 20 neurons in the hidden layer was developed to characterize the flow behavior of the alloy. The relative errors were limited in the range of −0.004% to 0.016%, which suggested an excellent predicted performance of the ANN model. Moreover, the back-propagation neural network can be used to predict the true stresses in the whole strain range conveniently.

Acknowledgments: This work was sponsored by the National Natural Science Foundation of China (No. 51371038 and 51604034), the China Postdoctoral Science Foundation (No. 2014M551234), the Specialized Research Fund for the Doctoral Program of Higher Education (No. 20132302120002), the Fundamental Research Funds for the Central Universities (No. HIT. NSRIF. 2014006) and Heilongjiang Postdoctoral Fund (No. LBH-Z14096).

Author Contributions: Ying Han conceived and designed the experiments; Shun Yan performed the experiments; Ying Han and Yu Sun analyzed the data; Hua Chen contributed analysis tools; Ying Han and Yu Sun wrote the paper.

Conflicts of Interest: The authors declare no conflict of interest. The founding sponsors had no role in the design of the study; in the collection, analyses, or interpretation of data; in the writing of the manuscript, and in the decision to publish the results.

References

1. Deng, Y.; Yin, Z.M.; Huang, J.W. Hot deformation behavior and microstructural evolution of homogenized 7050 aluminum alloy during compression at elevated temperature. *Mater. Sci. Eng. A* **2011**, 528, 1780–1786. [CrossRef]

2. Zhang, H.; Li, L.X.; Yuan, D.; Peng, D.S. Hot deformation behavior of the new Al-Mg-Si-Cu aluminum alloy during compression at elevated temperatures. *Mater. Charact.* **2007**, 58, 168–173. [CrossRef]

3. Quan, G.Z.; Mao, Y.P.; Li, G.S.; Lv, W.Q.; Wang, Y.; Zhou, J. A characterization for the dynamic recrystallization kinetics of as-extruded 7075 aluminum alloy based on true stress–strain curves. *Comp. Mater. Sci.* **2012**, 55, 65–72. [CrossRef]

4. Shakiba, M.; Parson, N.; Chen, X.-G. Modeling the effects of Cu content and deformation variables on high-temperature flow behavior of dilute Al-Fe-Si using an artificial neural network. *Materials* **2016**, 9, 536. [CrossRef]

5. Sun, Y.; Hu, L.; Ren, J. Modeling the constitutive relationship of powder metallurgy Ti-47Al-2Nb-2Cr alloy 400 during hot deformation. *JMEPG* **2015**, 24, 1313–1321. [CrossRef]

6. El Mehtedi, M.; Gabrielli, F.; Spigarelli, S. Hot workability in process modeling of a bearing steel by using combined constitutive equations and dynamic material model. *Mater. Des.* **2014**, 53, 398–404. [CrossRef]

7. Lin, Y.C.; Li, K.K.; Li, H.B.; Chen, J.; Chen, X.M.; Wen, D.X. New constitutive model for high-temperature deformation behavior of inconel 718 superalloy. *Mater. Des.* **2015**, 74, 108–118. [CrossRef]

8. Ji, G.L.; Li, Q.; Ding, K.Y.; Yang, L.; Li, L. A physically-based constitutive model for high temperature deformation of Cu-0.36Cr-0.03Zr alloy. *J. Alloy. Compd.* **2015**, 648, 397–407. [CrossRef]

9. Sarebanzadeh, M.; Mahmudi, R.; Roumina, R. Constitutive analysis and processing map of an extruded Mg-3Gd-1Zn alloy under hot shear deformation. *Mater. Sci. Eng. A* **2015**, 637, 155–161. [CrossRef]

10. Xu, Y.; Hu, L.X.; Deng, T.Q.; Ye, L. Hot deformation behavior and processing map of as-cast AZ61 magnesium alloy. *Mater. Sci. Eng. A* **2013**, 559, 528–533. [CrossRef]

11. Deng, T.Q.; Ye, L.; Sun, H.F.; Hu, L.X.; Yuan, S.J. Development of flow stress model for hot deformation of Ti-47%Al alloy. *Trans. Nonferr. Met. Soc. China* **2012**, *21*, 308–314. [CrossRef]
12. Khamei, A.A.; Dehghani, K. Effects of strain rate and temperature on hot tensile deformation of severe plastic deformed 6061 aluminum alloy. *Mater. Sci. Eng. A* **2015**, *627*, 1–9. [CrossRef]
13. Han, Y.; Qiao, G.J.; Sun, J.P.; Zou, D.N. A comparative study on constitutive relationship of as-cast 904L austenitic stainless steel during hot deformation based on Arrhenius-type and artificial neural network models. *Comput. Mater. Sci.* **2013**, *67*, 93–103. [CrossRef]
14. Li, B.; Pan, Q.L.; Yin, Z.M. Microstructural evolution and constitutive relationship of Al-Zn-Mg alloy containing small amount of Sc and Zr during hot deformation based on Arrhenius-type and artificial neural network models. *J. Alloy. Compd.* **2014**, *584*, 406–416. [CrossRef]
15. Ji, G.L.; Li, F.G.; Li, Q.H.; Li, H.Q.; Li, Z. Prediction of the hot deformation behavior for Aermet100 steel using an artificial neural network. *Comput. Mater. Sci.* **2010**, *48*, 626–632. [CrossRef]
16. Li, H.Y.; Wang, X.F.; Wei, D.D.; Hu, J.D.; Li, Y.H. A comparative study on modified Zerilli-Armstrong, Arrhenius-type and artificial neural network models to predict high-temperature deformation behavior in T24 steel. *Mater. Sci. Eng. A* **2012**, *536*, 216–222. [CrossRef]
17. Quan, G.Z.; Lv, W.Q.; Mao, Y.P.; Zhang, Y.W.; Zhou, J. Prediction of flow stress in a wide temperature range involving phase transformation for as-cast Ti-6Al-2Zr-1Mo-1V alloy by artificial neural network. *Mater. Des.* **2013**, *50*, 51–61. [CrossRef]
18. Haghdadi, N.; Zarei-Hanzaki, A.; Khalesian, A.R.; Abedi, H.R. Artificial neural network modeling to predict the hot deformation behavior of an A356 aluminum alloy. *Mater. Des.* **2013**, *49*, 386–391. [CrossRef]
19. Wu, H.; Wen, S.P.; Huang, H.; Wu, X.L.; Gao, K.Y.; Wang, W.; Nie, Z.R. Hot deformation behavior and constitutive equation of a new type Al-Zn-Mg-Er-Zr alloy during isothermal compression. *Mater. Sci. Eng. A* **2016**, *651*, 415–424. [CrossRef]
20. Zhou, P.; Ma, Q.X.; Luo, J.B. Hot deformation behavior of as-cast 30Cr2Ni4MoV steel using processing maps. *Metals* **2017**, *7*. [CrossRef]
21. Xiang, S.; Liu, D.Y.; Zhu, R.H.; Li, J.F.; Chen, Y.L.; Zhang, X.H. Hot deformation behavior and microstructure evolution of 1460 Al–Li alloy. *Trans. Nonferrous Met. Soc. China* **2015**, *25*, 3855–3864. [CrossRef]
22. Rezaei Ashtiani, H.R.; Parsa, M.H.; Bisadi, H. Constitutive equations for elevated temperature flow behavior of commercial purity aluminum. *Mater. Sci. Eng. A* **2012**, *545*, 61–67. [CrossRef]
23. Yang, Q.; Yang, D.; Zhang, Z.; Cao, L.; Wu, X.; Huang, G.; Liu, Q. Flow behavior and microstructure evolution of 6A82 aluminium alloy with high copper content during hot compression deformation at elevated temperatures. *Trans. Nonferrous Met. Soc. China* **2016**, *26*, 649–657. [CrossRef]
24. Chen, L.; Zhao, G.; Yu, J. Hot deformation behavior and constitutive modeling of homogenized 6026 aluminum alloy. *Mater. Des.* **2015**, *74*, 25–35. [CrossRef]
25. Chen, L.; Zhao, G.; Yu, J.; Zhang, W. Constitutive analysis of homogenized 7005 aluminum alloy at evaluated temperature for extrusion process. *Mater. Des.* **2015**, *66*, 129–136. [CrossRef]
26. Liu, W.; Zhao, H.; Li, D.; Zhang, Z.; Huang, G.; Liu, Q. Hot deformation behavior of AA7085 aluminum alloy during isothermal compression at elevated temperature. *Mater. Sci. Eng. A* **2014**, *596*, 176–182. [CrossRef]
27. Hollomon, J.H.; Zener, C. Problems in fracture of metals. *J. Appl. Phys.* **1946**, *82*, 82–90. [CrossRef]
28. Zhao, J.W.; Ding, H.; Zhao, W.J.; Huang, M.L.; Wei, D.B.; Jiang, Z.Y. Modelling of the hot deformation behaviour of a titanium alloy using constitutive equations and artificial neural network. *Comput. Mater. Sci.* **2014**, *92*, 47–56. [CrossRef]

metals

Article

Design and Mechanical Properties Analysis of AA5083 Ultrafine Grained Cams

Daniel Salcedo, Carmelo J. Luis *, Rodrigo Luri, Ignacio Puertas, Javier León and Juan P. Fuertes

Mechanical, Energetics and Materials Engineering Department, Public University of Navarre, Campus Arrosadía s/n, 31006 Pamplona, Navarra, Spain; daniel.salcedo@unavarra.es (D.S.); rodrigo.luri@unavarra.es (R.L.); inaki.puerta@unavarra.es (I.P.); javier.leon@unavarra.es (J.L.); juanpablo.fuertes@unavarra.es (J.P.F.)
* Correspondence: cluis.perez@unavarra.es; Tel.: +34-948-169-301

Academic Editor: Myoung-Gyu Lee
Received: 14 February 2017; Accepted: 22 March 2017; Published: 28 March 2017

Abstract: This present research work deals with the development of ultrafine grained cams obtained from previously ECAP (Equal Channel Angular Pressing)-processed material and manufactured by isothermal forging. The design and the manufacturing of the dies required for the isothermal forging of the cams are shown. Optimization techniques based on the combination of design of experiments, finite element and finite volume simulations are employed to develop the dies. A comparison is made between the mechanical properties obtained with the cams manufactured from material with no previous deformation and with those from previously SPD (Severe Plastic Deformation)-processed material. In addition, a comparative study between the experimental results and those obtained from the simulations is carried out. It has been demonstrated that it is possible to obtain ultrafine grained cams with an increase of 10.3% in the microhardness mean value as compared to that obtained from material with no previous deformation.

Keywords: isothermal forging; design; FEM; ECAP; DOE

1. Introduction

One of the most interesting aspects of dealing with cam design is the material employed in its manufacturing as this is directly related to the tribological behaviour of the mechanical component [1]. Inertia loads may cause some undesirable behaviour in the case of mechanical devices of type cam-follower, as is stated in Lee and Lee [2], and one of the most practical solutions is to reduce mass. This solution is in line with the use of nanostructured materials in order to manufacture the cam, where this leads to higher values of mechanical strength at lower weight values.

As is well-known, the most widespread Severe Plastic Deformation (SPD) process is Equal Channel Angular Pressing (ECAP). This process, which was initially proposed by V.M. Segal and his co-workers in the former Soviet Union [3], consists in compressing a material through a die with two channels with practically the same cross-section and at an angle which varies from 90° to 120° [4]. There have been a large number of research papers which deal with the ECAP process of different metallic materials, especially aluminium alloys [5–11].

Nevertheless, the number of practical applications to the manufacturing of mechanical components is still scant. Some of the most remarkable mechanical components manufactured from SPD materials in the existing bibliography are mentioned below.

In relation to the manufacturing of gears from ECAP-processed materials, research work from both Kim et al. [12] and Luis Pérez et al. [13] are noteworthy. From ECAP-processed AA6061 (twelve times using route Bc), Kim et al. [12] manufacture a micro-gear by extrusion at temperature values of 443 K and 553 K. One of their most remarkable conclusions is that the extrusion process may be carried out with no significant loss of the mechanical properties now improved by the previous ECAP process. Luis Pérez et al. [13] manufacture two different gears from ECAP-processed AA5083 by isothermal forging: one with a module of 2 and another with a module of 4. In the case of the 4 module gear, after having been forged from ECAP-processed AA5083, its microhardness value is between 8.6% and 13.6% higher than that from the gear forged with annealed AA5083.

As far as is known, most of the practical applications of mechanical components deal with the manufacturing of bolts. Choi et al. [14] propose the manufacturing of high strength bolts from ECAP-processed AA1050, where, in order to obtain this ultra-fine grained material, AA1050 is subjected to the ECAP process at room temperature up to three passages with routes A, Bc and C. Three stages are used in the bolt forming process as well as a final thread-machining stage. In this way, these authors achieve an ultimate tensile strength for the ultra-fine grained bolts which is two times higher than that of the conventional ones. In the research work from Yanagida et al. [15], the formability of four different ECAP-processed carbon steel materials is studied. This is carried out by means of the manufacturing of M1.6 micro-bolts. One of the most important conclusions is that the formability of the ECAP-processed carbon steel undergoes a low reduction compared to the non-processed material.

There have been several attempts made to improve the industrial applicability of the ECAP process (above all, in terms of continuity) in the manufacturing of bolts. For instance, in research work from Jin et al. [16] and Jin et al. [17], a spring-loaded ECAP system is proposed. In this newly proposed system, a wire rod is cut into the starting billet and this is automatically transferred to the ECAP die, where there is a sliding tool to form the die channels. A commercial multi-stage former is also installed to manufacture the bolts in four stages: extrusion, upsetting, pre-heading and finish heading. As shown in Jin et al. [17], the bolts are manufactured from ECAP-processed AA6061 and their resulting ultimate tensile strength is 7.9% higher than those conventionally manufactured.

Another attempt to develop a continuous process in order to manufacture bolts is carried out by Kim et al. [18]. These authors use a two-pass hybrid process, which consists of a series of pinch rolls, an ECAP die and a wire-drawing die. AA6061 wire is passed twice through the hybrid process in a continuous manner with routes A and C. Then, a forming process, which is composed of three stages, is used to manufacture M4.5 bolts from this ECAP-processed wire. Four cases are studied and compared: two-pass hybrid process with route A (Case 1), two-pass hybrid process with route C (Case 2), two-pass wire-drawing (Case 3) and three-pass wire-drawing (Case 4). It is found that the two-pass hybrid process with route C turns out to be the most uniform and it achieves the highest mechanical property values in terms of microhardness and ultimate tensile strength.

Other examples of mechanical components that have been manufactured from ECAP-processed materials are blades and impellers. Puertas et al. [19] study the manufacturing of Francis turbine blades from ECAP-processed AA1050. Once ECAP-processed AA1050 is obtained, blades are manufactured by isothermal forging at temperature values from 25 to 200 °C. A considerable increase in the microhardness of the previously ECAP-processed blades is achieved compared to that obtained from the annealed material, where this is approximately of some 25% for all the temperature range considered. Lee et al. [20] develop a forging process in order to manufacture an impeller with twisted blades of micro-thickness (from 400 to 600 μm). The impeller is forged from an AZ31 magnesium alloy which is previously ECAP-processed with four initial passages at 400 °C followed by five additional passages at 250 °C with route Bc. The forging process is carried out at a temperature value of 300 °C and at a strain rate value of 0.001 s^{-1}. Under these conditions, a complete forging die filling is achieved.

Another noteworthy application by Luis et al. [21] is the manufacturing of nanostructured rings from ECAP-processed AA5083. The AA5083 billets are previously ECAP-processed twice with route C and then they are subjected to two stages of isothermal forging after a specific heat treatment. The optimum forging temperature is at 250 °C as, at 200 °C, cracks appear in the forged rings. The specific heat treatment consists of an increase in temperature from room temperature to 340 °C at a heating rate of 12 °C/min.

Cisar et al. [22] manufacture a knuckle arm, which forms part of the steering system of an automobile, from ECAP-processed AZ31 magnesium alloy. AZ31 is subjected to up to four ECAP passages at a temperature of 250 °C and then, it is forged at a temperature of 150 °C. One of the aims of this research work is to evaluate the applicability of ECAP-processed materials to the manufacturing of industrial components. To this end, a comparison is made between the forging of ECAP-processed AZ31, non-ECAP-processed AZ31 and AA6061 subjected to a subsequent aging treatment at 175 °C during 8 h.

Fuertes et al. [23] study the mechanical properties of isothermally forged connecting rods which are manufactured from previously ECAP-processed AA1050 and AA5083. This research work encompasses the design stage by finite element simulations, the experimental tests and the use of metallographic techniques for the required properties to be analysed. It is observed that there is an improvement in the mechanical properties when the starting material is ECAP-processed before carrying out the isothermal forging.

In this present research work, the manufacturing of an ultrafine grained cam is dealt with, where the initial material used is ECAP-processed AA5083 which is subsequently isothermally forged. The design of the forging dies is also carried out, with the help of techniques such as finite element (FEM) and finite volume (FV) simulations, as well as design of experiments (DOE). It is shown that the ECAP-processed cams have better mechanical properties than those obtained from non-ECAP-processed AA5083.

The aim of this present study is to compare the mechanical properties of the cams manufactured from two different starting materials; one of these following conventional manufacturing processes by isothermal forging and the other from previously ECAP-processed material.

2. Optimization of the Die Design for Manufacturing the Cams

In this present section, the process of design and optimization for the dies to be used in the manufacturing of the cams is outlined. Taking into consideration the dimensions of the initial material, which are limited by the ECAP press used, a cam with a difference of 10 mm between the maximum and the minimum radius is designed. These radii refer to the distance between the cam profile and the centre of the rotation axis. In addition to this, another design restriction is that the geometry of the cavity allows different cam profile geometries such as cycloid, harmonic and sinusoidal ones (see Figure 1), among others, to be machined. Due to the fact that it is likely that buckling and cracks will occur, the design is carried out in two forging stages: preform and final cam.

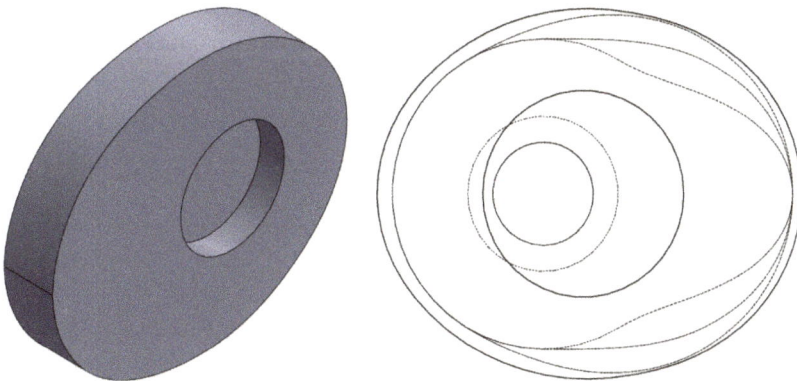

Figure 1. Basic geometry and the three types of profile.

The design process may be observed in Figure 2. Two sets of design of experiments are planned in order to optimize the geometry of both the preform and the final cam by performing FV simulations with the software Simufact. forming 12™ (MSC Software Company, Hamburg, Germany). Subsequently, different experimental tests and FEM simulations are carried out in order to optimize the final geometry for the manufactured cam that will be shown below.

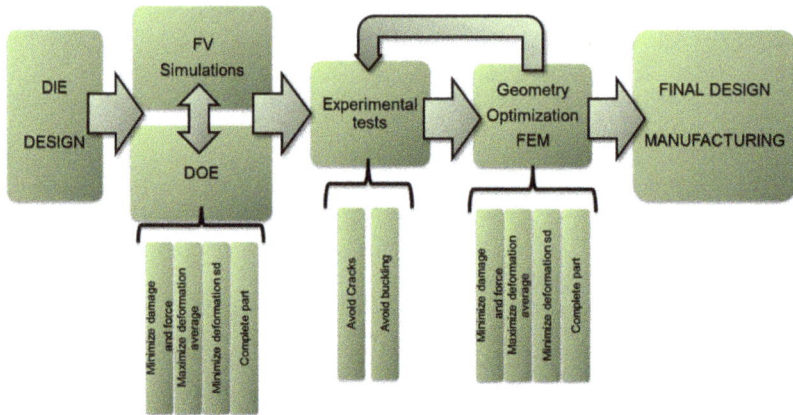

Figure 2. Optimization process for the cam design.

As previously-mentioned, the forging process is carried out in two stages. In the first of the two stages, a preform is obtained that will be subsequently forged in order to obtain the final geometry.

2.1. Cam Preform (DOE)

The simulations carried out in order to get the design employ a flow rule obtained from the isothermal compression of AA5083 at N2 state and at room temperature (Figure 3), as this is the most demanding case in terms of forging force and it is not possible to surpass a force value of 3000 kN, which is the maximum value the equipment allows. This flow rule is obtained from a series of compression tests over cylindrical billets of this material while temperature is kept constant, as is shown in Salcedo et al. [24].

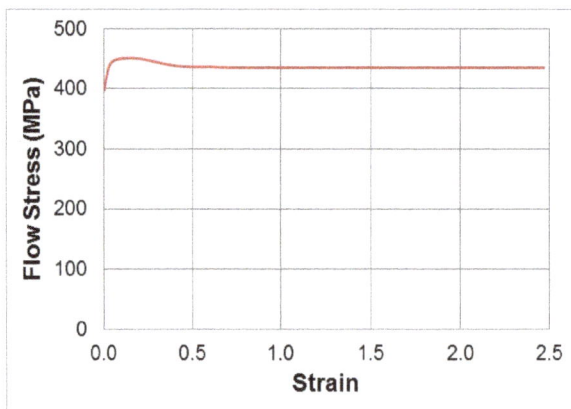

Figure 3. Flow rule fitted for AA5083 at N2 [24].

In order to perform the simulations, a cylindrical billet is taken with a starting diameter of 18 mm and an initial length of 55 mm. Rigid dies and a meshing size for the finite volume of 1 mm are employed, as is shown in Figure 4. All the simulations performed have the same characteristics so that their results are not affected. Conversion of deformation energy into heat is also taken into consideration and a friction coefficient (Shear's model) of 0.3 is considered. The element employed is a triangle with three nodes at the vertices as the outer surface of the part is meshed. The minimum size of element is 0.25 mm, whereas the cube size which models the finite volume is 1 mm. In addition, re-meshing in the simulations is carried out every 10% of the total stroke of the upper die.

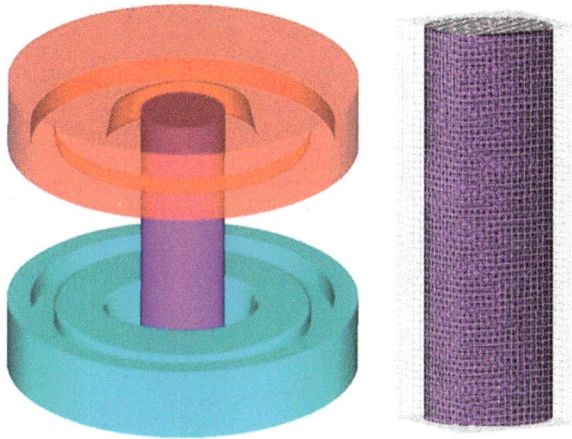

Figure 4. Simulation images from DOE Case 1.

A total of nine finite volume simulations are performed for the factorial design of experiments 2^3 with the following design factors: draft angle for all the vertical surfaces, flash thickness and fillet radii in all the die edges. DOEs are planned for forging force, damage, material efficiency, mean value of strain and standard deviation value of strain. The material efficiency parameter evaluates if the die filling is appropriate or not. In relation to the mean and to the standard deviation values for strain, these are calculated from the central section of the part in the different simulations. The results obtained are shown in Table 1.

Table 1. Design of experiments for the preform, where the experiment (exp.) in bold (Case 6) is the optimum configuration.

Exp.	Factors			Results				
	Angle (°)	Flash Thickness (mm)	Radii (mm)	Force (kN)	Damage	Strain (Mean)	Strain (sd)	Material Efficiency
1	10	2	2	230	0.13	0.17	0.13	1.40
2	30	2	2	202	0.04	0.10	0.07	1.00
3	10	4	2	214	0.12	0.15	0.11	1.20
4	30	4	2	177	0.04	0.09	0.06	1.00
5	10	2	4	214	0.08	0.14	0.12	1.10
6	**30**	**2**	**4**	**168**	**0.03**	**0.08**	**0.04**	**0.90**
7	10	4	4	163	0.05	0.11	0.07	1.00
8	30	4	4	163	0.03	0.08	0.04	0.90
9	20	3	3	182	0.06	0.11	0.07	1.00

A statistical analysis is carried out with the software Statgraphics™ (Centurion XVI, Statpoint Technologies, Inc., Warrenton, VA, USA) and it is observed from the Pareto chart and the main effects plot that there are no significant factors for the forging force. In the case of the strain homogeneity, the most significant factor is draft angle, whereas in the case of damage, the factors that have significant importance turn out to be draft angle and fillet radii, as can be observed in Figures 5–7.

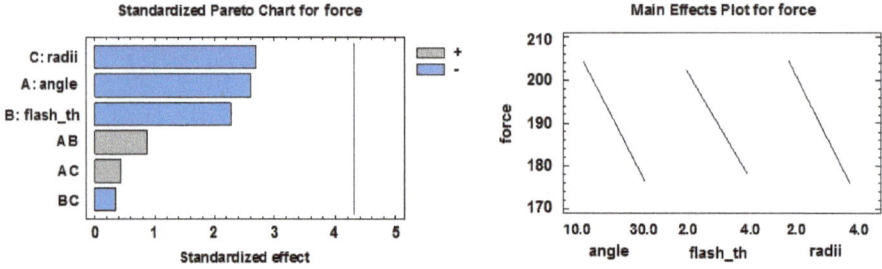

Figure 5. Design of experiments for forging force.

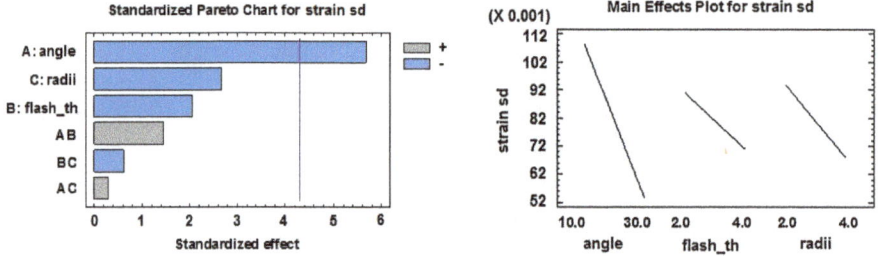

Figure 6. Design of experiments for strain homogeneity.

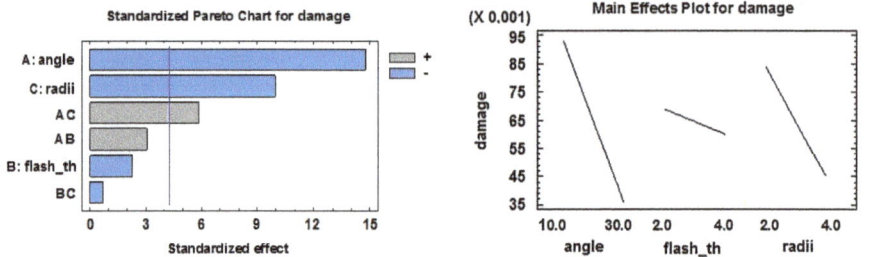

Figure 7. Design of experiments for damage.

From the data obtained, it is concluded that the optimum configuration is Case 6. The FEM results of strain, damage and die contact for the case under consideration (Case 6) are shown in Figure 8. The zone in blue colour from Figure 8c is not in contact with the die, which gives an idea about the filling of the die cavity. In the case of the preform, it is not interesting for flash to appear so that there is more material volume for the second forging stroke and the material is able to flow until the die cavity is completely filled.

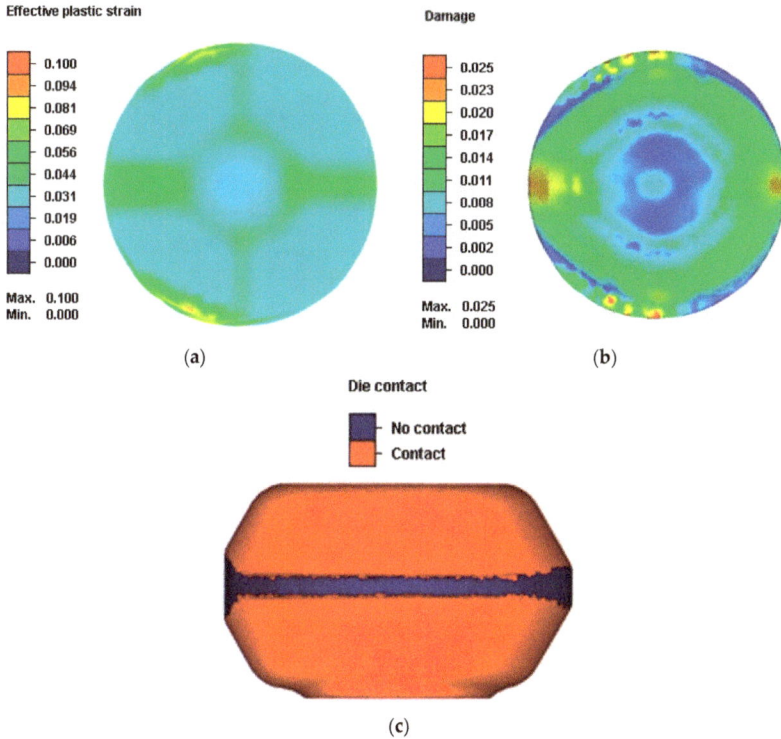

Figure 8. Strain value (a); damage value (b); and die filling (c) for the optimum case in the simulations of the cam preform.

2.2. Cam Initial Design

In the case of the second forging stage (see Figure 9), a procedure similar to that of the preform is followed, with the particularity that the geometry of the starting billet is obtained from the last increment in the calculation of the optimum case for the first forging stage, where this includes the accumulated plastic strain and the damage value. The rest of the simulation parameters coincide with those from the previous section.

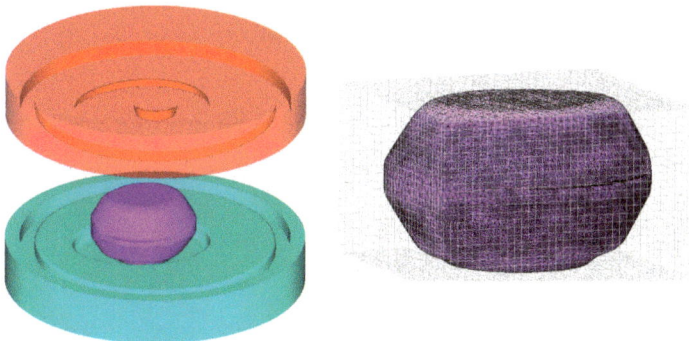

Figure 9. Simulation images from DOE Case 1.

Table 2 shows the results for the fifteen finite volume simulations carried out for the central composite design (CCD), which consists of a 2^3 factorial, six star points and one central point. The design factors and the response variables are the same as in the previous case.

Table 2. Design of experiments for the final cam, where the experiment (exp.) in bold (Case 6) is the optimum configuration.

Exp.	Factors			Results			
	Angle (°)	Flash Thickness (mm)	Radii (mm)	Force (kN)	Damage	Strain (Mean)	Strain (sd)
1	10	2	2	787	0.33	0.64	0.70
2	30	2	2	597	0.28	0.83	0.88
3	10	4	2	629	0.33	0.69	0.71
4	30	4	2	578	0.29	0.81	0.83
5	10	2	4	586	0.27	0.72	0.74
6	**30**	**2**	**4**	**874**	**0.26**	**0.89**	**0.78**
7	10	4	4	737	0.29	0.74	0.72
8	30	4	4	634	0.28	0.82	0.88
9	10	3	3	773	0.31	0.73	0.70
10	30	3	3	620	0.28	0.85	0.80
11	20	2	3	835	0.30	0.86	0.78
12	20	4	3	628	0.30	0.77	0.73
13	20	3	2	625	0.30	0.80	0.83
14	20	3	4	782	0.28	0.78	0.79
15	20	3	3	695	0.29	0.81	0.83

In a similar way to the previous case, there are no significant factors in the case of forging force, whereas in the case of damage, the most significant factors turn out to be draft angle and fillet radii. When these two factors (draft angle and fillet radii) are increased, damage decreases, as may be observed in Figures 10 and 11.

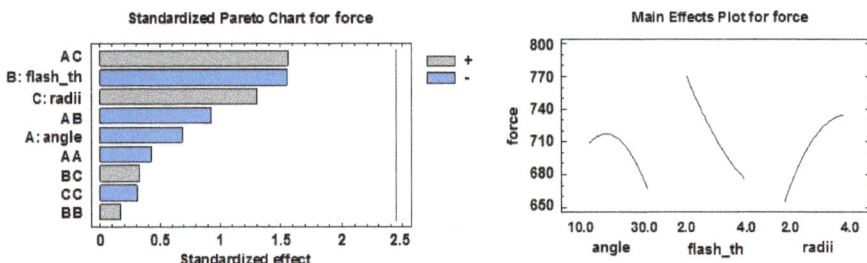

Figure 10. Pareto chart and main effects plot for forging force.

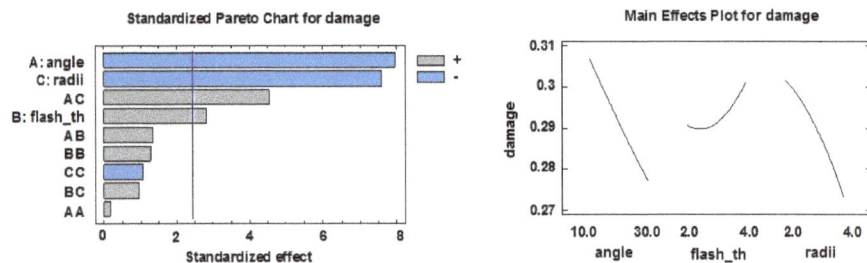

Figure 11. Pareto chart and main effects plot for damage.

From the data obtained, it is concluded that again, the optimum configuration is Case 6, which has a draft angle of 30°, a flash thickness of 2 mm and fillet radii of 4 mm. Several simulation images of this specific case are shown in Figure 12. Having a minimum damage value, a good die filling, a high strain mean value and a good strain homogeneity are all considered to be crucial. As in the case of the preform, the zone in blue colour from Figure 12c indicates that the material is not in contact with the die. It may be stated that the only zone that is not in contact corresponds to the flash zone, which predicts a correct filling of the die cavity.

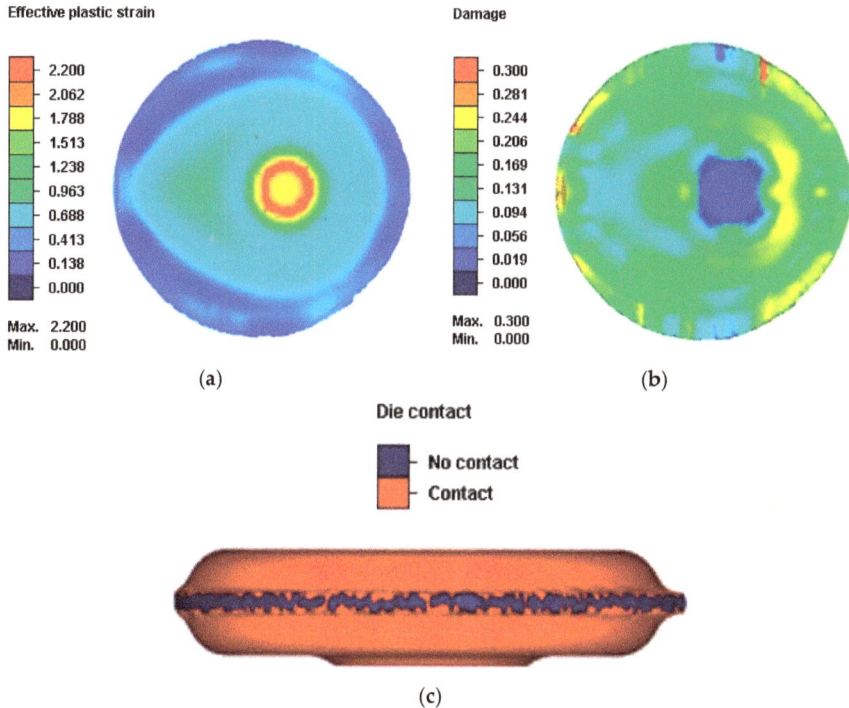

Figure 12. Strain value (**a**); damage value (**b**); and die filling (**c**) for the optimum case in the simulations of the cam initial design.

Prior to the manufacturing of the forging dies, diverse experimental tests were carried out in order to verify if buckling took place because of the slenderness of the preform as this phenomenon had occurred in some of the simulations. These tests showed the buckling of the preform with the initial dimensions as a result, as can be observed in Figure 13. Therefore, its length was reduced to a proportion of two to one in relation to its diameter and thus, the buckling, which might be caused by an inappropriate filling and internal cracks, was avoided. In conclusion, the preforms used in this research work were finally taken with a length of 36 mm and a diameter of 18 mm. Due to the smaller amount of initial material, the cam design also had to be redefined with smaller dimensions so that the preform volume was able to fill the die cavity completely.

Figure 13. Billets with buckling and final test without buckling.

From the optimum design obtained by the design of experiments (which turned out to be draft angle of 30°, flash thickness of 2 mm and fillet radii of 4 mm) and with the above-mentioned reduction in the size of the die cavity, a new stage of optimization in the geometry of the die cavity is carried out by FEM simulations. FEM simulations are now used instead of FV simulations as the former give a better precision than the latter, but at a higher computational cost. The element employed is a tetrahedron with five nodes, four at the vertices and one at the centre for the volume conservation. The minimum element size is 0.25 mm, which ensures a high number of elements at the fillet radii. A coarsening factor of 2.0 is used so that the size of the inner elements is bigger and thus, the computational cost is reduced. The friction coefficient is 0.3 with a Shear's model and the type of contact employed is type node to segment. Remeshing is carried out in case of node penetration inside the die and every 20 calculation increment, where the solver used is the so-called multifrontal sparse.

After several iterations modifying both the initial position of the preform and the dimensions of the cam, a compromise solution is achieved in order to reduce the forging force, improve strain mean values and strain homogeneity and reduce the damage exerted to the cam but, at the same time, filling the die cavity completely. The cavity of the forging dies has a geometry which allows different cam profiles to be machined with a minimum radius value of 15 mm and a maximum one of 25 mm.

2.3. Cam Final Design

Figure 14 shows the results of the final simulation for the preform, where the correct die filling, the damage value, the required forging force (700 kN) and the plastic strain value introduced in the material may be verified.

Similarly, Figure 15 shows the results of the final simulation for the cam in terms of the die filling, the damage value, the required forging force (2000 kN) and the plastic strain value introduced in the material. The damage value is calculated using the Crockroft–Latham criterion and it is concentrated at the flash. The required forging force is lower than the maximum capacity of the hydraulic press used in these experimental tests. In addition, the plastic strain is rather homogeneous and it has an approximate value of 1 at the zone where the profile is to be machined, which will lead to a higher surface hardness at this zone.

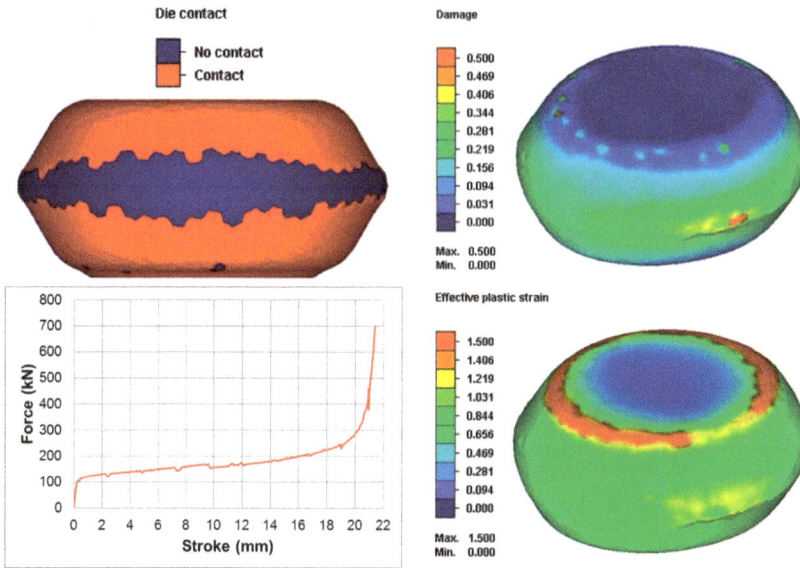

Figure 14. Cam preform.

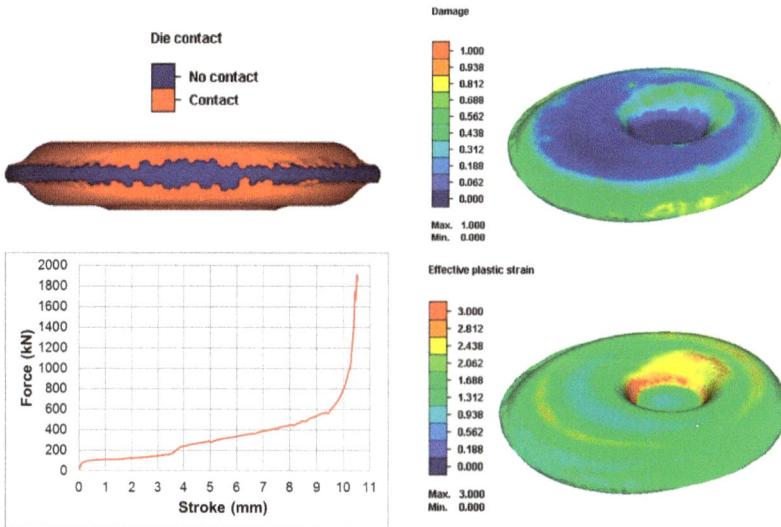

Figure 15. Second stroke for the cam.

3. Set-Up of the Experimentation

As was previously mentioned, the cams that are to be manufactured have a minimum radius of 15 mm and a maximum one of 25 mm and they are to be forged through two stages. Two different initial materials are used. The first of these (named as N0) consists of AA5083 as cast which is subjected to a heat treatment of annealing following the recommendations by ASM [25]. The second material (named as N2F) is obtained from N0 material which is subjected to equal channel angular pressing

(ECAP) twice and with route C [4]. A set of ECAP dies with a channel diameter of 20 mm and an intersection angle between the two channels of 90° is used. Moreover, the fillet radii between the channels have the same values and they are equal to 2.5 mm. Subsequently, this N2F material is subjected to a heat treatment of recovery, which is named as flash treatment and which consists in increasing the material temperature at a velocity of 12 °C/min inside a furnace until a temperature of 340 °C is reached and then the material is cooled in water.

The manufacturing of cams with both starting materials is carried out by a forging process at a temperature of 200 °C and from a preform of 18 mm in diameter and 36 mm in length. In order to manufacture them, a hydraulic press with a heating system for the set of die-holders is used, where the forging dies designed in this present research work are inserted in the previously-mentioned die-holders [13]. A Five-minute wait is required with the cam preform placed on the forging die before carrying out the first forging stage so that the temperature of the preform reaches 200 °C. In the case of the second forging stage, this is carried out just after the first one. In both cases, the descent velocity of the upper plate of the press is 1 mm/s and aqueous-based polytetrafluoroethylene (Teflon®) is used as a lubricant.

The set of forging dies designed from the optimization process described in this present study and used to manufacture the above-mentioned cams can be seen in Figure 16. They were made of F522 steel as this may be quenched with no practical change in its geometry and with final hardness values of around 60 HRC.

Figure 16. Manufactured forging dies.

The cams finally manufactured may be observed in Figure 17. In the figure on the left, the different steps (initial billet of the starting material, ECAP-processed billet and machined preform) in order to obtain the preform are shown, whereas in the figure on the right, several final parts obtained after the second forging stroke are shown.

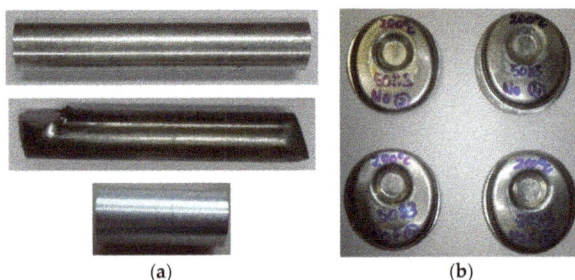

(a) (b)

Figure 17. (**a**,**b**) Preform obtained from starting material and cams manufactured after the second forging stroke.

The preparation of samples in order to obtain optical microscopy micrographs requires a metallographic saw, which does not modify their microstructure, and their subsequent mounting in a transparent resin. After encapsulating the samples, they are prepared with several abrasive papers and polishing cloths and then, they are subjected to Barker's reagent.

Once this previous process is carried out, optical microscopy micrographs at 100× and 20× are taken with the use of polarized filters. In addition, Scanning Electron Microscopy (SEM) is also utilised. For this case, the samples are electropolished with perchloric acid, ethanol and glycerine. All the SEM micrographs are taken at 1000×, 2500×, 10,000× and 25,000× with backscattered electrons.

4. Discussion of Results

In this section, the mechanical properties of the manufactured cams are discussed. In order to do this, microhardness measurements are taken from the cams and their microstructure is observed through optical microscopy and SEM, as was previously mentioned.

Regarding microhardness measurements, three different zones are selected at the outer part of the cam, as this zone undergoes the highest value of wear and thus, a higher value of hardness is required. As can be observed in Figure 18, zone A corresponds with the bottom zone of maximum radius, zone B with the upper zone of minimum radius and zone C with the lateral zone of minimum radius. Five microhardness measurements are taken in each zone in order to have more accurate data along with the calculation of the mean and the standard deviation values.

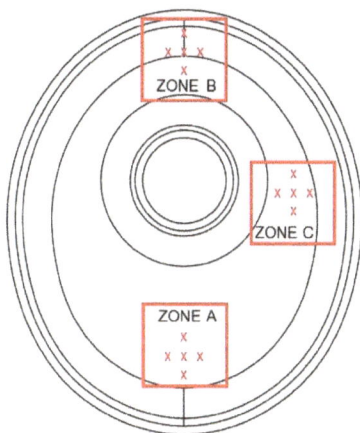

Figure 18. Zones selected for the measurements of microhardness.

A microhardness tester Mitutoyo HM-200® (Mitutoyo Corporation, Kawasaki, Japan) is used in order to carry out these measurements. Vickers microhardness tests are performed with a load value of 3 N. In this load range, the uncertainty value for the equipment is ±1% if room temperature is between 22 and 24 °C. The approach time taken is 3 s, the load maintenance time is 10 s and the withdrawal time is 3 s. With this procedure, the results obtained in the measurements are shown in Tables 3 and 4.

In both cases (N0 material and N2F material), the highest microhardness value occurs at zone C. It is observed that a better homogeneity is achieved in the case of the samples with N2F material. In addition, an increase of 10.3% in the mean value of microhardness is achieved for N2F material in relation to N0 material, as is shown in Figure 19.

Table 3. Microhardness measurements of the cam with N0 material.

AA5083 N0	Position (x, y) (mm)	Hardness Vickers (Hv)	Mean (Hv)	St Deviation (Hv)
A	(0, 0)	101.8	107.3	3.4
	(2.5, 0)	109.4		
	(−2.5, 0)	106.2		
	(0, 2.5)	108.7		
	(0, −2.5)	110.3		
B	(0, 0)	103.0	105.7	2.6
	(2.5, 0)	107.8		
	(−2.5, 0)	103.2		
	(0, 2.5)	108.8		
	(0, −2.5)	105.9		
C	(0, 0)	109.0	109.5	2.6
	(2.5, 0)	113.1		
	(−2.5, 0)	105.8		
	(0, 2.5)	110.1		
	(0, −2.5)	109.7		
Total	-	-	107.5	3.1

Table 4. Microhardness measurements of the cam with N2F material.

AA5083 N2F	Position (x, y) (mm)	Hardness Vickers (Hv)	Mean (Hv)	St Deviation (Hv)
A	(0, 0)	117.4	117.3	1.8
	(2.5, 0)	114.4		
	(−2.5, 0)	118.2		
	(0, 2.5)	119.2		
	(0, −2.5)	117.5		
B	(0, 0)	118.2	117.8	2.7
	(2.5, 0)	113.4		
	(−2.5, 0)	120.1		
	(0, 2.5)	119.7		
	(0, −2.5)	117.7		
C	(0, 0)	121.4	120.7	1.2
	(2.5, 0)	121.8		
	(−2.5, 0)	120.4		
	(0, 2.5)	118.7		
	(0, −2.5)	121.2		
Total	-	-	118.6	2.4

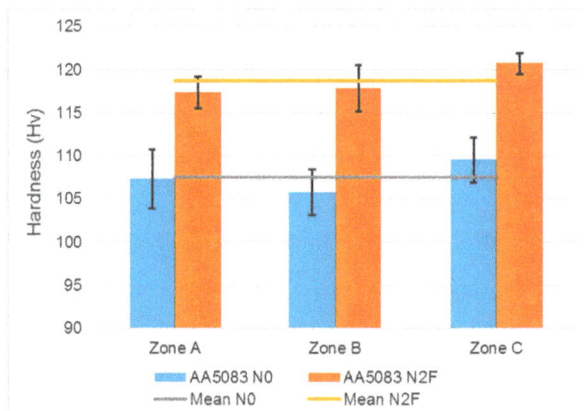

Figure 19. Comparison of microhardness values (Hv) for N0 and N2F initial materials.

Figure 20 shows optical micrographs taken from the cams with the two initial materials: N0 and N2F. In both cases, stretched grains are observed because of the forging process. In the micrographs from N2F material, it is observed that there are a higher number of deformation bands.

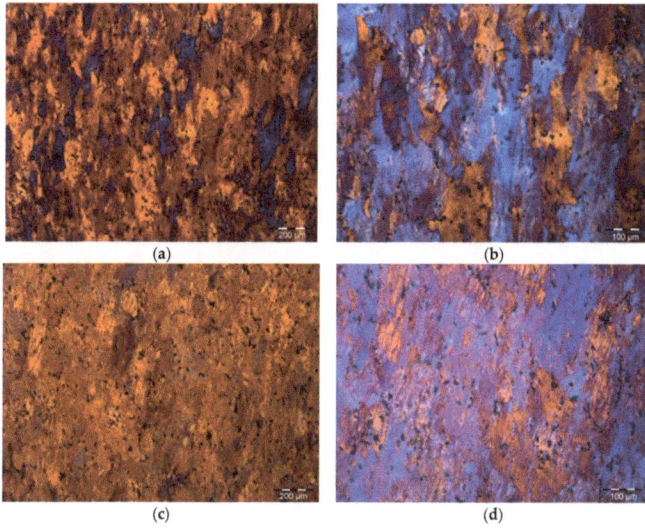

Figure 20. Optical micrographs of the cams with N0 and N2F materials: (**a**) N0 at 100×; (**b**) N0 at 200×; (**c**) N2F at 100×; and (**d**) N2F at 200×.

Figure 21 shows SEM micrographs for the cams with N0 initial material whereas Figure 22 shows those corresponding with the cams with N2F initial material. In both cases, the grain size turns out to be lower than 1 μm and even grains with a size of around 250 nm may be found.

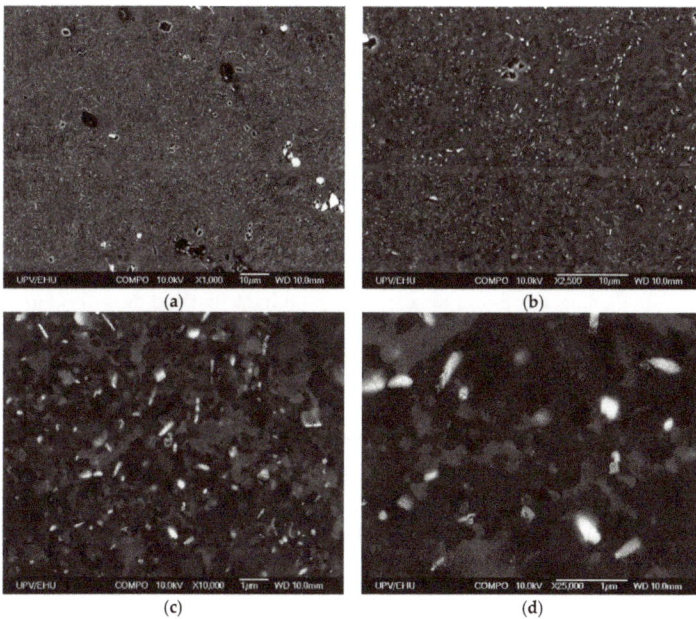

Figure 21. SEM micrographs of the cams with N0 material: (**a**) N0 at 1000×; (**b**) N0 at 2500×; (**c**) N0 at 10,000×; and (**d**) N0 at 25,000×.

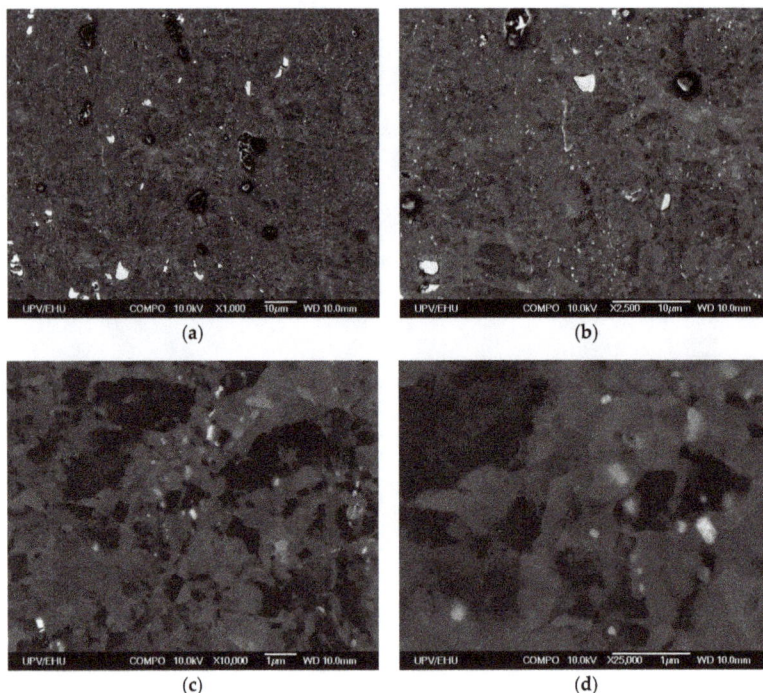

Figure 22. SEM micrographs of the cams with N2F material: (**a**) N2F at 1000×; (**b**) N2F at 2500×; (**c**) N2F at 10,000×; and (**d**) N2F at 25,000×.

Furthermore, a comparative study is made between the experimental results and the FEM simulations. For the FEM simulations, flow rules of N0 and N2F initial materials are used, where these are obtained from isothermal compression tests and by the application of a new constitutive model developed by these present authors [26].

With regard to the forging force, Figure 23 shows a comparison between the experimental forging force and that calculated from the simulations. As can be observed in Figure 23, the experimental force turns out to be a little higher than the simulated one but the approach of the latter is good. The maximum value for the forging force is lower than 3000 kN, which is the maximum capacity of the hydraulic press, and the final maximum force values for both the first and the second forging stage are smaller in the case of N2F material, which leads to a lower wear value for the forging dies. Taking this into account along with the fact that no cracks appear in any of the tests, it may be affirmed that N2F material has a better forgeability.

Figure 24 shows the plastic strain value, the damage value and the die contact obtained in the FEM simulations for N0 initial material. The plastic strain values obtained in the simulations may be compared to the microhardness values from Tables 3 and 4. For example, it may be observed in Figure 24 that, in the case of the second forging stage, the strain value is higher at zone C, which is the lateral zone (in red) close to the guide from the upper die, and this agrees quite well with the experimental microhardness measurements.

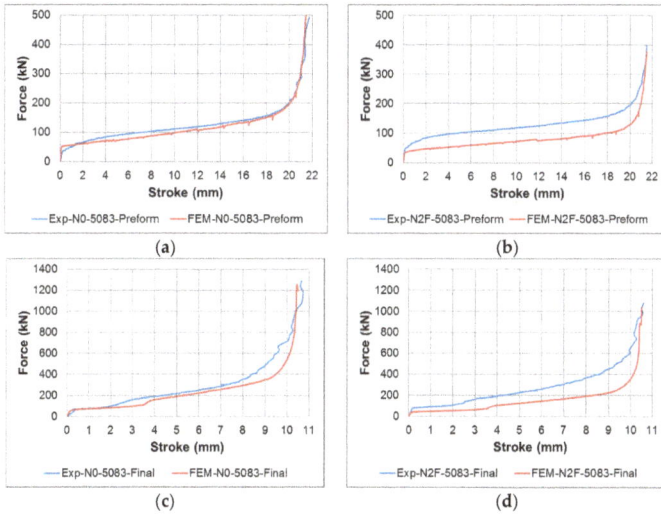

Figure 23. Forging force values obtained in the experiments: (**a**) first stage for N0; (**b**) first stage for N2F; (**c**) second stage for N0; and (**d**) second stage for N2F.

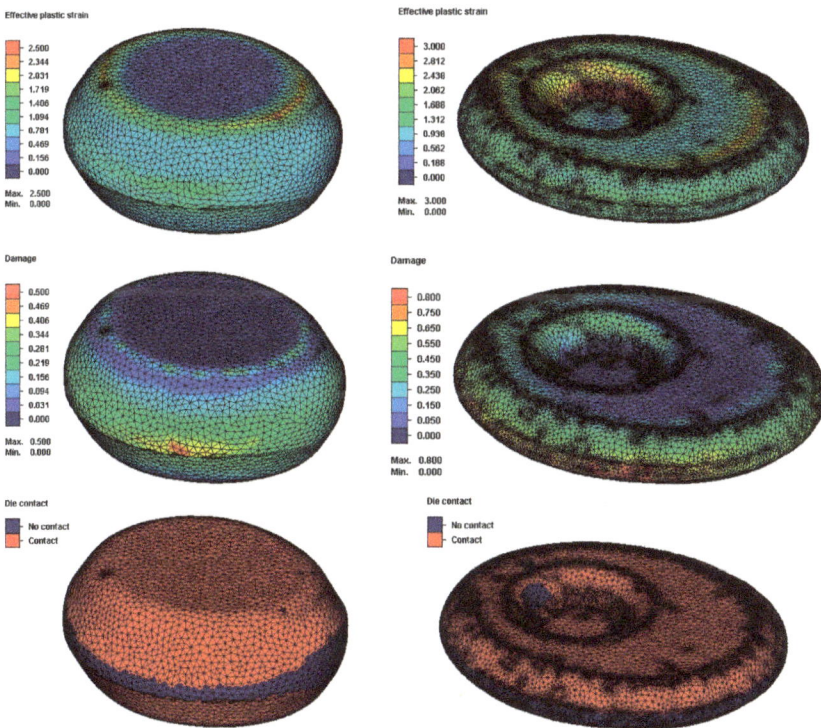

Figure 24. Plastic strain, damage and die contact values obtained in the FEM simulations for N0 starting material.

In relation to the damage value, as no cracks appear, it is assumed that the material can undergo a value of 0.8 (according to the Crockroft–Latham criterion) because this is the maximum value achieved in the simulations at the zone of the flash.

It may also be pointed out that there is a good approach between the geometry of the manufactured cams and the results obtained in the FEM simulations. In addition, Figure 24 shows the small amount of material waste in the form of flash, which leads to the conclusion that this design makes good use of the ECAP-processed material.

5. Conclusions

This research work demonstrates that it is feasible to manufacture cams from AA5083 using the ECAP processing and a subsequent isothermal forging. An increase of 10.3% in the microhardness mean value has been verified for the material previously processed by ECAP compared to that obtained from material with no previous deformation.

From the data obtained, it may be concluded that the optimum configuration for the geometry shown in this work is that given by Case 6, which has a draft angle of 30°, a flash thickness of 2 mm and fillet radii of 4 mm. This geometry has a minimum damage value compared to the different geometries shown in this study, a good die filling, a high strain mean value and a good strain homogeneity.

In both cases (N0 material and N2F material), the highest microhardness value occurs at zone C. It can be observed that a better homogeneity is achieved in the case of the samples with N2F material. A higher number of deformation bands and subgrains have been observed in the cams with N2F material, which may lead to dynamic recrystallization processes during the forging process and thus, to an improvement in forgeability.

The maximum value for the forging force is lower than 3000 kN, which is the maximum capacity of the hydraulic press employed in the experiments, and the final maximum force values for both the first and the second forging stage are smaller in the case of N2F material, which leads to a lower wear value for the forging dies. Taking this into account, along with the fact that no cracks appear in any of the tests, it may be affirmed that N2F material has a better forgeability.

Acknowledgments: The authors of this present research work acknowledge the support given by the Spanish Ministry of Economy and Competitiveness through the Research Project DPI2013-41954-P.

Author Contributions: Daniel Salcedo, Carmelo J. Luis, Rodrigo Luri, Ignacio Puertas, Javier León and Juan P. Fuertes have approximately equally contributed to most of the research tasks.

Conflicts of Interest: The authors declare no conflict of interest.

References

1. Priest, M.; Taylor, C.M. Automobile engine tribology—Approaching the surface. *Wear* **2000**, *241*, 193–203. [CrossRef]
2. Lee, S.W.; Lee, D.G. Composite hybrid valve lifter for automotive engines. *Compos. Struct.* **2005**, *71*, 26–33. [CrossRef]
3. Segal, V.M. Materials processing by simple shear. *Mater. Sci. Eng. A* **1995**, *197*, 157–164. [CrossRef]
4. Valiev, R.Z.; Langdon, T.G. Principles of equal-channel angular pressing as a processing tool for grain refinement. *Prog. Mater. Sci.* **2006**, *51*, 881–981. [CrossRef]
5. El-Danaf, E.A. Mechanical properties, microstructure and texture of single pass equal channel angular pressed 1050, 5083, 6082 and 7010 aluminum alloys with different dies. *Mater. Des.* **2011**, *32*, 3838–3853. [CrossRef]
6. El-Danaf, E.A. Mechanical properties, microstructure and micro-texture evolution for 1050AA deformed by equal channel angular pressing (ECAP) and post ECAP plane strain compression using two loading schemes. *Mater. Des.* **2012**, *34*, 793–807. [CrossRef]
7. Baig, M.; El-Danaf, E.; Mohammad, J.A. Thermo-mechanical responses of an aluminum alloy processed by equal channel angular pressing. *Mater. Des.* **2014**, *57*, 510–519. [CrossRef]

8. Zha, M.; Li, Y.-J.; Mathiesen, R.; Bjørge, R.; Roven, H.J. Microstructure, hardness evolution and thermal stability of binary Al-7Mg alloy processed by ECAP with intermediate annealing. *Trans. Nonferr. Met. Soc. China* **2014**, *24*, 2301–2306. [CrossRef]

9. Luri, R.; Fuertes, J.P.; Luis, C.J.; Salcedo, D.; Puertas, I.; León, J. Experimental modelling of critical damage obtained in Al–Mg and Al–Mn alloys for both annealed state and previously deformed by ECAP. *Mater. Des.* **2016**, *90*, 881–890. [CrossRef]

10. Fakhar, N.; Fereshteh-Saniee, F.; Mahmudi, R. High strain-rate superplasticity of fine- and ultrafine-grained AA5083 aluminum alloy at intermediate temperatures. *Mater. Des.* **2015**, *85*, 342–348. [CrossRef]

11. Fakhar, N.; Fereshteh-Saniee, F.; Mahmudi, R. Significant improvements in mechanical properties of AA5083 aluminum alloy using dual equal channel lateral extrusion. *Trans. Nonferr. Met. Soc. China* **2016**, *26*, 3081–3090. [CrossRef]

12. Kim, W.J.; Sa, Y.K.; Kim, H.K.; Yoon, U.S. Plastic forming of the equal-channel angular pressing processed 6061 aluminum alloy. *Mater. Sci. Eng. A* **2008**, *487*, 360–368. [CrossRef]

13. Luis Pérez, C.J.; Salcedo Pérez, D.; Puertas Arbizu, I. Design and mechanical property analysis of ultrafine grained gears from AA5083 previously processed by equal channel angular pressing and isothermal forging. *Mater. Des.* **2014**, *63*, 126–135. [CrossRef]

14. Choi, J.S.; Nawaz, S.; Hwang, S.K.; Lee, H.C.; Im, Y.T. Forgeability of ultra-fine grained aluminum alloy for bolt forming. *Int. J. Mech. Sci.* **2010**, *52*, 1269–1276. [CrossRef]

15. Yanagida, A.; Joko, K.; Azushima, A. Formability of steels subjected to cold ECAE process. *J. Mater. Process. Technol.* **2008**, *201*, 390–394. [CrossRef]

16. Jin, Y.G.; Baek, H.M.; Im, Y.-T.; Jeon, B.C. Continuous ECAP process design for manufacturing a microstructure-refined bolt. *Mater. Sci. Eng. A* **2011**, *530*, 462–468. [CrossRef]

17. Jin, Y.G.; Baek, H.M.; Hwang, S.K.; Im, Y.T.; Jeon, B.C. Continuous high strength aluminum bolt manufacturing by the spring-loaded ECAP system. *J. Mater. Process. Technol.* **2012**, *212*, 848–855. [CrossRef]

18. Kim, J.H.; Hwang, S.K.; Im, Y.-T.; Son, I.-H.; Bae, C.M. High-strength bolt-forming of fine-grained aluminum alloy 6061 with a continuous hybrid process. *Mater. Sci. Eng. A* **2012**, *552*, 316–322. [CrossRef]

19. Puertas, I.; Luis Pérez, C.J.; Salcedo, D.; León, J.; Fuertes, J.P.; Luri, R. Design and mechanical property analysis of AA1050 turbine blades manufactured by equal channel angular extrusion and isothermal forging. *Mater. Des.* **2013**, *52*, 774–784. [CrossRef]

20. Lee, J.H.; Kang, S.H.; Yang, D.Y. Novel forging technology of a magnesium alloy impeller with twisted blades of micro-thickness. *CIRP Ann. Manuf. Technol.* **2008**, *57*, 261–264. [CrossRef]

21. Luis, C.J.; Salcedo, D.; León, J.; Puertas, I.; Fuertes, J.P.; Luri, R. Manufacturing of nanostructured rings from previously ECAE-processed AA5083 alloy by isothermal forging. *J. Nanomater.* **2013**, *2013*, 613102. [CrossRef]

22. Cisar, L.; Yoshida, Y.; Kamado, S.; Kojima, Y.; Watanabe, F. Microstructures and tensile properties of ECAE-processed and forged AZ31 magnesium alloy. *Mater. Trans.* **2003**, *44*, 476–483. [CrossRef]

23. Fuertes, J.P.; Luis, C.J.; Luri, R.; Salcedo, D.; León, J.; Puertas, I. Design, simulation and manufacturing of a connecting rod from ultra-fine grained material and isothermal forging. *J. Manuf. Processes* **2016**, *21*, 56–68. [CrossRef]

24. Salcedo, D.; Luis, C.J.; Puertas, I.; León, J.; Luri, R.; Fuertes, J.P. FEM Modelling and Experimental Analysis of an AA5083 Turbine Blade from ECAP Processed Material. *Mater. Manuf. Processes* **2014**, *29*, 434–441. [CrossRef]

25. ASM Handbook Committee. *ASM HandBook, Heat Treating*; ASM International: Materials Park, OH, USA, 1991; Volume 4.

26. León, J.; Luis, C.J.; Fuertes, J.P.; Puertas, I.; Luri, R.; Salcedo, D. A proposal of a constitutive description for aluminium alloys in both cold and hot working. *Metals* **2016**, *6*, 244. [CrossRef]

metals

MDPI

Article

Advanced Plasticity Modeling for Ultra-Low-Cycle-Fatigue Simulation of Steel Pipe

Rongting Li [1,2,*]**, Philip Eyckens** [2]**, Daxin E** [1,*]**, Jerzy Gawad** [3]**, Maarten Van Poucke** [4]**, Steven Cooreman** [4] **and Albert Van Bael** [2,*]

[1] School of Materials Science and Engineering, Beijing Institute of Technology, Beijing 100081, China
[2] Department of Materials Engineering, KU Leuven, Kasteelpark Arenberg 44-Box 2450, 3001 Heverlee, Belgium; wen.qian@student.kuleuven.com
[3] Department of Computer Science, KU Leuven, Celestijnenlaan 200A, 3001 Leuven, Belgium; yangxi.wow@gmail.com
[4] OnderzoeksCentrum voor de Aanwending van Staal, Technologiepark 935, 9052 Zwijnaarde, Belgium; ava818ting@163.com (M.V.P.); stevencooreman@outlook.com (S.C.)
* Correspondence: lirongting818@163.com (R.L.); daxine@bit.edu.cn (D.E.); albert.vanbael@kuleuven.be (A.V.B.); Tel.: +86-10-6891-3947 (R.L. & D.E.); +32-16-37-3453 (A.V.B.)

Academic Editor: Myoung-Gyu Lee
Received: 5 March 2017; Accepted: 13 April 2017; Published: 14 April 2017

Abstract: Pipelines and piping components may be exposed to extreme loading conditions, for instance earthquakes and hurricanes. In such conditions, they undergo severe plastic strains, which may locally reach the fracture limits due to either monotonic loading or ultra-low cycle fatigue (ULCF). Aiming to investigate the failure process and strain evolution of pipes enduring ULCF, a lab-scale ULCF test on an X65 steel pipeline component is simulated with finite element models, and experimental data are used to validate various material modeling assumptions. The paper focuses on plastic material modeling and compares different models for plastic anisotropy in combination with various hardening models, including isotropic, linear kinematic and combined hardening models. Both isotropic and anisotropic assumptions for plastic yielding are considered. As pipes pose difficulty for the measurement of plastic properties in mechanical testing, we calibrate an anisotropic yield locus using advanced multi-scale simulation based on texture measurements. Moreover, the importance of the anisotropy gradient across thickness is studied in detail for this thick-walled pipeline steel. It is found that the usage of a combined hardening model is essential to accurately predict the number of the cycles until failure, as well as the strain evolution during the fatigue test. The advanced hardening modeling featuring kinematic hardening has a substantially higher impact on result accuracy compared to the yield locus assumption for the studied ULCF test. Cyclic tension-compression testing is conducted to calibrate the kinematic hardening models. Additionally, plastic anisotropy and its gradient across the thickness play a notable, yet secondary role. Based on this research, it is advised to focus on improvements in strain hardening characteristics in future developments of pipeline steel with enhanced earthquake resistance.

Keywords: plasticity modeling; kinematic hardening; plastic anisotropy; finite element simulation; ultra-low cycle fatigue; failure; strain evolution

---

1. Introduction

The finite element (FE) method has been widely used in the design and implementation of metal forming to predict the distribution of stress and strain in the formed part. Metal forming is usually a complex process in which material properties, i.e., microstructure, crystal orientation, flow stress and plastic anisotropy, will change due to the accumulation of the local plastic deformation. In general, the constitutive plasticity modelling significantly influences the attainable accuracy in FE simulations. As discussed next, the constitutive model is composed of two aspects, i.e., a hardening model and a yield locus model.

Isotropic hardening is often assumed in numerical simulations of sheet metal forming processes. It has the advantage of simplicity, but only approximates real material behavior, as for instance, it is unable to present the Bauschinger effect, which is encountered in many metals. The Bauschinger effect is characterized by a reduced yield stress upon load reversal after initial plastic loading. For instance, in uniaxial tension followed by compression, the flow stress in compression is significantly lower than the flow stress at the end of tensile prestraining. To enable modeling of the Bauschinger effect, the linear kinematic hardening rule has been proposed [1], which translates the yield locus by the kinematic stress tensor or back stress tensor with a single constant hardening modulus. Combined hardening models have also been introduced, which integrate isotropic and kinematic hardening, e.g., as illustrated in Wu [2] and Chung et al. [3]. Combined hardening models may reproduce the Bauschinger effect more accurately, at the cost of a more elaborate hardening parameter identification procedure.

The earliest yield locus for anisotropic plastic deformation was derived by Hill [4] as a straightforward extension of the isotropic von Mises yield locus [5]. The anisotropy parameters of the classical Hill′48 yield locus are usually identified from Lankford coefficients (*r*-values) measured in tensile test in three directions. More recent yield criteria require more effort in terms of the experimental calibration. Barlat's criterion Yld2000-2D [6] requires eight measurements: three directional yield stresses obtained from the uniaxial tensile tests, σ_0, σ_{45}, σ_{90}, and corresponding *r*-value, r_0, r_{45} and r_{90}. Additionally, it requires the equibiaxial yield stress σ_b and the equibiaxial *r*-value r_b, which are typically obtained from either disk compression or bulge test. To provide more accurate predictions, yield locus expressions with more coefficients have been introduced, which necessitate more material tests for calibration. For example, the Yld2004-18p [7] yield function includes 18 parameters; the criterion BBC2008 [8] needs 16 or 24 parameters; and the CPB06 [9] yield locus may contain 28 anisotropy coefficients.

As mentioned above, the coefficients in these yield criteria are typically identified based on the results of various mechanical tests that probe particular points on the yield locus. This entails much technical complexity and methodological issues in ensuring consistency among the tests, which in turn increase the cost of experiments needed for calibration of anisotropy coefficients. To circumvent this issue, one may employ virtual experiments that either replace or complement mechanical testing [6]. The virtual experiments often use multilevel polycrystal plasticity models that derive macroscopic plastic behavior from microstructural features via a homogenization procedure applied on the material response at the micro-scale level. Embedding crystal plasticity models directly in large-scale FE simulations is computationally extremely demanding. Therefore, Van Houtte et al. [10] have established the facet method using homogeneous polynomials to describe the plastic potential in either stress or strain rate space, which comprises much more coefficients than the conventional phenomenological constitutive law. These coefficients have been readily derived from a series of virtual experiments performed on the basis of, e.g., the Taylor or the ALAMEL (Advanced Lamel model) models [11–16].

Ultra-low cycle fatigue (ULCF) testing typically applies cyclic loads that induce a small amount (a few percent) of plastic deformation, which leads to failure after applying a low number of cycles, usually less than a hundred. Kanvinde and Deierlein [17] consider that ULCF is in the range of 10–20 cycles, and Xue [18] put the limit to 100 for ULCF. Despite these differences, there is a general agreement that the failure is driven by the plastic response of the material attributed to ULCF [19]. However, the characterization of the parameters driving ULCF was not found until numerous experimental research works were carried out to calibrate the material constants for various metals in the 1950s. The classical constitutive model of predicting material failure due to ULCF is the Barcelona plastic damage model initially proposed by Lubliner et al. [20], which is based on an internal variable-formulation of plasticity theory for the non-linear analysis of concrete. Martinez and Oller et al. [21] developed a new plastic-damage formulation to simulate the mechanical response and failure due to ULCF specially for steels. It is based on the Barcelona plastic model, but provides enhancements by adding a non-linear kinematic hardening law coupled with a new isotropic hardening law. Van Poucke et al. [22] performed simulations of a large-scale bending test under cyclic loading and validated it with experimental ULCF on pipes with or without defect. Different hardening models are considered (isotropic hardening, non-linear kinematic hardening and combined hardening) in combination with isotropic yield locus (von Mises). Assuming combined hardening, the buckle may be accurately predicted for the ULCF test with a buckle-initiating defect. Without such a defect, however, failure occurs near clamping in the test, which could not be accurately predicted, attributed to insufficient modelling of the clamping conditions.

As each combination of the homogenization response of polycrystalline plasticity models and hardening models will lead to a yield criterion [14], the aim of the paper is to investigate the capability of advanced plastic anisotropic yield criteria to represent the behavior of steel pipe during the ULCF forming process including the material failure and the strain evolution, on the basis of the study of Van Poucke et al. and Van Wittenberghe [22]. The advanced plastic anisotropic yield criteria adopted in this research are:

(1)   isotropic hardening model combined with von Mises, Hill and facet yield loci.
(2)   linear kinematic hardening model in combination with von Mises and Hill yield loci.
(3)   combined hardening model combined with von Mises and Hill yield loci.

The adopted anisotropic yield loci (Hill and facet) are calibrated by the ALAMEL polycrystal plasticity model. To investigate the accuracy of the above yield criteria on the prediction of cycles before ULCF failure and strain evolution during the process, finite element simulations were carried out using ABAQUS (6.13, DS SIMULIA, Paris, France) and validations were made with large-scale ULCF experimental results.

2. ULCF Testing and FE Simulation Setup

2.1. Experimental Test Setup

In the full-scale ULCF test of straight pipe (Figure 1), the pure-bending method was selected. The tube is welded to both tube holders to prevent any relative rotation and translation during the test. The test input was a cylinder displacement, and the bending moment in the bending setup was delivered by hydraulic cylinders with a capacity of 500 kN on a moment arm of 2 m, giving the setup a bending capacity of 1000 kNm [22]. The test sample was filled with water and slightly pressurized (to 0.27 MPa). During the test, the water pressure value was monitored to serve as a leak detection where a sudden pressure drop indicates a through-thickness crack. The tube holder was fixed to the tube by using a small weld to avoid a rotational and longitudinal movement of the tube inside the tube holder.

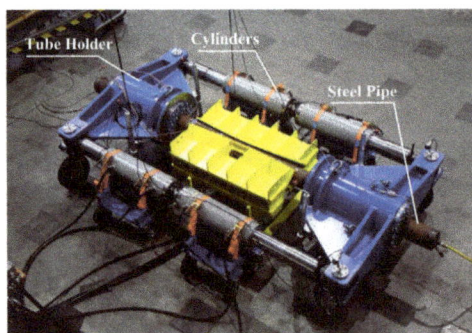

Figure 1. The experimental bending setup of ultra-low cycle fatigue (ULCF) [22].

2.2. Finite Element Simulation Setup

The simulations of a four-point bending test were used to validate the experimental ultra-low cycle fatigue (ULCF) tests on pipes as shown in Figure 2. In the Finite Element (FE) simulation, the pipe displacement was imposed in two points along the pipe length (Reference Points (RP) 3 and 4 in Figure 2) with the cyclic loading schedule in the *X*-direction as shown in Figure 3. In the simulations, the hydraulic cylinder displacement was calculated out of RP 5, 6, 7 and 8 displacements. A single cycle comprises bending in one way followed by unloading, then bending in the other way and finally unloading.

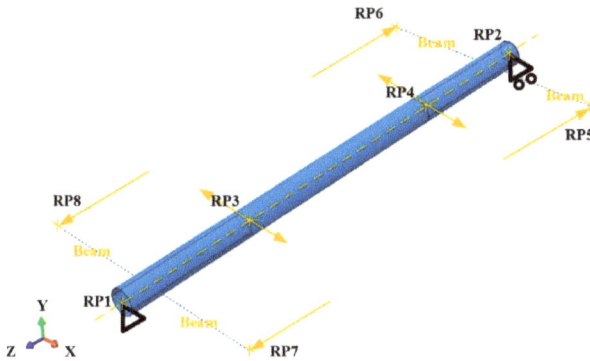

Figure 2. The geometry of the ULCF in ABAQUS FE model, with four reference points (RP) for defining the boundary conditions.

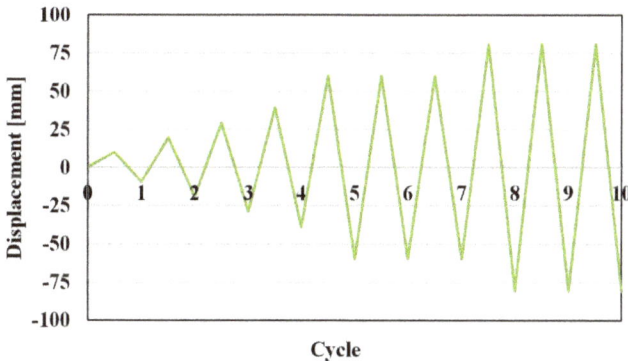

Figure 3. Schematic of the cyclic load pattern.

A kinematic coupling interaction was defined in Reference Points 3 and 4 to generate a rigid connection of the reference point with the pipe surface, similarly as the welded connections between the tube and tube holders in the experiment. One support joint (RP1) was axially restrained, and another support (RP2) was only allowed to move freely in the tube axial direction. Like in the experimental test, an internal pressure was imposed on the inner surface of the tube in the simulations, and an axial force in RP2 was defined as the result of pressure on the pipe end section.

Shell elements with four nodes and reduced integration (S4R) were selected for the deformable tube. Having an element size of approximately 10 mm by 12 mm, it resulted in 20,104 elements in total for the tube mesh.

3. Material Model Description

3.1. Pipe Material

The large-scale ULCF tests were performed using an on-shore pipeline of X65 steel. Geometric parameters describing the considered tube are illustrated in Figure 4. To initiate buckling at the pipe center, 1 mm of thickness was removed in a central zone of 200 mm length and 60 mm width at either side (0° and 180° around circumference). The initial steel sheet was formed from a hot-rolled coil by uncoiling and levelling in a cold forming process. Next, the tube was rolled by a series of non-cylindrical rolls and high frequency welded along the axial direction.

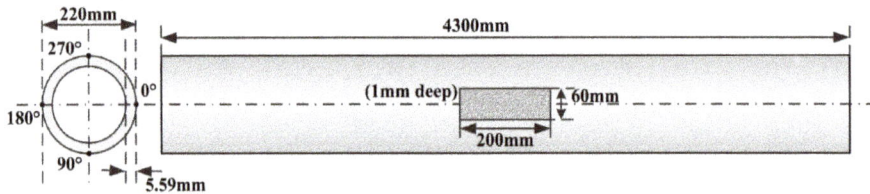

Figure 4. Geometry and dimensions of the experimental pipe.

The optical microstructure analysis of the tube was performed in multiple sections. As shown in Figure 5, the microstructure is composed of ferritic (bcc) grains with a small fraction of bainite along ferritic grain boundaries. The average grain diameter is about 5 μm.

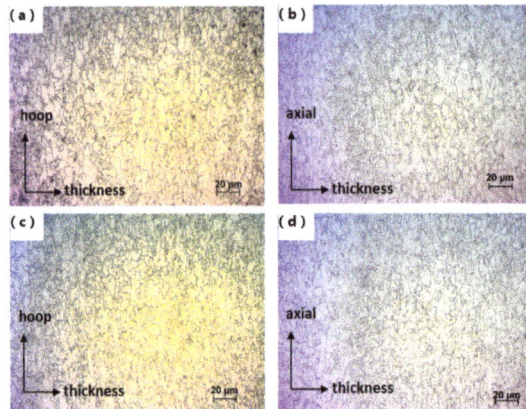

Figure 5. The optical microstructure of X65 steel: (**a,b**) in the weld zone; (**c,d**) 90° to welding.

3.2. Plastic Strain Hardening

3.2.1. Mechanical Tests

To test the material properties under different stress and strain paths, monotonic tensile tests and a cyclic tension-compression test were performed.

In the monotonic tensile tests, full-thickness tensile specimens of dog-bone geometry, oriented along tube axial direction, have been tested (without prior flattening of the sample). Three tests were performed to guarantee repeatability. The obtained stress-strain curve is presented in Figure 6 for the X65 steel pipe for the monotonic tensile test. The average yield stress and ultimate tensile strength are respectively about 595 MPa and 692 MPa.

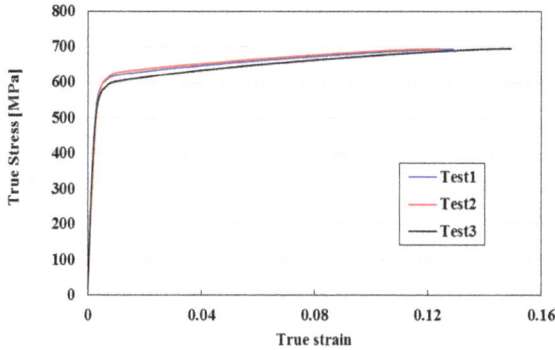

Figure 6. The applied stress-strain behavior obtained from monotonic tensile tests along the tube axial direction.

In cyclic tension-compression tests, flat samples of a 3-mm thickness were extracted with face mill and subsequently polished to avoid premature failure. Samples were oriented along the tube axial direction. The tests were performed with strain amplitudes of 1% and 1.5% (Figure 7a). It was observed in Figure 7b that the stress value decreased gradually as the number of cycles increased under all conditions.

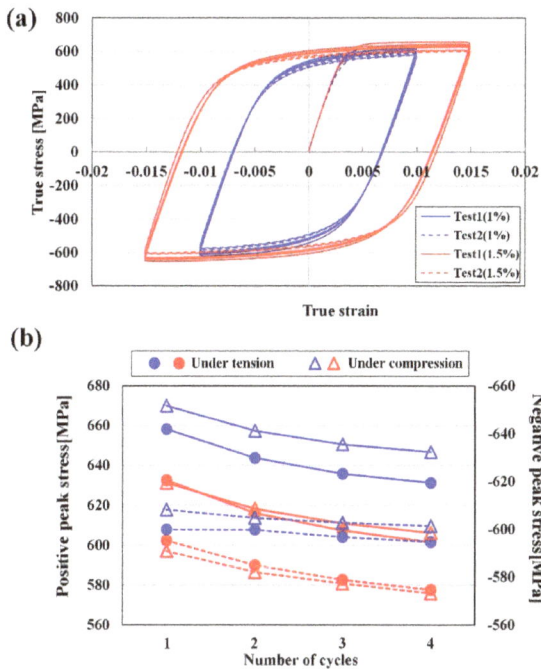

Figure 7. Results of cyclic tension-compression tests: (**a**) the true stress-strain curve under strain different amplitudes and (**b**) changes in stress at the end of tension and compression as a function of the number of cycles.

3.2.2. Strain Hardening Models

The differences between the isotropic, kinematic and combined hardening models lies in the relationship between yield function and flow stress, as described by continuum mechanics. Neglecting temperature dependence and assuming rate-insensitive behavior, the following general yield criterion holds:

$$\Phi(\sigma - \alpha) = \sigma^0\left(\bar{\varepsilon}^{pl}\right) \tag{1}$$

where the yield function Φ is a positive, homogeneous function of one degree, σ is the stress tensor, α is the back stress tensor and σ^0 represents the flow stress (yield stress), which is a function of the equivalent plastic strain $\bar{\varepsilon}^{pl}$.

(1) For the isotropic hardening, the Voce hardening law is used:

$$\sigma^0 = \sigma|_0 + Q_\infty\left(1 - e^{-b\bar{\varepsilon}^{pl}}\right) \tag{2}$$

which defines the evolution of the flow stress σ^0, as a function of the equivalent plastic strain $\bar{\varepsilon}^{pl}$. Here, $\sigma|_0$ is the yield stress; Q_∞ is the change of yield surface size at saturation; and b is the rate at which the yield surface grows as plastic straining develops. The stress-strain is calibrated from the quasi-static tension tests presented in Figure 6. For isotropic hardening, the back stress $\alpha = 0$.

(2) For the linear kinematic hardening, the evolution law (3) describes the translation of the yield surface in stress space through the backstress α with constant hardening modulus C:

$$\dot{\alpha} = C\frac{1}{\sigma^0}(\sigma - \alpha)\dot{\bar{\varepsilon}}^{pl} \tag{3}$$

The flow stress is a piece-wise linear function of equivalent strain; cf. Table 1. All parameters are calibrated by fitting to uniaxial tensile and tension-compression data.

Table 1. Material parameters corresponding to the isotropic hardening law, linear kinematic hardening law and combined hardening law.

Hardening Law		Material Parameters		
		σ^0 (MPa)	$\bar{\varepsilon}^{pl}$	Tabular Data
Isotropic hardening		—	—	$Q_\infty = 760$ MPa
				$B = 0.05$
Linear kinematic hardening		594	0	$C = 656$ MPa
		678	0.128	
Combined hardening	Nonlinear kinematic hardening component	—	—	$\sigma_0 = 600$ MPa
				$C_1 = 49{,}376$ MPa
				$\gamma_1 = 234.351$
	Isotropic hardening component	600	0	—
		444	0.044	
		512	1	

(3) The evolution law of combined hardening model consists of two components: (i) a nonlinear kinematic hardening component as:

$$\dot{\alpha} = \frac{C_1}{\gamma_1}\left(1 - e^{-\gamma_1\bar{\varepsilon}^{pl}}\right) \tag{4}$$

in which the kinematic hardening modulus C_1 and kinematic hardening exponent γ_1 are calibrated from cyclic test data and (ii) an isotropic hardening component Equation (2).

Material parameters are given in Table 1 for different hardening models.

3.3. Plastic Anisotropy

3.3.1. Texture Measurements

The non-random orientation of crystal lattice planes, or (crystallographic) texture, is the main microstructural data to calibrate the ALAMEL multi-scale model [23]. The texture at different locations along the tube hoop direction and across the tube wall thickness direction have been measured by X-ray diffraction experiments, from which orientation distribution functions (ODFs) were calculated. A discrete set of 5000 orientations, representative for the texture, was extracted from the continuous ODFs using the STAT algorithm in the MTM-FHM software [24].

Figure 8 shows $\phi_2 = 45°$ sections of the ODFs of the measured texture for the investigated steel tubes at seven different depths (from outer to inner tube surface: 0%/16.7%/25%/50%/75%/83.3%/100%) and three locations (90°/180°/270°) along the tube hoop direction. A significant texture gradient across thickness is observed. The texture intensity or texture sharpness is quantitatively represented by the texture index (TI) integrated by the square of the ODF function $f(g)$ over the entire Euler space:

$$\text{TI}(g) = \int f^2(g) dg \qquad (5)$$

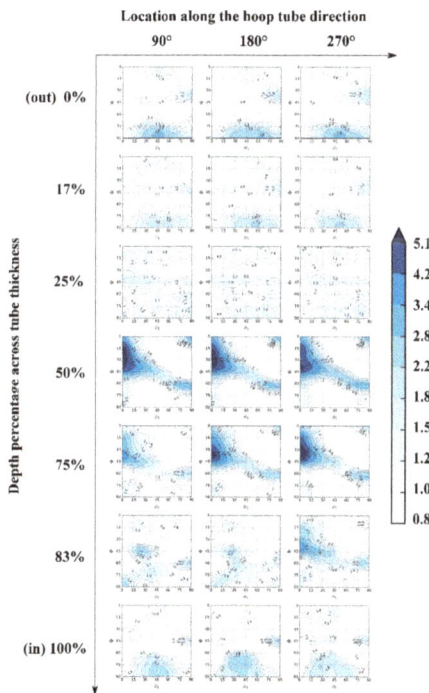

Figure 8. $\phi_2 = 45°$ orientation distribution function (ODF) sections of the initial texture of the material at different locations.

It is shown in Figure 9 that the texture at 50% depth across thickness has the sharpest texture (highest TI) and that the texture at 25% depth possesses the lowest TI value, which is close to a random texture (i.e., TI = 1).

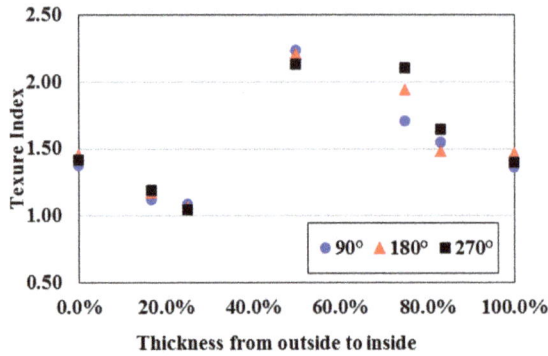

Figure 9. The texture index at different locations.

Figure 10 shows $\phi_2 = 45°$ sections of the merged texture at equal depths across the thickness of three locations along the hoop direction. Further results are merged in the ODF of the overall merged texture of both hoop and thickness direction, as presented. These merged texture data are used as the input data of the multi-scale virtual experiments for the Hill anisotropy calibration, as detailed next.

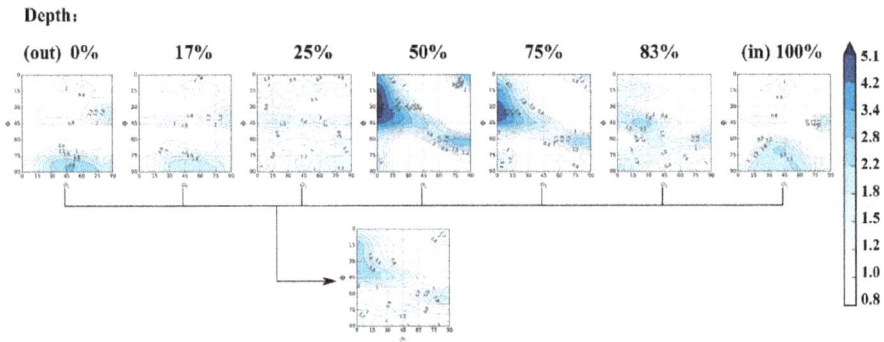

Figure 10. $\phi_2 = 45°$ ODF sections of the merged textures along the hoop (positions 90°, 180° and 270°) (first row) and the ODF section of the overall merged texture (second row).

3.3.2. Advanced Plasticity Modeling

In this study, the following isotropic and anisotropic yield loci have been employed in the FE simulations in order to check their influence on the prediction of ULCF:

(1) von Mises: the isotropic von Mises yield function calibrated by giving the value of the uniaxial yield stress as a function of uniaxial equivalent plastic strain.
(2) Hill-average: the anisotropic Hill yield function calibrated by r-values in three directions. The r-values are obtained from the ALAMEL multi-scale model on the basis of the overall average texture.

(3) Hill-gradient: the Hill yield function with different *r*-value parameters for the different integration points across tube thickness. They are calibrated by the ALAMEL model from the corresponding texture gradient.

(4) Facet: the anisotropic facet yield function with 360 parameters, calibrated to closely reproduce the behavior of ALAMEL under all conceivable deformation conditions.

The Hill-average, Hill-gradient and facet criteria are detailed in the next paragraphs.

3.3.3. The Hill-Average Yield Function

The quadratic yield function proposed by Hill [4] is a widely used anisotropic yield function for orthotropic materials. It is given by:

$$\sigma^0 = \sqrt{F(\sigma_{22} - \sigma_{33})^2 + G(\sigma_{33} - \sigma_{11})^2 + H(\sigma_{11} - \sigma_{22})^2 + 2L\sigma_{23}^2 + 2M\sigma_{31}^2 + 2N\sigma_{12}^2} \tag{6}$$

in which the stress tensor σ is expressed in a coordinate system aligned with the main directions of orthotropic symmetry, in this case being a cylindrical Cartesian coordinate system having its first reference direction aligned with the tube axial direction, second with the hoop direction and third with the radial (tube thickness) direction. If the equation holds, the material is yielding at the considered material point.

It is commonly assumed that the yield point in pure shear is independent of the plane and direction, which means that $L = M = N$. Defining the flow stress σ^0 as the uniaxial tensile stress in the uniaxial tensile test along rolled direction (RD), it follows that $G + H = 1$. Therefore, out of six, there remain three independent Hill parameters, which can be associated with the *r*-values in uniaxial tensile tests at $0°$, $45°$ and $90°$ to RD, as follows:

$$F = \frac{r_{0°}}{r_{90°}(r_{0°} + 1)} \tag{7}$$

$$H = \frac{r_{90°}}{r_{90°}(r_{0°} + 1)} \tag{8}$$

$$G = 1 - H = \frac{r_{0°} r_{90°}}{r_{90°}(r_{0°} + 1)} \tag{9}$$

$$L = M = N = \frac{(2r_{45°} + 1)(r_{0°} + r_{90°})}{2r_{90°}(r_{0°} + 1)} \tag{10}$$

For the simulations with the Hill-average yield function, only one material definition and one shell section are defined for the whole tube part. As pipes pose difficulty for the measurement of plastic properties (*r*-values) in tensile testing along directions other than the axial direction, the *r*-values in the Hill-average yield criterion are calibrated based on an advanced identification algorithm: the crystal plasticity virtual experiment framework (VEF) [13,15]. The VEF is a software suite that manages high-level stress-based virtual testing capabilities (e.g., uni- and multi-axial tensile/compressive testing, yield locus and *r*-value calculation) for multi-scale plasticity models that have been developed for strain-rate-based input of plasticity, such as ALAMEL.

The above overall merged texture (in Section 3.3.1) is used as input data of virtual experiments to calibrate the anisotropy coefficients: r_0, r_{45} and r_{90} in Table 2. From Equations (7)–(10), the Hill-average normalized σ_{11}–σ_{22} yield locus section is readily generated as shown in Figure 11 ("Hill-average by ALAMEL"). The recalibrated normalized σ_{11}–σ_{22} yield locus section by VEF based on ALAMEL is also obtained as shown in Figure 11 ("ALAMEL"). Whereas the quadratic Hill yield criterion generates an elliptic yield locus shape, the yield locus generated directly by ALAMEL deviates from elliptic shape. It can also be seen that the slope of yield locus contours coincide for the point on the horizontal and vertical axes, corresponding to calibration by r_0 and r_{90}, respectively.

Table 2. *r*-values calibrated by VEF based on the ALAMEL model.

	r Value of Overall Merged Texture		
ALAMEL	r_0	r_{45}	r_{90}
	0.858	1.250	1.070

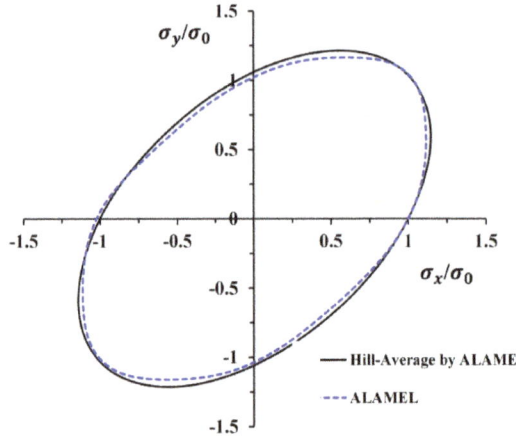

Figure 11. Two-dimensional sections of the yield surfaces calibrated by VEF.

3.3.4. The Hill-Gradient Yield Function

For the simulations with the Hill-gradient yield locus, Figure 12 illustrates the shell section definition. For the region without thickness reduction (Shell Section 1), five material definitions across thickness are imposed, while for the region with reduced thickness (Shell Section 2), four material definitions are used.

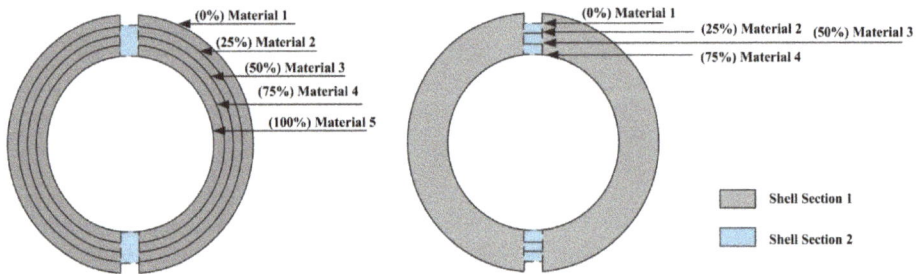

Figure 12. The diagram of the material definition in ABAQUS with the Hill-gradient yield locus.

Five sets of *r*-values corresponding to the merged texture data at five different depths across tube thickness are also calibrated by means of VEF algorithm as shown in Figure 13. Note that at 16% and 25%, the behaviour is near-isotropic (all the *r*-values close to one), which corresponds to the very weak texture found at these depths. Two-dimensional sections of the yield surfaces, as calculated by ALAMEL, are presented for these for textures at different depths in Figure 14. A significant gradient in plastic properties (*r*-values and yield surfaces) across thickness is observed.

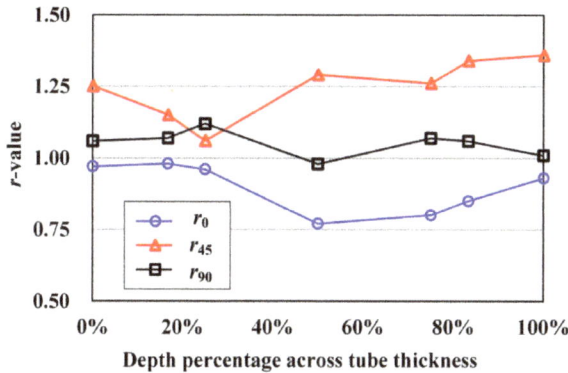

Figure 13. The *r*-values corresponding to the merged texture data at five different depths across tube thickness calibrated by VEF.

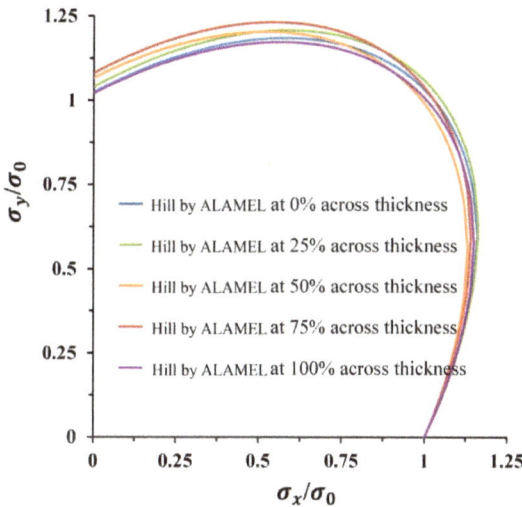

Figure 14. Two-dimensional sections of the yield surfaces calibrated by VEF at 0%, 25%, 50%, 75% and 100% across the thickness.

3.3.5. The Facet Yield Function

Van Houtte et al. [10] have proposed the facet expression, describing the plastic potential by means of a homogeneous polynomial. This mathematical formulation can be used in either strain rate space or stress space. The following is a brief introduction of the method.

Using the description of plastic potential accounting for the plastic anisotropy of textured polycrystalline materials [13], ψ is defined as the plastic potential in strain rate space scaled by the equivalent stress and be expressed as:

$$\psi(D) = \frac{\Psi(D)}{\sigma^0} \tag{11}$$

where D is the plastic strain rate tensor, $\Psi(D)$ represents the rate of plastic work per unit volume and σ^0 is the equivalent stress. Assuming plastic incompressibility and strain rate insensitivity, the deviatoric stress tensor S can be derived as:

$$S(D) = \frac{\partial \Psi(D)}{\partial D} \tag{12}$$

The facet expression defines the scaled plastic potential $\psi(d)$ in the strain rate space as:

$$\psi(d) = [G_n(d)]^{\frac{1}{n}} \tag{13}$$

in which d is a normalized plastic strain rate, n is an even natural number and $G_n(d)$ is a homogeneous polynomial of degree n, expressed as:

$$G_n(d) = \sum_{k=1}^{M} \lambda_k (S_k^m \cdot d)^n \tag{14}$$

where the deviatoric stresses S_k^m contribute λ_k weights to the plastic potential. The superscript m denotes a quantity derived from multi-scale modeling. In the facet method, $M \approx 200$ is required to reproduce the plastic surface. Combing Equations (11)–(14), a normalized deviatoric stress s can be derived as:

$$s(d) = \frac{\partial \Psi(d)}{\partial d} = \overline{\sigma}[G_n(d)]^{(\frac{1}{n}-1)} \sum_{k=1}^{M} \lambda_k (S_k^m \cdot d)^{n-1} S_k^m \tag{15}$$

The ALAMEL model was employed to obtain the stress parameters S_k^m in Formula (14). In this paper, the anisotropic yield locus by the facet method was calibrated on the basis of the ALAMEL multi-scale model. The anisotropic plastic potential of the pipe was calibrated from the average of the measured texture data. The facet normalized $\sigma_{11} - \sigma_{22}$ yield locus section, yield locus "Hill-average" and "ALAMEL" are presented together in Figure 15. It is clear that the facet yield locus approximates the multi-scale model ALAMEL much better than "Hill-average".

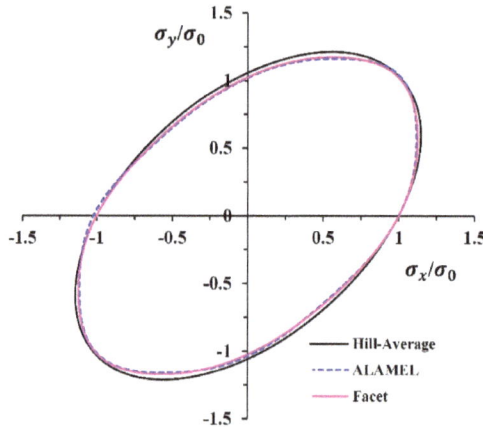

Figure 15. Two-dimensional sections of the yield surfaces respectively calibrated by the VEF/ facet method.

4. Numerical Results of ULCF

4.1. Onset of Buckling

As illustrated in Figures 4 and 16, the wall thickness of the sample was reduced by 1 mm at the pipe center over a surface of 200 mm by 60 mm on both sides of the pipe to induce buckling at the pipe center. In the experimental test, a strain gauge was attached next to the zone of reduced thickness, at 25 cm from the pipe center (Figure 16). The axial strain at this so-called reference position has been used to detect the onset of buckling in both experiment and simulation.

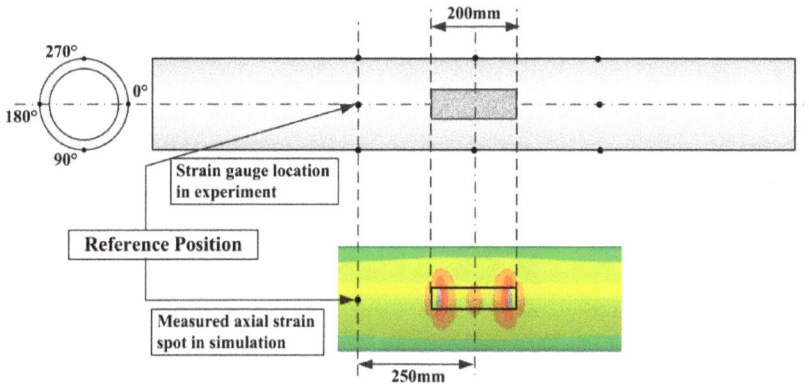

Figure 16. The reference position for tracking strain evolution in the experiment and simulation.

The adopted criterion for onset of buckling in the large-scale ULCF experiments [22] is a significant decrease of oscillation amplitude in axial strain from one half-cycle to the previous one in the reference position, as illustrated in Figure 17.

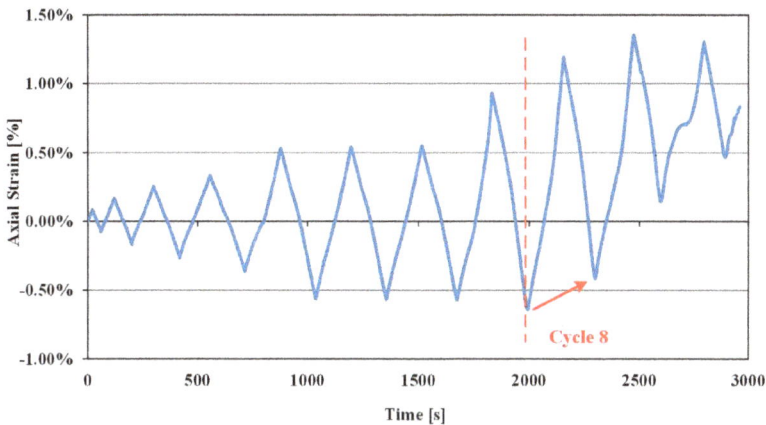

Figure 17. Illustration of the criterion for determining the onset of buckling. Experimental measurement data of axial strain in the reference position [22].

To test the influence of various plasticity modeling assumptions, simulations of the ULCF are performed using ABAQUS, and the results are compared to the stable experimental results with good repeatability in former work [22]. As the calculation with the facet method is available in explicit time integration schemes, in order to compare all of the results with different yield criteria in the same time integration scheme, both explicit and implicit time integration schemes are adopted in the FE simulations to firstly exclude the influence of time integration schemes on the results. A comparison of the number of cycles at onset of buckling is presented in Figure 18. It can be seen that for identical constitutive models, the onset of buckling is in nearly all cases predicted at the same cycle for both explicit (Figure 18a) and implicit (Figure 18b) FE integration schemes. This confirms that the physical phenomenon of ULCF buckling is captured, independent of the numerical implementation in the finite element simulation scheme. Comparing results for the various constitutive models, it is seen that the assumption of combined kinematic hardening leads to close agreement to the experiment, with only a small influence of the adopted yield function. Isotropic and linear kinematic hardening models give systematic under-prediction of the onset of buckling.

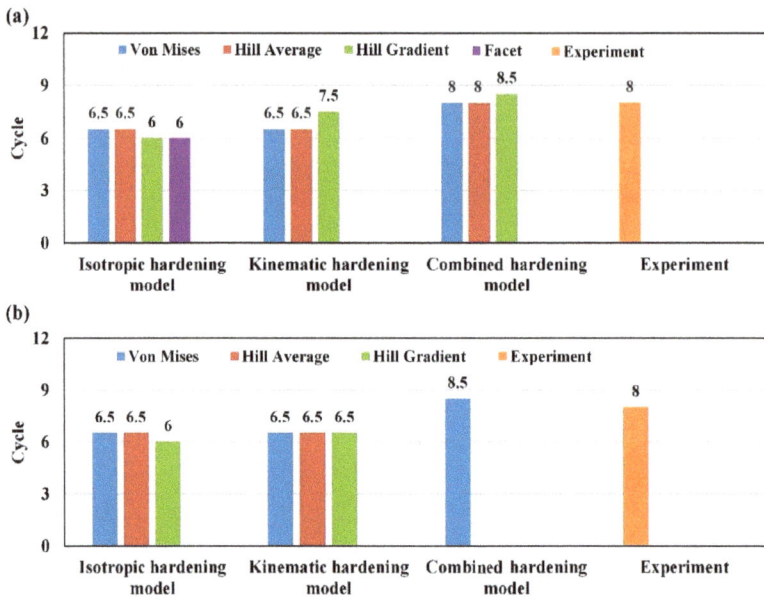

Figure 18. Comparison of prediction of the onset of buckling to the experiment for (**a**) explicit and (**b**) implicit FE simulations.

4.2. Strain Evolution

4.2.1. Influence of Anisotropic Yield Locus on Strain Evolution and Buckling Subjected to ULCF

Figures 19–23 compare the axial strain evolution of the experimental results at the reference position with those of the numerical simulations with different constitutive models. In all simulations, the occurrence of plastic deformation was captured all around the fourth cycle by the equivalent plastic strain values, resulting in the nearly same strain evolution before the fourth cycle. The differences of axial strain evolution with respect to various plastic models were discussed in terms of the plastic deformation before buckling.

Simulations with different yield criteria taking into account combined hardening all resulted in buckling at the center of the pipe, respectively at the 8.5th cycle for Hill-gradient and the eight cycle for both von Mises and Hill-average. Accurate predictions of strain evolution for these models are obtained (Figure 19).

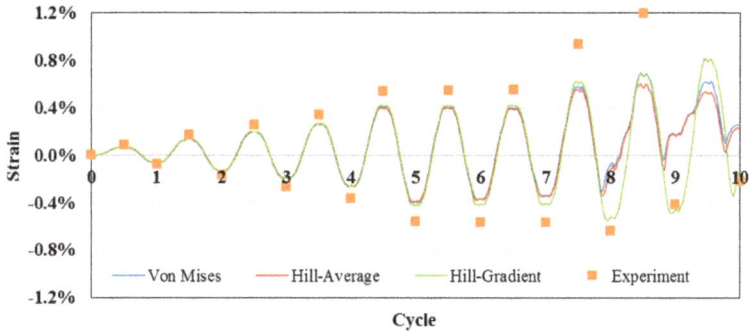

Figure 19. Axial strain evolution in explicit time integration FE simulation with the combined hardening model.

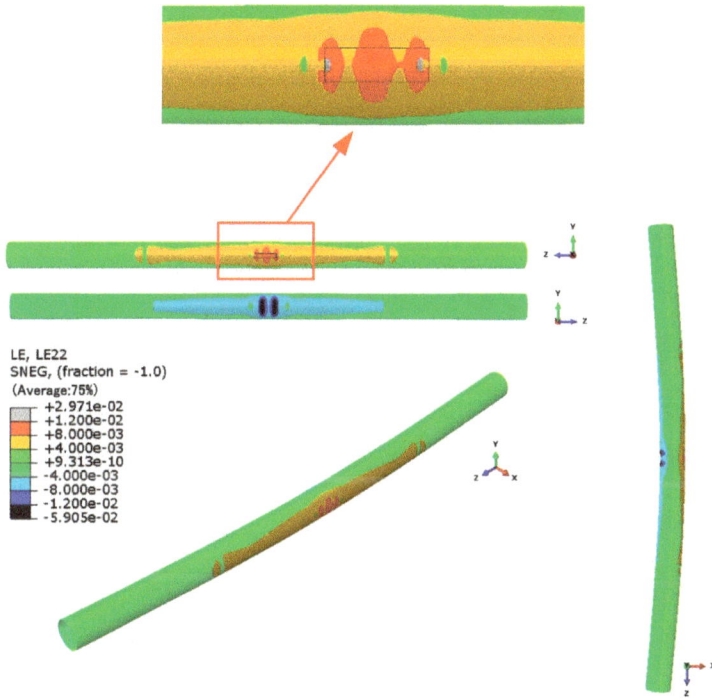

Figure 20. Axial strain and buckling overview at the 8.5th cycle with combined hardening and the Hill-gradient yield locus.

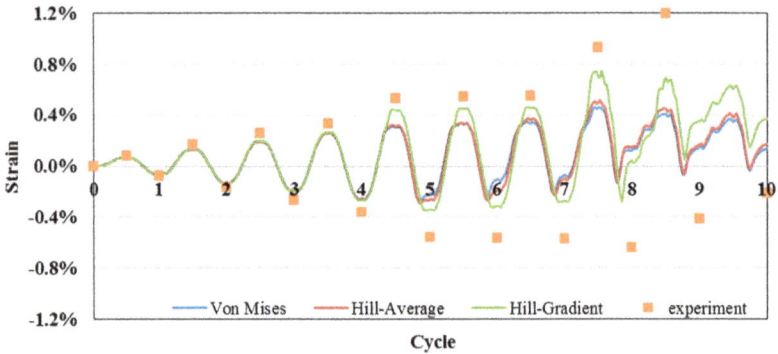

Figure 21. Axial strain evolution in explicit calculation with the linear kinematic hardening model.

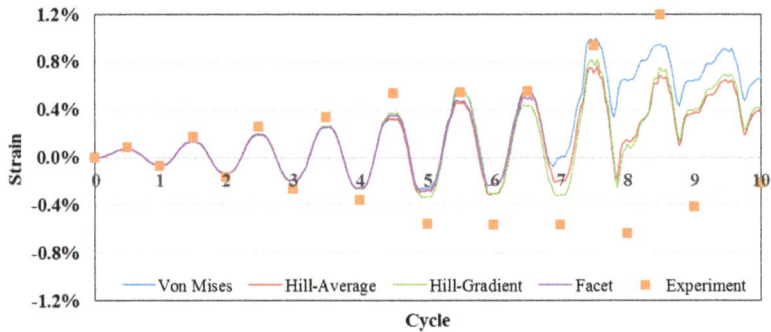

Figure 22. Axial strain evolution in explicit calculation with the isotropic hardening model.

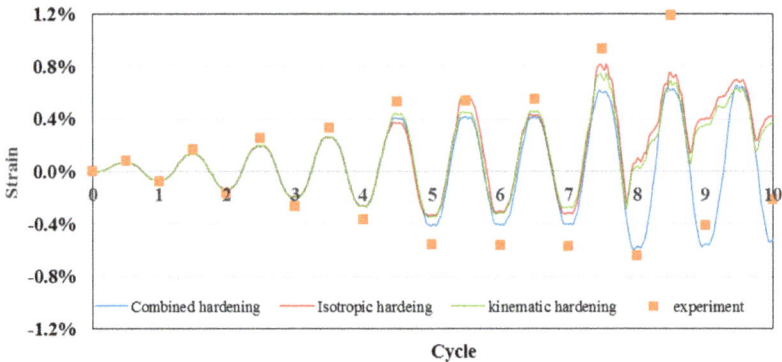

Figure 23. Axial strain evolution in explicit calculation with different hardening models.

The experimentally-measured axial strain at 25 cm from the center was as a consequence a bit higher than in the simulation, and the strain-evolution predictions during the ULCF process assuming the Hill anisotropic model with gradient properties over thickness (Hill-gradient) are largest in magnitude and closer to the experiment values.

In BS8010 [25] and Gresnight [26], relationships are proposed to calculate the critical buckling strain from the pipe outer diameter (OD) and wall thickness (WT), independently of the strength of the tube material. Applied to the current geometry (OD = 220 mm and WT = 5.56 mm), the critical buckling strain amplitude is found to be 1.2%. Figure 20 shows the axial strain distribution and buckling with the Hill-gradient yield locus taking into account combined hardening at the 8.5th cycle, which coincides with the occurrence of buckling at the center of the pipe. It can be seen that in the central region with reduced thickness, the strain locally exceeds 1.2% and reaches 2.97%, and this validates the critical strain criterion for buckling proposed in [22].

Figure 21 compares the experimental axial strain at the reference position and the predicted ones for linear kinematic hardening with von Mises, Hill-average and Hill-gradient. Furthermore, in this case, the most accurate prediction of strain evolution before onset of buckling is realized by the Hill-gradient. A similar conclusion can be drawn comparing the different yield loci in combination with the isotropic hardening law in Figure 22.

4.2.2. Influence of Strain Hardening on Strain Evolution and Buckling Subjected to ULCF with Hill-Gradient Anisotropic Yield Locus

Considering the comparatively accurate prediction of strain evolution with respect to the Hill-gradient anisotropic yield locus, the axial strain evolution under the Hill-gradient yield locus with different strain hardening models is presented again in Figure 23. The result confirms that the assumption of a combined hardening model delays the onset of buckling in ULCF, compared to isotropic and linear kinematic hardening models.

5. Summary and Conclusions

In this paper, a detailed experimental and numerical study of advanced plasticity modeling for ultra-low cycle fatigue simulation of steel pipe was presented. From the results of numerical simulations and experiments, the following conclusions are drawn:

1   In terms of simulations with different strain-hardening models, the combined hardening model enables predicting accurately the onset of buckling. Compared to isotropic and linear kinematic hardening assumptions, the prediction of buckling is delayed with two bending cycles, resulting in eight or 8.5 cycles in total for the ULCF test setup under consideration, whereas eight cycles are experimentally found.

2   Regarding the yield function assumption, it is systematically found that strain evolution predictions during the ULCF process are closest in agreement with the experiment for the Hill anisotropic model that accounts for the gradient in properties over the tube wall thickness.

3   A significant texture gradient across thickness is observed. The texture anisotropy gradient has an obvious effect on the strain evolution of ULCF simulation and to some degree also on the buckling failure process.

4   Hardening modeling matters more than modeling assumptions regarding the (an-)isotropy (i.e., the choice and calibration of the yield locus) in the simulation of pipeline steel undergoing ULCF.

Acknowledgments: Authors Philip Eyckens and Albert Van Bael gratefully acknowledge the financial support of the Industrieel Onderzoeksfonds of KU Leuven (Project C32/15/017). The author Rongting Li wishes to acknowledge the scholarship of China Scholarship Council to study at KU LEUVEN for one year, gratefully thanks Prof. Van Bael, Philip Eyckens and Jerzy Gawad for all of the guidance in the one year's study and acknowledges Maarten Van Poucke and Steven Cooreman for the former experimental work and simulation research of the ULCF test.

Author Contributions: Rongting Li performed the sample preparation, finite element simulation data analysis and manuscript writing and editing. Maarten Van Poucke and Steven Cooreman contributed to the lab ultra-low cycle fatigue (ULCF) test and the test analysis. Jerzy Gawad contributed to the computational technology support. Philip Eyckens, Daxin E and Albert Van Bael contributed to program supervision and manuscript revision.

Conflicts of Interest: The authors declare no conflicts of interest.

References

1. Prager, W. Recent developments in the mathematical theory of plasticity. *J. Appl. Phys.* **1949**, *20*, 235–241. [CrossRef]

2. Wu, H.-C. Anisotropic plasticity for sheet metals using the concept of combined isotropic-kinematic hardening. *Int. J. Plast.* **2002**, *18*, 1661–1682. [CrossRef]

3. Chung, K.; Lee, M.-G.; Kim, D.; Kim, C.; Wenner, M.L.; Barlat, F. Spring-back evaluation of automotive sheets based on isotropic-kinematic hardening laws and non-quadratic anisotropic yield functions. *Int. J. Plast.* **2005**, *21*, 861–882. [CrossRef]

4. Hill, R. A theory of the yielding and plastic flow of anisotropic metals. *Proc. R. Soc. Lond.* **1948**, *193*, 281–297. [CrossRef]

5. Mises, R.V. Mechanik der plastischen formänderung von kristallen. *ZAMM J. Appl. Math. Mech./Z. Angew. Math. Mech.* **1928**, *8*, 161–185. [CrossRef]

6. Barlat, F.; Ferreira, D.J.M.; Gracio, J.J.; Lopes, A.B.; Rauch, E.F. Plastic flow for non-monotonic loading conditions of an aluminum alloy sheet sample. *Int. J. Plast.* **2003**, *19*, 1215–1244. [CrossRef]

7. Yoon, J.W.; Barlat, F.; Dick, R.E.; Karabin, M.E. Prediction of six or eight ears in a drawn cup based on a new anisotropic yield function. *Int. J. Plast.* **2006**, *22*, 174–193. [CrossRef]

8. Comsa, D.S; Banabic, D. *Plane-Stress Yield Criterion for Highly-Anisotropic Sheet Metals*; Numisheet 2008: Interlaken, Switzerland, 2008; pp. 43–48.

9. Plunkett, B.; Cazacu, O.; Barlat, F. Orthotropic yield criteria for description of the anisotropy in tension and compression of sheet metals. *Int. J. Plast.* **2008**, *24*, 847–866. [CrossRef]

10. Van Houtte, P.; Yerra, S.K.; Van Bael, A. The facet method: A hierarchical multilevel modelling scheme for anisotropic convex plastic potentials. *Int. J. Plast.* **2009**, *25*, 332–360. [CrossRef]

11. Van Houtte, P.; Kanjarla, A.K.; Van Bael, A.; Seefeldt, M.; Delannay, L. Multiscale modelling of the plastic anisotropy and deformation texture of polycrystalline materials. *Eur. J. Mech. A/Solids* **2006**, *25*, 634–648. [CrossRef]

12. Van Houtte, P.; Gawad, J.; Eyckens, P.; Van Bael, B.; Samaey, G.; Roose, D. Multi-scale modelling of the development of heterogeneous distributions of stress, strain, deformation texture and anisotropy in sheet metal forming. *Procedia IUTAM* **2012**, *3*, 67–75. [CrossRef]

13. Gawad, J.; Van Bael, A.; Eyckens, P.; Samaey, G.; Van Houtte, P.; Roose, D. Hierarchical multi-scale modeling of texture induced plastic anisotropy in sheet forming. *Comput. Mater. Sci.* **2013**, *66*, 65–83. [CrossRef]

14. Eyckens, P.; Mulder, H.; Gawad, J.; Vegter, H.; Roose, D.; van den Boogaard, T.; Van Bael, A.; Van Houtte, P. The prediction of differential hardening behaviour of steels by multi-scale crystal plasticity modelling. *Int. J. Plast.* **2015**, *73*, 119–141. [CrossRef]

15. Gawad, J.; Banabic, D.; Van Bael, A.; Comsa, D.S.; Gologanu, M.; Eyckens, P.; Van Houtte, P.; Roose, D. An evolving plane stress yield criterion based on crystal plasticity virtual experiments. *Int. J. Plast.* **2015**, *75*, 141–169. [CrossRef]

16. Gawad, J.; Banabic, D.; Comsa, D.S.; Gologanu, M.; Van Bael, A.; Eyckens, P.; Van Houtte, P.; Roose, D. Evolving Texture-Informed Anisotropic Yield Criterion for Sheet Forming. *AIP Conf. Proc.* **2013**, *1567*, 350–355.

17. Kanvinde, A.M. Micromechanical Simulation of Earthquake-Induced Fracture in Steel Structures. Available online: http://adsabs.harvard.edu/abs/2004PhDT.......249K (accessed on 5 March 2017).

18. Xue, L. A unified expression for low cycle fatigue and extremely low cycle fatigue and its implication for monotonic loading. *Int. J. Fatigue* **2008**, *30*, 1691–1698. [CrossRef]

19. Campbell, F.C. *Elements of Metallurgy and Engineering Alloys*; ASM International: Geauga County, OH, USA, 2008.

20. Lubliner, J.; Oliver, J.; Oller, S.; Oñate, E. A plastic-damage model for concrete. *Int. J. Solids Struct.* **1989**, *25*, 299–326. [CrossRef]

21. Martinez, X.; Oller, S.; Barbu, L.G.; Barbat, A.H.; Jesus, A.M.P.D. Analysis of ultra low cycle fatigue problems with the barcelona plastic damage model and a new isotropic hardening law. *Int. J. Fatigue* **2015**, *73*, 132–142. [CrossRef]

22. European Commission. Ultra Low Cycle Fatigue of Steel under Cyclic High-Strain Loading Conditions (ULCF). Available online: http://bookshop.europa.eu/en/ultra-low-cycle-fatigue-of-steel-under-cyclic-high-strain-loading-conditions-ulcf--pbKINA27731/ (accessed on 5 March 2017).

23. Vanhoutte, P. Deformation texture prediction: From the taylor model to the advanced lamel model. *Int. J. Plast.* **2005**, *21*, 589–624. [CrossRef]
24. Van Houtte, P. *The MTM-FHM Software System Version 2*; MTM-KU: Leuven, Belgium, 1995.
25. Institution, British Standards. Code of Practice for Pipelines. Pipelines Subsea: Design, Construction and Installation. Available online: http://shop.bsigroup.com/ProductDetail/?pid=000000000000289228 (accessed on 5 March 2017).
26. Gresnigt, A.M.; Foeken, R.J.V. Local Buckling of Uoe and Seamless Steel Pipes. In Proceedings of the Eleventh International Offshore and Polar Engineering Conference, Stavanger, Norway, 17–22 June 2001.

metals

MDPI

Article

Design of U-Geometry Parameters Using Statistical Analysis Techniques in the U-Bending Process

Wiriyakorn Phanitwong *, Untika Boochakul and Sutasn Thipprakmas

Department of Tool and Materials Engineering, King Mongkut's University of Technology Thonburi, Bangkok 10140, Thailand; Untika.bo@gmail.com (U.B.); sutasn.thi@kmutt.ac.th (S.T.)
* Correspondence: wiriyakorn.wp@gmail.com; Tel.: +66-2-470-9218

Received: 16 March 2017; Accepted: 10 June 2017; Published: 26 June 2017

Abstract: The various U-geometry parameters in the U-bending process result in processing difficulties in the control of the spring-back characteristic. In this study, the effects of U-geometry parameters, including channel width, bend angle, material thickness, tool radius, as well as workpiece length, and their design, were investigated using a combination of finite element method (FEM) simulation, and statistical analysis techniques. Based on stress distribution analyses, the FEM simulation results clearly identified the different bending mechanisms and effects of U-geometry parameters on the spring-back characteristic in the U-bending process, with and without pressure pads. The statistical analyses elucidated that the bend angle and channel width have a major influence in cases with and without pressure pads, respectively. The experiments were carried out to validate the FEM simulation results. Additionally, the FEM simulation results were in agreement with the experimental results, in terms of the bending forces and bending angles.

Keywords: U-bending; pad; finite element method; spring-back; analysis of variance (ANOVA)

1. Introduction

In recent years, with more severe requirements for industrial sheet-metal parts, complicated geometrical parts with high dimensional accuracy and difficult-to-form materials are increasingly needed. To fabricate these bent parts, owing to the merits of the die-bending process, which are that several kinds of shapes and sizes can be fabricated more quickly with low production-costs, the bending process is a common and widely-applied sheet-metal forming process. Spring-back is the the principal problem faced in formation and is the main barrier faced in product quality improvement in precision-bent parts. Various geometry and process parameters, including bend angle, material thickness, tool radius, and material properties, result in processing difficulties in the control of the spring-back feature. Therefore, the bending process has also been developed using different methods to achieve precision bent parts [1–7]. Zong et al. [1] studied spring-back evaluation in hot v-bending of Ti-6Al-4V alloy sheets. Their results showed the effects of punch radius and bending temperature on spring-back characteristics. By applying a smaller punch, a greater plastic deformation was generated. The results also illustrated that the shape of the bent parts more closely matched those of the required bent-shape parts when holding times were increased. Leu [2] investigated the effects of process parameters, including, lubrication, material properties, and process geometries, on the position deviation and spring-back of high-strength steel sheets, in order to develop process design guidelines for the asymmetric V-die-bending process. Kumar et al. [3] carried out experiments to determine the spring-back and thinning effects of aluminum sheet metal during L-bending operations. As clearance increased, the spring-back characteristics and fracture propagation increased. Lim [4] observed time-dependent spring-back in advanced high-strength steel (AHSS). The results showed that Young's moduli significantly affected both the initial spring-back and the time-dependent spring-back. Thipprakmas studied the punch height effect and coined-bend technique in the V-die-bending process [5,6], as well as proposing the side coined-bead technique to prevent the bead mark on the bend radius in the V-die-bending process [7].

As an industrial part, the U-shape of the channel, beam, and frame increasingly requires the aforementioned complicated geometry; high dimension precision is also needed. As the main problem in the bending process, spring-back is the key factor that affects the quality of complicated U-shaped parts. A great deal of research has been conducted on this in the past [8–18]. Jiang and Dai [8] proposed a novel model to predict U-bend spring-back and time-dependent spring-back for a high-strength low-alloy (HSLA) steel plate. Using this novel model, prediction was done at the stage of tool design and resulted in increased accuracy of the drawn part. Thipprakmas and Boochakul [9] investigated the effects of asymmetrical U-bent shapes, including bend angle and tool radius, on the spring-back characteristics and compared them with those in the case of a symmetrical U-bent shape. Phanitwong and Thipprakmas [10] proposed the centered coined-bead technique for precise fabrication of U-bent parts. Li et al. [11] investigated delamination using U-channel formation in the bending mode. Their results indicated that increases in the forming speed somewhat decreased the tendency towards delamination, and increases in the blank holding force significantly diminished the occurrence of delamination. The predictions of spring-back in cases of the U stretch-bending process [12], and in the pre-strained U-draw/bending process [13], were investigated. Thipprakmas and Phanitwong [14] used the finite element method to analyze the bending mechanisms and spring-back/spring-go characteristics in various U-bending processes. Seong et al. [15] analyzed the core shear stress in welded deformable sandwich plates to prevent de-bonding failure during U-bending. To prevent failure during the bending phase of sandwich plates, it was recommended that the clearance be three times larger than the thickness of the sandwich plate. Marretta and Lorenzo [16] studied the influence of material property variability on spring-back and thinning in aluminum alloy sheet stamping of S-shaped U-channel processes. Tang et al. [17] proposed the mixed hardening rule, coupled with the Hill48' yielding function, to predict the spring-back of sheet U-bending. Zhang et al. [18] predicted spring-back and side wall curl after the U-bending process. The increases in the blank holding force and the friction between the sheet and die reduced the spring-back characteristics.

However, the geometry parameter design for controlling the spring-back characteristic has not yet been researched. In the present research, the effects of U-geometry parameters, including bend angle, tool radius, material thickness, channel width, and workpiece length were investigated, using finite element method (FEM) simulations. In addition, U-geometry parameter design was also examined using a combination of FEM simulations and statistical analysis techniques to determine the degree of importance of each U-geometry parameter. Experiments were carried out to validate the FEM simulation results. The FEM simulation results were in agreement with the experimental results, in terms of the bending forces and bending angles. Based on stress distribution analysis, the FEM simulation results clearly identified the different bending mechanisms and effects of the U-geometry parameters on the spring-back characteristic between U-bending U-geometry, with and without pressure pads. The statistical analysis of a central composite design (CCD) and analysis of variance (ANOVA) techniques were used out to examine the degree of importance of U-geometry parameters in relation to the spring-back feature. The statistical analysis results were able to specify the U-geometry parameters that markedly influence the spring-back characteristic, and yielded information about the degree of importance of each pro U-geometry parameter on the U-bending process. The results showed that the bend angle and channel width have a major influence on the spring-back feature, in cases with and without pressure pads, respectively.

2. Finite Element Method (FEM) Simulations and Experimental Procedures

2.1. FEM Simulation Model

The U-bending process is typically subdivided into two types of U-die-bending: with and without pressure pads [19]. Diagrams of the U-bending models investigated in the present study, with and without pressure pads, are shown in Figure 1. In the present research, to avoid the effects of pressure pads on the spring-back characteristics, a fixed pressure pad mode was set. Specifically, in the FEM simulations, a pressure pad was set by moving downward at the same velocity of the punch. This resulted in the workpiece being completely clamped by the punch and pressure pad during the bending phase. These U-bending models with three levels of channel width (W), workpiece thickness (t), bend angle (θ), punch radius (R_p), and workpiece length (WP_L), were examined. The details of the U-geometry parameter conditions investigated are listed in Table 1. The two-dimensional, implicit quasi-static finite element method, as implemented in the commercial analytical code DEFORM-2D, was used for the FEM simulations. One half of the simulation model, with a two-dimensional plane strain simulation, was applied. The solution algorithm of the Newton-Raphson iteration was utilized in this FEM model. To prevent the divergence calculation due to excessive deformation of the elements, the adaptive remeshing technique was applied by setting every three steps. In accordance with past research [9,10], the punch and die were set as rigid types, and the workpiece material was set as an elasto-plastic type. The workpiece was meshed and used approximately 4000 rectangular elements. An element size of approximately of 0.6 mm was generated overall for the workpiece, with the seven elements over the workpiece thickness. The fine element region was also generated on the bending allowance zone. In this zone, element sizes of approximately 0.2 mm, and the 27 elements over the workpiece thickness, were generated. The workpiece material was A1100-O (Japanese Industrial Standards (JIS)) aluminum and the corresponding properties were obtained from tensile testing data. On the basis of the U-die-bending process, in the present research, the phenomenon of the cyclic loading of bending-unbending was not generated, and the Bauschinger effect could be neglected. The plastic properties of the workpiece were assumed to be isotropic and were described using the von Mises yield function. In the present research, based on the plane strain model of the U-die-bending process, which only focused on the longitudinal bend direction, the effects of material anisotropy were small and could be neglected. The elasto-plastic, power-exponent, isotropic hardening model was used, and the constitutive equation was determined from the stress–strain curve, as shown in Figure 2. The strength coefficient and the strain hardening exponent values were 153.5 MPa and 0.2, respectively. In addition, a plastic strain ratio in the rolling direction (r_0) of 0.521 was observed. The other material properties are given in Table 1, where E, σ_u and ν denote, respectively, the Young modulus, ultimate tensile stress, and the Poisson's ratio. To define the accuracy of the friction coefficient, as per past research [4,8,19], the contact surface model, defined by Coulomb friction law, was applied. By comparing the FEM simulation results obtained from each friction coefficient (0.08, 0.10 and 0.12) with the experimental results, the contact interfaces between the sheet and the tool, defined with a friction coefficient value of 0.10, were in good agreement, which corresponded well with the literature [10,13,14,16]. The other process parameter conditions were designed as shown in Table 1.

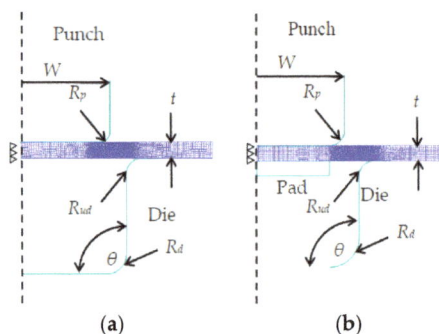

Figure 1. Finite element method (FEM) simulation model: (**a**) without pressure pad and (**b**) with pressure pad.

Table 1. FEM simulation and experimental conditions.

Simulation Model	Plane Strain Model	
Object types	Workpiece: Elasto-plastic Punch/Die/Pad: Rigid	
Workpiece material	A1100-O (JIS) Ultimate tensile stress (σ_u): 102.5 MPa Elongation (δ): 43.5% Plastic anisotropy (r_0): 0.521 Young's modulus (E): 69000 MPa Poisson's ratio (ν): 0.33	
Friction coefficient (μ) Flow curve equation Punch velocity Pressure pad velocity	0.1 $\bar{\sigma} = 153.5\bar{\varepsilon}^{0.20} + 88$ 30 mm/min 30 mm/min	
Workpiece geometries	Thickness (t)	With pad: 3 mm, 4 mm, 5 mm Without pad: 1 mm, 2 mm, 3 mm
	Length (WP_L): 120 mm, 125 mm, 130 mm	
U-die geometries	Tool radius (R_p)	With pad: 5 mm, 6 mm, 7 mm Without pad: 3 mm, 4 mm, 5 mm
	Bend angle (θ): 90°, 105°, 120° Channel width (W): 30 mm, 45 mm, 60 mm Upper die radius (R_{ud}): 5 mm Depth of U-die (D): 35–55 mm	

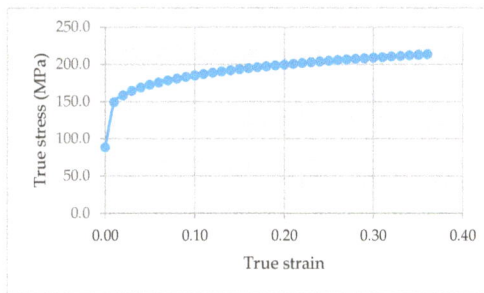

Figure 2. True stress-strain curve obtained from the tensile test.

2.2. Experimental Procedures

Validation of the FEM simulation results were done using laboratory U-bending experiments. As per the experiments in previous research [9,10], Figure 3 shows examples of the punch and die set for the U-bending experiments. A five-tonne universal tensile testing machine (Lloyd instruments Ltd., West Sussex, UK) was used as the press machine for the laboratory experiments. In the case with pressure pads, as shown in Figure 3a, the workpiece was clamped by the pad plate that was screwed on the punch. A geometrical comparison between the experimental geometry and the simulation-based geometry was performed, based on the obtained bend angles and workpiece thickness. Five samples from each bending condition were used to inspect the obtained bend angles and the workpiece thickness at the leg, bend radius, and bottom surface. The bend angle and workpiece thickness, after unloading, were measured using a profile projector (Model PJ-A3000, Mitutoyo, Kawasaki, Japan). The amount of spring-back and the workpiece thickness were calculated based on the obtained bend angles and the workpiece thickness. The average spring-back and workpiece thickness values, as well as the standard deviation (SD), were reported. The bending force was also recorded and compared with the bending force analyzed using FEM simulations.

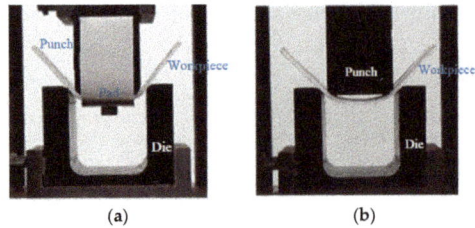

Figure 3. U-die sets for the experiments: (**a**) without pressure pad and (**b**) with pressure pad.

2.3. Statistical Analysis Techniques

To examine the degree of importance of the U-geometry parameters in relation to the spring-back characteristic, first, the central composite design technique was used to plan the experimental design for the FEM simulations. The three levels of the five parameters, channel width (*W*), workpiece thickness (*t*), bend angle (θ), punch radius (R_p), and workpiece length (WP_L), were applied, as shown in Table 2. The ANOVA technique was also applied to illustrate the degree of importance of each parameter that markedly influenced the spring-back characteristic, as depicted in the Equation (1):

$$\% \text{ Contributions}_{\text{treatment}} = [SS_{\text{treatment}} / SS_{\text{total}}] \times 100 \tag{1}$$

where $SS_{\text{treatment}}$ and SS_{total} represent the treatment sum of squares and the total sum of squares, respectively.

Table 2. U-geometry parameters and their levels.

Parameters		Parameter Levels			Units
		Low	Medium	High	
Channel width (*W*)	With pad	30	45	60	mm
	With no pad				
Bend angle (θ)	With pad	90	105	120	°
	With no pad				
Workpiece length (WP_L)	With pad	120	125	130	mm
	With no pad				
Tool radius (R_p)	With pad	5	6	7	mm
	With no pad	3	4	5	
Workpiece thickness (*t*)	With pad	3	4	5	mm
	With no pad	1	2	3	

3. Results and Discussion

3.1. Effects of U-Geometry Parameters on the Spring-Back Characteristic

3.1.1. Channel Width

Figure 4 shows a comparison of the stress distribution analyzed in the workpiece, before unloading, with respect to the various channel widths in the cases of a U-bending process with and without pressure pads. By using a pressure pad, a bending characteristic could not be formed across the punch radius during the bending phase. This resulted in bending stress being generated in the bending allowance zone (bend radius zone) and a low reversed bending stress being generated on the legs. Reversed bending stress is the stress generated by the bending phenomena, where the generated compressive and tensile stresses are formed on the opposite side of those formed in the bending allowance zone. These stress distribution analyses generally agree with those reported in the literature [14]. For the aforementioned stress characteristics, the same level of generated bending stress on the bending allowance zone and the reversed bending stress on legs could be obtained as the channel width increases; the measured length of the generated stresses are shown in Figure 4a. This manner of the stress distribution analysis corresponded well with bending theory and the literature; the changes in bending stress generated in the bending allowance zone did not depend on the channel width, but rather, depended on the bend radius and bend angle [10,20]. In contrast, in the case where no pressure pad was used, the bending characteristic was formed across the punch radius during bending phase, as reported in the literature [9,10]. Based on the bending moment theory, the more the channel width increased, the more the bending characteristics formed across the punch radius increased. This resulted in the bending stress characteristics being generated on the bottom surface, as well as the reversed bending stress characteristic generated on the legs, as shown in Figure 4b. Specifically, the generated reversed bending stress on the bottom increased, but the generated reversed bending stress on the legs decreased slightly as the channel width increased; the measured length of the generated stress is listed in Figure 4b. These stress distribution analyses corresponded well with bending theory and the literature [9,10]. After compensating these stress distribution analyses, as per previous research [5,7,9,10], the amount of spring-back was predicted, and is depicted in Figure 5. The results showed that, as the channel width increased, the amount of spring-back was reasonably consistent in the case of a pressure pad being used. The results of spring-back prediction agreed well with the aforementioned stress distribution analysis, and also corresponded well with bending theory and the literature [20]. Specifically, in the case where a pressure pad was used, the stress distributions were analyzed at the same level, with respect to channel widths, and resulted in the same level of predicted spring-back characteristics. On the other hand, with the aforementioned stress distribution analysis in the case of no pressure pad, the results showed that the spring-go characteristic increased as the channel width increased. This behavior also generally agrees with that which is reported in the literature [10].

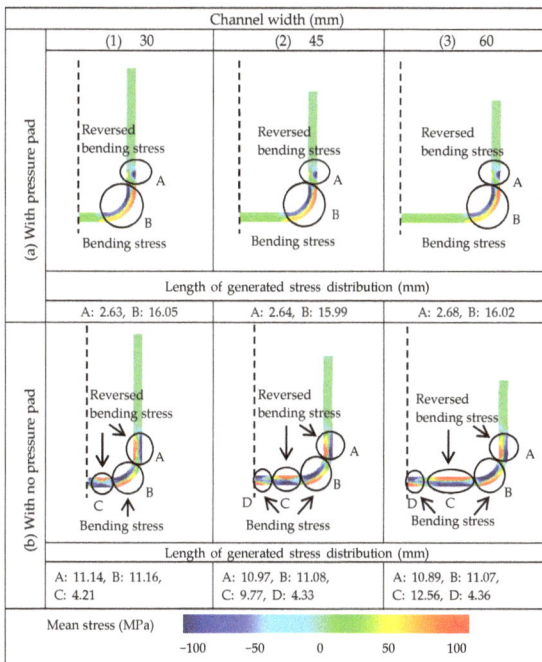

Figure 4. Comparison of the stress distribution analyzed in the workpiece, before unloading, with respect to various channel widths (θ: 90°, R_p: 7 mm, t: 3 mm, WP_L: 120 mm, R_{ud}: 5 mm).

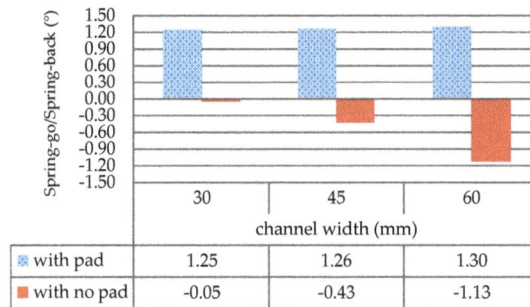

	30	45	60
with pad	1.25	1.26	1.30
with no pad	-0.05	-0.43	-1.13

Figure 5. Comparison of the predicted spring-back/spring-go in cases of U-bending, with and without pressure pads, with respect to channel widths (θ: 90°, R_p: 7 mm, t: 3 mm, WP_L: 120 mm, R_{ud}: 5 mm).

3.1.2. Workpiece Thickness

A comparison of the stress distribution analyzed in the workpiece, before unloading, with respect to workpiece thicknesses in the cases of the U-bending process, with and without pressure pads, is shown in Figure 6. In the case where a pressure pad was used, the workpiece thickness affected the bending stress and reversed the bending stress characteristics in the bending allowance and the leg zones. This could be explained by the fact that changes in workpiece thickness resulted in increases in the outer bend radius, as well as in increases in the length of bending allowance zone over the bend radius. This manner of stress distribution analysis, again, corresponded well with bending theory and the literature, where changes in bending stress generated in the bending allowance zone depended on the bend radius [10,20]. Specifically, as the workpiece thickness increased, the bending stress characteristics in the bending allowance zone increased. In addition, the reversed bending stress characteristics on the legs increased as the workpiece thickness increased. However, with the measured length of the generated stress shown in Figure 6a, the increase in bending stress characteristic in the bending allowance zone was smaller than the increase in reversed bending characteristic on the legs. After compensating the stress distribution analyses, the amount of spring-back decreased as the workpiece thickness increased (depicted in Figure 7). This behavior generally agrees with that which is reported in the literature, that spring-back characteristics decreased as workpiece thickness increased [20]. On the other hand, in the case where no pressure pad was used, the results showed that there was a decrease in bending stress characteristics, on the bottom and in the bending allowance zone as the workpiece thickness increased. As per previous research [9,10], the bending characteristic had difficulty forming across the punch radius during the bending phase as the workpiece thickness increased. This resulted in the bending stress characteristic not being formed, on the bottom surface, and the reversed bending characteristic was more easily formed on the legs in the case where a large workpiece thickness was applied. Therefore, the results showed an increase in the reversed bending stress characteristics on the legs, as measured by the length of the generated stress (Figure 6b). Again, after compensating these stress distribution analyses, the amount of spring-back decreased as the workpiece thickness increased, as shown in Figure 7. It was also observed, that the decrease in the spring-back characteristic in the case where no pressure pad was used, was larger than that in the case of a pressure pad being used.

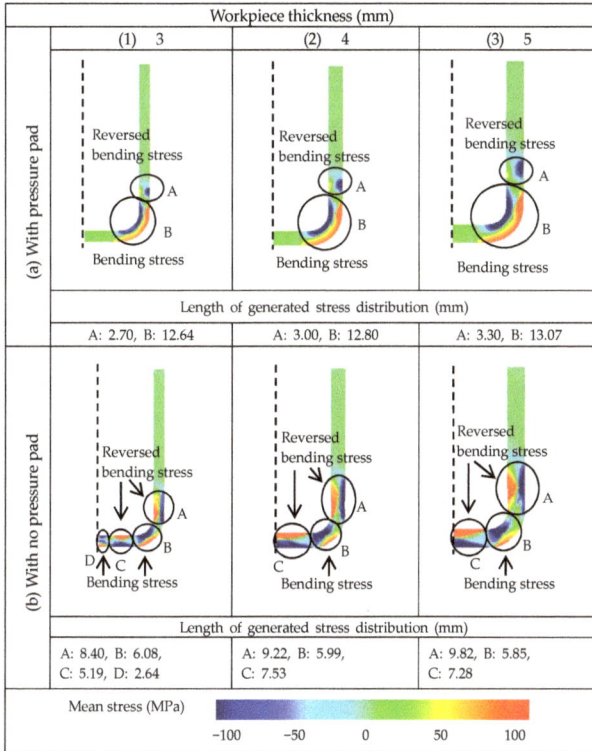

Figure 6. Comparison of the stress distribution analyzed in the workpiece, before unloading, with respect to various workpiece thicknesses (θ: 90°, R_p: 5 mm, W: 30 mm, WP_L: 120 mm, R_{ud}: 5 mm).

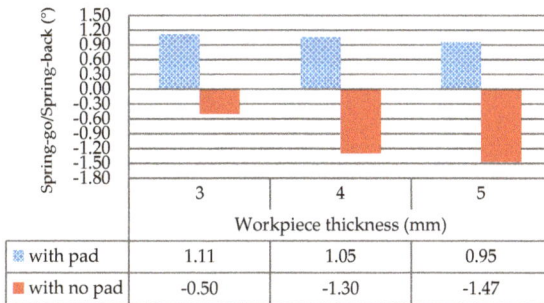

	3	4	5
with pad	1.11	1.05	0.95
with no pad	-0.50	-1.30	-1.47

Figure 7. Comparison of the predicted spring-back/spring-go between the cases of U-bending with and without pressure pads, with respect to workpiece thicknesses (θ: 90°, R_p: 5 mm, W: 30 mm, WP_L: 120 mm, R_{ud}: 5 mm).

3.1.3. Bend Angle

Figure 8 shows a comparison of the stress distribution analyzed in the workpiece, before unloading, with respect to the various bend angles in the cases of the U-bending process, with and without pressure pads. With the use of pressure pads, according to past research [14], the bending and reversed bending stress characteristics were generated in the bending allowance and in the leg zones. It could be explained that the changes in bend angle directly resulted in changes in the bending allowance zone over the bend radius. Specifically, the larger the bend angle applied, the smaller the bending allowance zone obtained. This change resulted in the bending and reversed bending stress characteristics being decreased and increased as the bend angle increased, respectively, as measured by the length of generated stress (Figure 8a); these corresponded well with bending theory and the literature [10,20]. After compensating these stress distribution analyses, the amount of spring-back decreased as the bend angle increased, as depicted in Figure 9; these results corresponded well with bending theory and the literature [20]. In the case where no pressure pad was used, according to previous research [9,10] and based on the bending moment theory, the bend angle affected the bending characteristic that formed across the punch radius during bending phase, and affected the bending stress and reversed bending stress characteristics in the bottom surface and leg zones. This resulted in reversed bending stress characteristics in the leg zone decreasing, and the bending stress characteristic on the bottom increasing, as shown in Figure 8b, which corresponded well with bending theory and the literature [9,10,14]. After compensating for these stress distribution analyses, the increase in the bending stress characteristics was larger than the increase in the reversed bending characteristics; the measured length of the generated stress is listed in Figure 8b. This resulted in the increase in the amount of spring-back as the bend angle increased, as shown in Figure 9. These amounts of spring-back, again, corresponded well with bending theory and the literature [9].

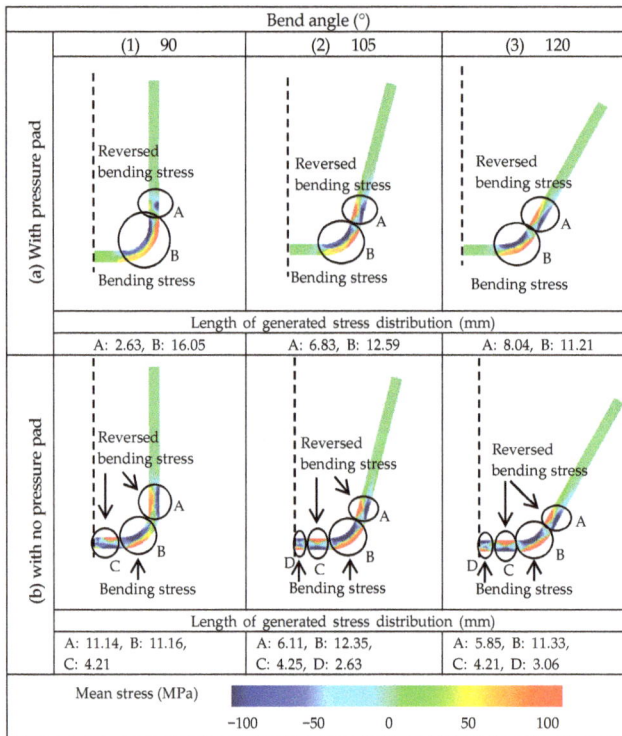

Figure 8. Comparison of the stress distribution analyzed in the workpiece, before unloading, with respect to various bend angles (R_p: 7 mm, W: 30 mm, t: 3 mm, WP_L: 120 mm, R_{ud}: 5 mm).

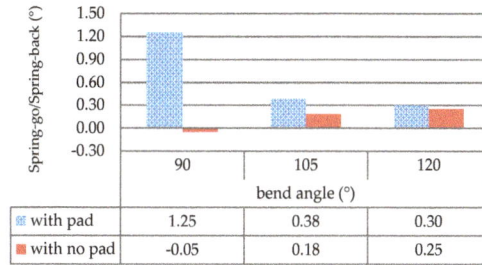

Figure 9. Comparison of the predicted spring-back/spring-go characteristics between the cases of U-bending, with and without pressure pads, with respect to bend angles (R_p: 7 mm, W: 30 mm, t: 3 mm, WP_L: 120 mm, R_{ud}: 5 mm).

3.1.4. Tool Radius

Figure 10 shows a comparison of the stress distribution analyzed in the workpiece, before unloading, with respect to various punch radii, in the cases of the U-bending process, with and without pressure pads. With the use of a pressure pad, as per past research [14,20], the bending stress characteristic was generated over the tool radius and a small reversed bending stress was generated on the legs, as shown in Figure 10a. The results showed that, as the punch radius increased, the bending stress characteristic increased; however, the reversed bending stress remained somewhat constant. These results corresponded well with bending theory and the literature, in that the tool radius directly affected the bending allowance zone and was increased as the tool radius increased [20]. After compensating these stress distribution analyses, the amount of spring-back increased as the tool radius increased, as shown in Figure 11. These results corresponded well with bending theory and the literature [20]. In the case where no pressure pad was used, the tool radius affected the bending stress and the reversed bending stress characteristics in the bottom surface and leg zones, which corresponded well with bending theory and the literature [9,10,14]. Specifically, the tool radius, not only affected the bending allowance zone, but it also affected the bending characteristics formed across the punch radius during the bending phase, and resulted in the change in bending stress and the reversed bending stress characteristics in the bottom surface and leg zones. It was observed that, as the tool radius increased, the increase in the bending stress characteristic was larger than the increase in the reversed bending characteristic; the measured length of the generated stress is listed in Figure 10b. This resulted in the increase in the amount of spring-back as the tool radius increased, as shown in Figure 11. These stress distribution analyses, and the amount of spring-back, corresponded well with bending theory and the literature [9,10].

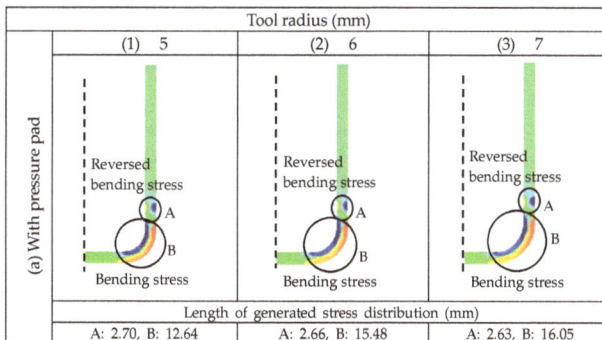

Figure 10. *Cont.*

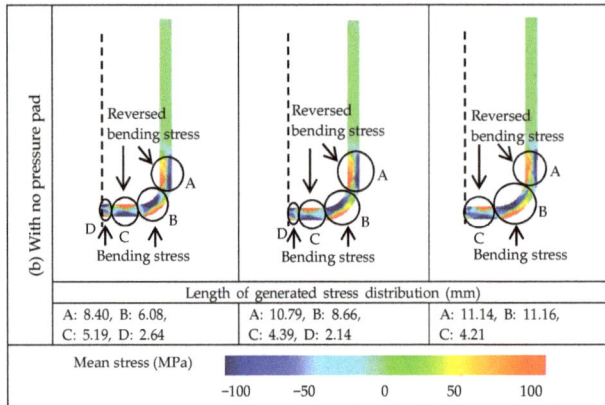

Figure 10. Comparison of the predicted spring-back/spring-go characteristics between the cases of U-bending, with and without pressure pads, with respect to tool radius (θ: 90°, W: 30 mm, t: 3 mm, WP_L: 120 mm, R_{ud}: 5 mm).

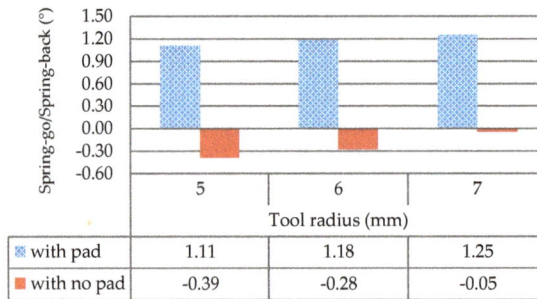

	with pad	with no pad
Tool radius (mm)		

	5	6	7
with pad	1.11	1.18	1.25
with no pad	-0.39	-0.28	-0.05

Figure 11. Comparison of the predicted spring-back/spring-go characteristics between the cases of U-bending, with and without pressure pads, with respect to tool radii (θ: 90°, W: 30 mm, t: 3 mm, WP_L: 120 mm, R_{ud}: 5 mm).

3.1.5. Workpiece Length

Figure 12 shows a comparison of stress distribution, analyzed in the workpiece, before unloading, with respect to various workpiece lengths, in the cases of the U-bending process with and without pressure pads. As the workpiece length increased, the bending stress and reversed bending characteristics were somewhat constant in the case of pressure pads being used; additionally, the bending stress and reversed bending stress characteristics were somewhat constant in the case where no pressure pad was used. It could be explained that, based on the bending moment theory, the workpiece length rarely had any effect on the bending characteristics formed across the punch radius during the bending phase, and this resulted in the same level of bending and reversed bending stress characteristics being generated on the workpiece. In cases where pressure pads and no pressure pads were used, the amounts of spring-back and spring-go were also somewhat constant, respectively, as the workpiece length increased (Figure 13). The aforementioned effects of U-geometry parameters on spring-back/spring-go characteristics, including channel width, workpiece thickness, bend angle, tool radius, and workpiece length, were clearly identified; in the case of pressure pads being used, the channel width and workpiece length did not have any effect on the spring-back/spring-go characteristics, but only workpiece length did not have any effects on spring-back characteristics in the case where no pressure pad was used. Although the results clearly elucidated the effects of the U-geometry parameters on the spring-back/spring-go characteristics, the degree of importance of the U-geometry parameters, in relation to the spring-back/spring-go characteristics, could not be determined. The degree of importance of the U-geometry parameters is very important for die and process designs. Therefore, the statistical technique was needed to examine the degree of importance of the U-geometry parameters, in relation to the spring-back/spring-go characteristics.

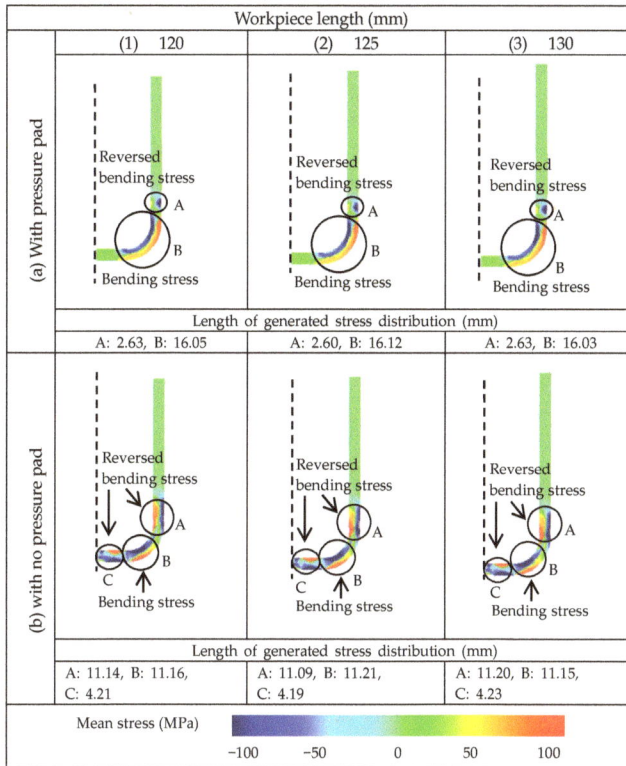

Figure 12. Comparison of the stress distribution analyzed in the workpiece, before unloading, with respect to various workpiece lengths (θ: 90°, R_p: 7 mm, W: 30 mm, t: 3 mm, R_{ud}: 5 mm).

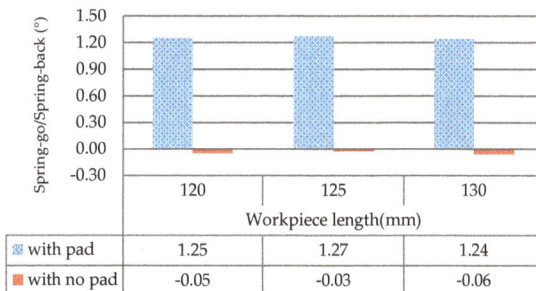

	120	125	130
with pad	1.25	1.27	1.24
with no pad	-0.05	-0.03	-0.06

Figure 13. Comparison of the predicted spring-back/spring-go characteristics between the cases of U-bending with and without pressure pads, with respect to workpiece lengths (θ: 90°, R_p: 7 mm, W: 30 mm, t: 3 mm, R_{ud}: 5 mm).

3.2. The Use of FEM Simulation and Its Validation

To reduce the number of experiments for the examination of the degree of importance of the U-geometry parameters, FEM simulations were used. Therefore, validation of the accuracy of the FEM simulation results was necessary. Figure 14 shows examples of the comparison between bent parts, with and without pressure pads, obtained by FEM and experiments. The results showed the spring-back and spring-go characteristics formed where pressure pads and no pressure pads were used, respectively. In addition, as shown in Figure 15, additional bending conditions were also investigated. The analyzed bending angle was also compared with those obtained by the experiments. FEM simulation results showed the formation of the spring-back and spring-go characteristics, which corresponded well with the experiments, and the error, compared with that of the experimental results, was approximately 1%. Next, workpiece thickness of the bent parts was also examined and compared with that which was obtained from the FEM simulation results. As the results in Figure 14 show, the FEM simulation results showed that the workpiece thickness, which corresponded well with the experiments and the error, compared with that from the experimental results, was approximately 1%. The bending force was also compared with that obtained by the experiments, as shown in Figure 16. The bending force increased as the bending stroke increased due to the workpiece being largely bent. Next, after a bending stroke of approximately 12 mm, the legs of the workpiece were pushed into the die using a punch and causing a small bending characteristic; this resulted in the decreases in the bending force. After the entire workpiece was moved to the die, the bending force was decreased to nearly zero, and it, again, increased when the bottom surface made contact with the die, and increased sharply when it was completely compressed. The FEM simulations were in good agreement with the experimental results, wherein the error was approximately 1%.

FEM simulation result	Experimental result	Bend angle (°)			Thickness (mm)		
			Left	Right	P1	P2	P3
(a)		1.	91.35	91.37	3.011	2.990	3.006
		2.	91.39	91.35	3.010	2.989	3.010
		3.	91.45	91.41	3.009	2.982	3.007
		4.	91.37	91.39	3.009	2.990	3.012
P1 — — P1		5.	91.51	91.41	3.012	2.985	3.015
		Ave.	91.40		3.010	2.987	3.010
P2 — — P2							
P3 P3		S.D.	0.05		0.001	0.004	0.004
bend angle 91.40°	bend angle 91.41°	% Error	0.00		0.33	0.43	0.33
Thickness (mm)							
P1: 3.000	P2: 2.990	P3: 3.000	P3: 3.009	P2: 2.982	P1: 3.007		
FEM simulation result	Experimental result	Bend angle (°)			Thickness (mm)		
			Left	Right	P1	P2	P3
(b)		1.	89.49	89.40	3.009	2.995	3.007
		2.	89.44	89.38	3.008	2.998	3.010
		3.	89.38	89.44	3.011	2.989	3.005
P1 — — P1		4.	89.35	89.49	3.010	2.990	3.010
		5.	89.40	89.35	3.010	2.992	3.008
P2 — — P2		Ave.	89.41		3.010	2.993	3.008
P3 P3		S.D.	0.05		0.001	0.004	0.002
bend angle 89.34°	bend angle 89.35°	% Error	0.08		0.33	0.23	0.27
Thickness (mm)							
P1: 3.000	P2: 2.998	P3: 3.000	P3: 3.010	P2: 2.992	P1: 3.008		

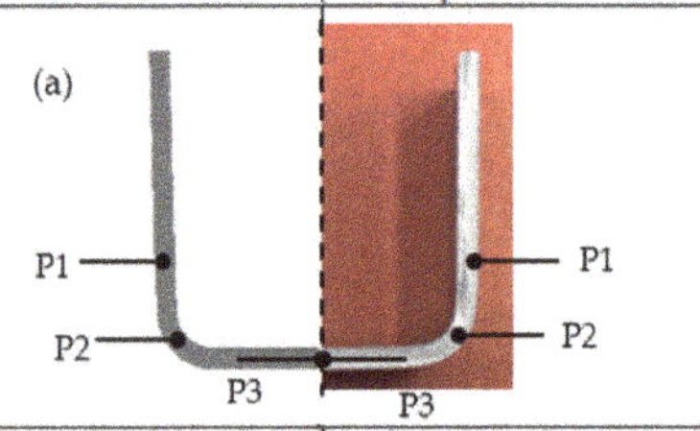

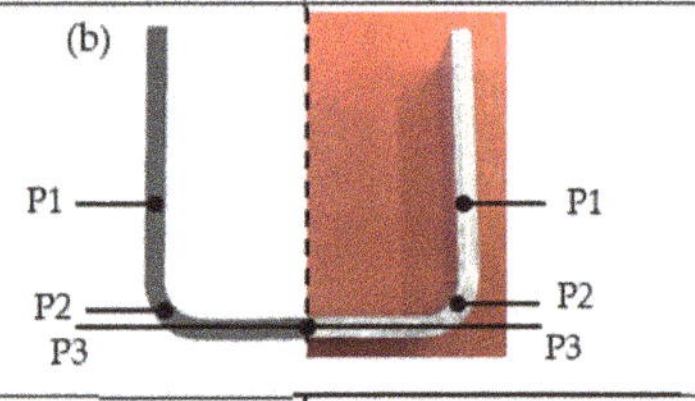

Figure 14. Comparison of the bent parts, obtained using FEM simulations and experiments: (**a**) with pressure pads; (**b**) without pressure pads (θ: 90°, t: 3 mm, WP_L: 120 mm, W: 40 mm, R_p: 5 mm, R_{ud}: 5 mm). SD: Standard deviation.

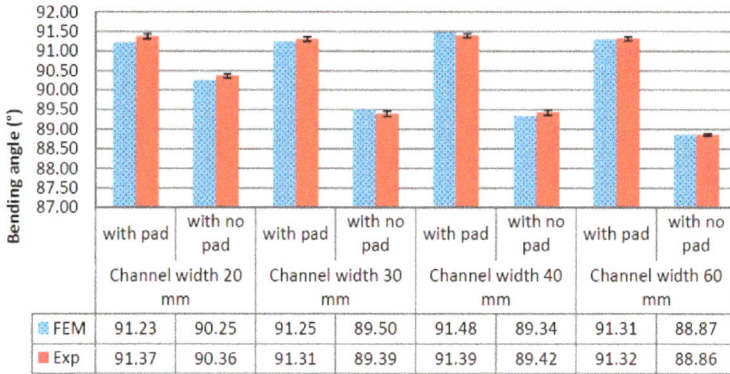

Remark:
Channel width 20 mm: $WP_L = 100$ mm, $R_p = 7$ mm
Channel width 30 mm: $WP_L = 120$ mm, $R_p = 7$ mm
Channel width 40 mm: $WP_L = 100$ mm, $R_p = 5$ mm
Channel width 60 mm: $WP_L = 100$ mm, $R_p = 7$ mm

	with pad	with no pad	with pad	with no pad	with pad	with no pad	with pad	with no pad
	Channel width 20 mm		Channel width 30 mm		Channel width 40 mm		Channel width 60 mm	
FEM	91.23	90.25	91.25	89.50	91.48	89.34	91.31	88.87
Exp	91.37	90.36	91.31	89.39	91.39	89.42	91.32	88.86

Figure 15. Comparison of the bending angle between the FEM simulations and the experimental results (θ: 90°, t: 3 mm, R_{ud}: 5 mm).

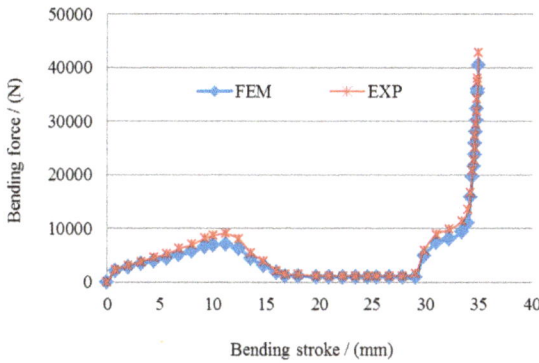

Figure 16. Comparison of bending forces between FEM simulations and experimental results (θ: 90°, t: 3 mm, WP_L: 130 mm, W: 60 mm, R_p: 5 mm, R_{ud}: 5 mm, without pad). EXP: Experiment.

3.3. Statistical Analysis

On the basis of the central composite design technique, Table 3 shows the amounts of spring-back, analyzed using FEM simulations for both pressure pads and no pressure pads. Based on these bending conditions, to confirm the accuracy of the FEM simulation results, some bending conditions were chosen for experimentation. As shown in Figure 17, the FEM simulation results, again, showed the formation of the spring-back characteristics, which corresponded well with the experiments; the error compared with that from the experimental results was approximately 1%.

Table 3. The amounts of spring-back in cases where pressure pads and no pressure pads were used, analyzed using FEM simulations.

| FEM No. | U-Geometry Parameters | | | | | | | | | | Spring-Back (°) | |
| | θ | | W_p | | R_p | | t | | WP_L | | | |
	with Pad	with no Pad	with Pad	with no Pad	with Pad	with no Pad	with Pad	with no Pad	with Pad	with no Pad	with Pad	with no Pad
1	90	90	30	30	5	3	3	1	120	120	1.11	0.33
2	120	90	30	30	5	3	3	1	120	130	0.29	0.80
3	90	90	60	30	5	3	3	3	120	120	1.56	0.67
4	120	90	60	30	5	3	3	3	120	130	0.34	0.93
5	90	90	30	30	7	5	3	1	120	120	1.25	0.69
6	120	90	30	30	7	5	3	1	120	130	0.32	0.93
7	90	90	60	30	7	5	3	3	120	120	1.62	0.50
8	120	90	60	30	7	5	3	3	120	130	0.49	0.78
9	90	90	30	60	5	3	5	1	120	120	0.95	0.57
10	120	90	30	60	5	3	5	1	120	130	0.06	0.52
11	90	90	60	60	5	3	5	3	120	120	1.31	0.95
12	120	90	60	60	5	3	5	3	120	130	0.04	0.96
13	90	90	30	60	7	5	5	1	120	120	1.45	0.36
14	120	90	30	60	7	5	5	1	120	130	0.33	0.42
15	90	90	60	60	7	5	5	3	120	120	1.39	0.69
16	120	90	60	60	7	5	5	3	120	130	0.37	0.55
17	90	120	30	30	5	3	3	1	130	120	1.52	1.07
18	120	120	30	30	5	3	3	1	130	130	0.24	1.74
19	90	120	60	30	5	3	3	3	130	120	1.54	0.25
20	120	120	60	30	5	3	3	3	130	130	0.36	0.16
21	90	120	30	30	7	5	3	1	130	120	1.93	0.87
22	120	120	30	30	7	5	3	1	130	130	0.27	1.56
23	90	120	60	30	7	5	3	3	130	120	1.65	0.63
24	120	120	60	30	7	5	3	3	130	130	0.47	0.53
25	90	120	30	60	5	3	5	1	130	120	1.30	1.00
26	120	120	30	60	5	3	5	1	130	130	0.08	1.87
27	90	120	60	60	5	3	5	3	130	120	1.33	0.42
28	120	120	60	60	5	3	5	3	130	130	0.05	0.38
29	90	120	30	60	7	5	5	1	130	120	1.45	1.71
30	120	120	30	60	7	5	5	1	130	130	0.33	1.31
31	90	120	60	60	7	5	5	3	130	120	1.39	0.73
32	120	120	60	60	7	5	5	3	130	130	0.34	0.45
33	105	105	45	45	6	4	4	2	125	125	0.30	2.71
34	90	90	45	45	6	4	4	2	125	125	1.52	2.53
35	120	120	45	45	6	4	4	2	125	125	0.38	2.62
36	105	105	30	30	6	4	4	2	125	125	0.41	1.50
37	105	105	60	60	6	4	4	2	125	125	0.47	2.16
38	105	105	45	45	5	3	4	2	125	125	0.42	2.15
39	105	105	45	45	7	5	4	2	125	125	0.41	2.68
40	105	105	45	45	6	4	3	1	125	125	0.33	1.65
41	105	105	45	45	6	4	5	3	125	125	0.47	1.18
42	105	105	45	45	6	4	4	2	120	120	0.46	2.16
43	105	105	45	45	6	4	4	2	130	130	0.45	2.37

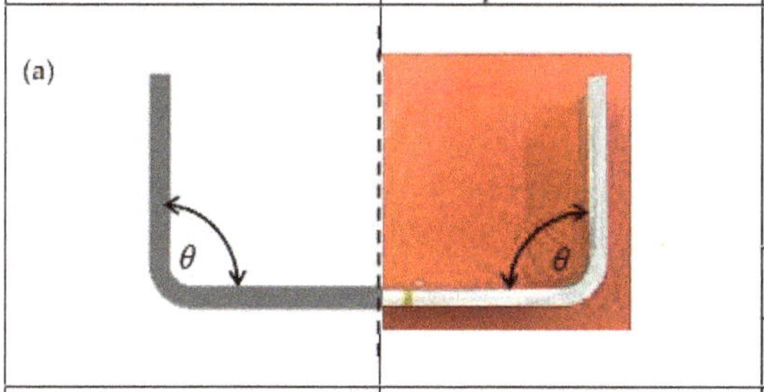

FEM simulation result	Experimental result	Bend angle (°)			
				Left	Right
(a)		Sample No.	1.	90.99	91.07
			2.	90.86	90.85
			3.	90.79	90.75
			4.	90.99	90.89
			5.	91.03	91.09
		Ave.		90.93	
		S.D.		0.12	
bend angle 90.96°	bend angle 90.89°	% Error		0.03	
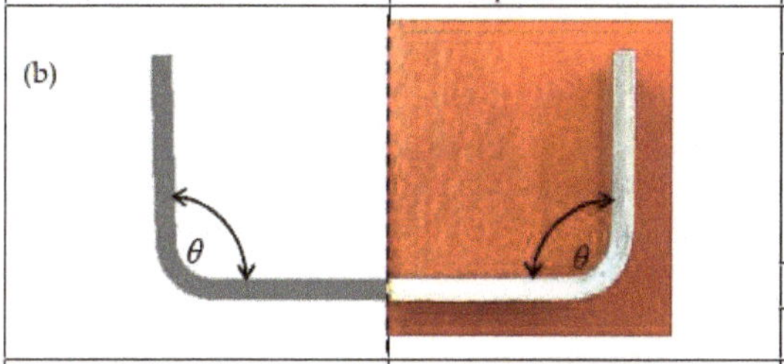 FEM simulation result	Experimental result	Bend angle (°)			
				Left	Right
(b)		Sample No.	1.	91.48	91.30
			2.	91.69	91.70
			3.	91.45	91.35
			4.	91.87	91.76
			5.	91.77	91.86
		Ave.		91.62	
		S.D.		0.21	
bend angle 91.54°	bend angle 91.35 °	% Error		0.09	
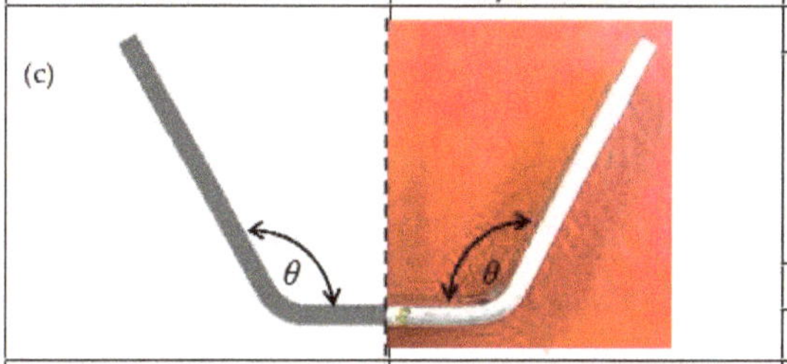 FEM simulation result	Experimental result	Bend angle (°)			
				Left	Right
(c)		Sample No.	1.	120.43	120.28
			2.	120.93	121.10
			3.	120.92	120.85
			4.	120.87	121.00
			5.	120.95	120.79
		Ave.		120.81	
		S.D.		0.26	
bend angle 120.63°	bend angle 120.85°	% Error		0.15	

Figure 17. Comparison of the bent parts, obtained using FEM simulations and experiments: (**a**) without pressure pads (θ: 90°, t: 3 mm, WP_L: 130 mm, W: 60 mm, R_p: 3 mm, R_{ud}: 5 mm); (**b**) with pressure pads (θ: 90°, t: 3 mm, WP_L: 120 mm, W: 60 mm, R_p: 5 mm, R_{ud}: 5 mm); (**c**) with no pressure pads (θ: 120°, t: 3 mm, WP_L: 120 mm, W: 30 mm, R_p: 5 mm, R_{ud}: 5 mm). SD: Standard deviation.

To investigate the degree of importance of the U-geometry parameters on the spring-back characteristics, based on the spring-back prediction amounts, ANOVA was carried out. The sums of squares due to the variations of the overall mean (SS_{total}) and regarding the mean of the U-geometry parameters ($SS_{treatment}$) in the cases where pressure pads and no pressure pads were used, were calculated and are listed in Table 4. The percentage contributions were calculated, according to Equation (1), and are listed in Table 4; these values were generally applied for considering the degree of importance of each parameter. The results showed that, in the case of using pressure pads, the percentage contribution to the bend angle was the largest, followed by workpiece thickness and tool radius, with their percentage contributions being 91.07%, 2.01% and 1.37%, respectively. These results indicated that the U-geometry parameters of the bend angle had the most influence on the spring-back characteristic, followed by workpiece thickness and tool radius, respectively. These results corresponded well with the stress distribution analyses, as previously mentioned; additionally they confirmed that channel width and workpiece length rarely had any effects on spring-back characteristics in the case where pressure pads were used. On the other hand, in the case where no pressure pads were used, the percentage contributions of channel width, workpiece thickness, and bend angle were 60.93%, 16.96% and 1.81%, respectively. These results showed that the U-geometry parameters of the channel width had the most influence on the spring-back characteristics, followed by workpiece thickness and bend angle, respectively. In the case of where no pressure pads were used, this result corresponded well with the stress distribution analyses, as previously mentioned; they again confirmed that tool radius and workpiece length rarely had any effects on spring-back characteristics in the case where no pressure pad was used. As these results indicated, the bend angle and channel width have a major influence in the cases where pressure pads and no pressure pads are used, respectively. These statistical results indicate that the U-geometry parameters affected spring-back characteristics differently in the case where pressure pads and no pressure pads were used. It could be explained that, owing to the different bending mechanisms between bending with pads and without pads, there were no bending characteristics formed across the punch radius during the bending phase in the case where pressure pads were used, but it was formed in the case where no pressure pads were used. Therefore, the U-geometry parameters, which directly affected these bending mechanisms, were different. Specifically, there were no stresses generated on the bottom surface in the case of pressure pads being used, resulting in the spring-back characteristics, though they only depended on the stress generated on the bending allowance and leg zones. Therefore, as mentioned previously, the U-geometry parameters of the bend angle, workpiece thickness, and tool radius, which greatly affected these zones, had a great influence on the spring-back characteristics. On the other hand, excluding the generated stress on the bending allowance and leg zones, the bending and reversed bending stresses commonly generated on the bottom surface, in the case of pressure pads being used, resulted in the spring-back characteristics not being dependent only on the stresses generated on the bending allowance and leg zones, but also dependent on the stresses generated on the bottom surface. These generated stresses on the bottom surface zone were based on the bending characteristics formed across the punch radius during the bending phase. Therefore, as mentioned previously, the U-geometry parameters of channel width, workpiece thickness, and bend angle, which greatly affected these zones, had a great deal of influence on the spring-back characteristics.

Table 4. Percentage contributions.

Parameters	$SS_{treatment}$		% Contributions	
	with Pad	with no Pad	with Pad	with no Pad
Bend angle (θ)	12.52	0.39	91.07	1.81
Workpiece thickness (t)	0.28	1.10	2.01	5.11
Tool radius (R_p)	0.19	-	1.37	-
Channel width (W)	-	0.01	-	0.05
Workpiece length (WP_L)	-	0.14	-	0.66
Bend angle (θ) × Bend angle (θ)	0.76	-	5.55	-
Channel width (W) × Channel width (W)	-	13.17	-	60.93
Workpiece thickness (t) × Workpiece thickness (t)	-	3.67	-	16.96
Bend angle (θ) × Workpiece thickness (t)	-	2.82	-	13.05
Workpiece thickness (t) × Workpiece length (WP_L)	-	0.31	-	1.43
Total	13.75	21.61	100.00	100.00

4. Conclusions

In the present study, in order to examine the degree of importance of U-geometry parameters, including channel width, workpiece thickness, bend angle, tool radius, and workpiece length, in relation to the spring-back characteristics, FEM simulations, in association with statistical analyses of a central composite design, and ANOVA, were applied. The effects of the U-geometry parameters in the U-bending process, with and without pressure pads, were investigated using FEM simulations. First, on the basis of stress distribution analyses, the FEM simulation results clearly identified the difference of the bending mechanism and the effects of the U-geometry parameters on the spring-back characteristics of the U-bending process, with and without pressure pads. Specifically, the FEM simulation results clearly revealed that channel width and workpiece length rarely had any effects on the change in bending and reversed bending stresses in the case where pressure pads were used. Again, in the case where no pressure pads were used, the workpiece length rarely had any effects on the change in the bending and reversed bending stresses. However, the FEM simulation results could only elucidate the tendency of effects for each U-geometry parameter on the spring-back characteristics. Next, by using the central composite design and ANOVA, the degree of importance of the U-geometry parameters, in relation to the spring-back characteristics, could be obtained. The ANOVA results illustrated the influence of each U-geometry parameter on the spring-back characteristics, together with their calculated percentage contributions. The bend angle had the most influence in the case of pressure pads being used, followed by workpiece thickness and tool radius, respectively. On the other hand, the ANOVA results illustrated the influence of each U-geometry parameter on the spring-back characteristics, together with their calculated percentage contributions; channel width had the most influence in the case of no pressure pads being used, followed by workpiece thickness and bend angle, respectively. Laboratory experiments were carried out and FEM simulation results, as validated by laboratory experiments, showing a good agreement with the experimental results. The error in the bend angle, compared with the laboratory experimental results, was approximately 1%. The analyzed bending forces were also compared with that obtained by experiments, and showed a good agreement with the experimental results, in which the error was approximately 1%. Therefore, this technique could be applied as a tool for quality control of U-bent parts, based on spring-back characteristics, by optimizing the U-geometry parameter design.

Acknowledgments: This research was partially supported by a Grant from the Higher Education Research Promotion and National Research University Project of Thailand, Office of the Higher Education Commission, under Grant No. 56000519 and Grant No. 57000618. The authors thank Nopnarong Sirisatien for his advice in this study. The authors also thank graduate students, Pakkawat Komolruji, and Arkarapon Sontamino, for their help in this study.

Author Contributions: Wiriyakorn Phanitwong and Sutasn Thipprakmas conceived and designed the experiments; Wiriyakorn Phanitwong and Untika Boochakul performed the experiments; Wiriyakorn Phanitwong, Untika Boochakul and Sutasn Thipprakmas analyzed the data; Wiriyakorn Phanitwong and Sutasn Thipprakmas wrote the paper.

Conflicts of Interest: The authors declare no conflict of interest.

References

1. Zong, Y.Y.; Liu, P.; Guo, B.; Shan, D. Springback evaluation in hot V-bending of Ti-6Al-4V alloy sheets. *Int. J. Adv. Manuf. Technol.* **2015**, *76*, 577–585. [CrossRef]
2. Leu, D.K. Position deviation and springback in V-die bending process with asymmetric dies. *Int. J. Adv. Manuf. Technol.* **2015**, *79*, 1095–1108. [CrossRef]
3. Dilip Kumar, K.; Appukuttan, K.K.; Neelakantha, V.L.; Naik, P.S. Experimental determination of spring back and thinning effect of aluminum sheet metal during L-bending operation. *J. Mater. Des.* **2014**, *56*, 613–619. [CrossRef]
4. Lim, H.; Lee, M.G.; Sung, J.H.; Kim, J.H.; Wagoner, R.H. Time dependent springback of advanced high strength steels. *Int. J. Plast.* **2012**, *29*, 42–59. [CrossRef]
5. Thipprakmas, S. Finite element analysis on the coined-bead mechanism during the V-bending process. *Mater. Des.* **2011**, *32*, 4909–4917. [CrossRef]
6. Thipprakmas, S. Finite element analysis of punch height effect on V-bending angle. *Mater. Des.* **2010**, *32*, 4430–4436. [CrossRef]
7. Thipprakmas, S. Finite element analysis of sided coined-bead technique in precision V-bending process. *Int. J. Adv. Manuf. Technol.* **2013**, *65*, 679–688. [CrossRef]
8. Jiang, H.J.; Dai, H.L. A novel model to predict U-bending spring-back and time-dependent spring-back for a HSLA steel plate. *Int. J. Adv. Manuf. Technol.* **2015**, *81*, 1055–1066. [CrossRef]

9. Thipprakmas, S.; Boochakul, U. Comparison of spring-back characteristics in symmetrical and asymmetrical U-bending processes. *Int. J. Precis. Eng. Manuf.* **2015**, *16*, 1441–1446. [CrossRef]
10. Phanitwong, W.; Thipprakmas, S. Centered coined-bead technique for precise U-bent part fabrication. *Int. J. Adv. Manuf. Technol.* **2015**, *84*, 2139–2150. [CrossRef]
11. Li, H.; Chen, J.; Yang, J. Experiment and numerical simulation on delamination during the laminated steel sheet forming processes. *Int. J. Adv. Manuf. Technol.* **2013**, *68*, 641–649. [CrossRef]
12. Nanu, N.; Brabie, G. Analytical model for prediction of springback parameters in the case of U stretch-bending process as a function of stresses distribution in the sheet thickness. *Int. J. Mech. Sci.* **2012**, *64*, 11–21. [CrossRef]
13. Lee, J.Y.; Lee, J.W.; Lee, M.G.; Barlat, F. An application of homogeneous anisotropic hardening to springback prediction in pre-strained U-draw/bending. *Int. J. Solids Struct.* **2012**, *49*, 3562–3572. [CrossRef]
14. Thipprakmas, S.; Phanitwong, W. Finite element analysis of bending mechanism and spring-back/spring-go feature in various U-bending processes. *Steel Res. Int.* **2012**, 351–354.
15. Seong, D.Y.; Jung, C.G.; Yang, D.Y.; Ahn, J.; Na, S.J.; Chung, W.J.; Kim, J.H. Analysis of core shear stress in welded deformable sandwich plates to prevent de-bonding failure during U-bending. *J. Mater. Process. Technol.* **2010**, *210*, 1171–1179. [CrossRef]
16. Marretta, L.; Lorenzo, R.D. Influence of material properties variability on spring-back and thinning in sheet stamping processes: A stochastic analysis. *Int. J. Adv. Manuf. Technol.* **2010**, *51*, 117–134. [CrossRef]
17. Tang, B.; Zhao, G.; Wang, Z. A mixed hardening rule coupled with Hill48' yielding function to predict the spring-back of sheet U-bending. *Int. J. Mater. Form.* **2008**, *1*, 169–175. [CrossRef]
18. Zhang, D.; Cui, Z.; Ruan, X.; Li, Y. An analytical model for predicting spring back and side wall curl of sheet after U-bending. *Comput. Mater. Sci.* **2007**, *38*, 707–715. [CrossRef]
19. Phanitwong, W.; Thipprakmas, S. Development of anew spring-back factor for a wiping die bending process. *Mater. Des.* **2016**, *89*, 749–758. [CrossRef]
20. Lange, K. *Handbook of Metal Forming*; McGraw-Hill: New York, NY, USA, 1985; pp. 1–35.

![metals logo] *metals*

MDPI

Article

Particle Size and Particle Percentage Effect of AZ61/SiCp Magnesium Matrix Micro- and Nano-Composites on Their Mechanical Properties Due to Extrusion and Subsequent Annealing

Weigang Zhao [1], Song-Jeng Huang [2,*], Yi-Jhang Wu [2] and Cheng-Wei Kang [2]

[1] College of Material Engineering, Fujian Agricultural and Forestry University, No. 15 Shangxiadian Road, Cangshan District, Fuzhou 350002, China; weigang-zhao@fafu.edu.cn
[2] Department of Mechanical Engineering, National Taiwan University of Science and Technology, No. 43 Keelung Rd. Sec. 4, Taipei 10607, Taiwan; materials_fafu@163.com (Y.-J.W.); zhao610323@163.com (C.-W.K.)
*   Correspondence: sgjghuang@mail.ntust.edu.tw; Tel.: +886-2-2737-6485

Received: 4 June 2017; Accepted: 25 July 2017; Published: 1 August 2017

Abstract: Magnesium metal matrix composites (Mg MMCs) possess relatively more favorable mechanical properties than Mg alloys because they add reinforcements, such as small particles, short fibers, or continuous fibers, into the matrix. This study investigated the influence of adding different sizes and percentages of silicon carbide particles (SiCp) for manufacturing AZ61/SiCp Mg alloy composite extrusion plates on the mechanical properties of SiCp. We also examined the impact and discussed the evolution of microstructures, changes of material strength, ductility, formability, and other mechanical properties caused by a subsequent annealing treatment after plate extrusion. The results showed that the mechanical properties of plates can be improved by adding reinforcement particles. The effects of grain refinement were as follows: the smaller the size of the reinforcement particles, the greater the enhancement of mechanical properties. Among them, the AZ61/1 wt % SiCp/50 nm MMC plate had relatively excellent mechanical properties. Specifically, the ultimate tensile strength, yielding strength, ductility, hardness, and grain size of the plate were 331 MPa, 136.4 MPa, 43.1%, 62 HV, and 3.3 μm, respectively. Compared with SiCp-free Mg MMC plates, these properties of the AZ61/1 wt % SiCp/50 nm MMC plate were enhanced (or refined) by 6.4%, 3.4%, 83.4%, 2%, and 13.2%, respectively; by contrast, formability decreased by 9.1%.

Keywords: magnesium metal matrix composites (Mg MMCs); extrusion; annealing; mechanical properties

1. Introduction

Magnesium (Mg) alloys were widely used in the 1950s and 1960s in the aerospace and automotive industries (e.g., the B-37 airplane and Volkswagen "Beetle"). Since then, Mg alloys have gained more recognition as a structural material for lightweight applications because of their low density, high stiffness-to-weight ratio, favorable castability, shock absorption, and excellent damping. Nevertheless, Mg alloys have not been used for critical applications because of their inferior mechanical properties, high manufacturing cost, and unfavorable formability at room temperature compared with other engineering materials [1–4]. One of the most frequently used methods for enhancing the mechanical properties of alloys is through strengthening by means of second-phase particles [5–7].

Composite materials are vital engineering materials because of their outstanding mechanical properties [8,9]. Metal matrix composite (MMC) materials are some of the most widely known composites, but widespread engineering application of MMC materials has been met with resistance because of their unfavorable machining characteristics, in particular excessive tool wear and inferior surface finish, despite superior physical and mechanical properties [10,11]. Continuous fiber reinforcement was a key research objective for MMCs in the early stage. However, because of the complexity of manufacturing and the relatively high cost of fibrous MMCs, many researchers have instead focused on using discontinuous particles as reinforcement for MMCs [12]. The advantages of discontinuous particle-reinforced MMCs are low cost, being able to fabricate them through secondary processing (e.g., forging or extrusion), more favorable mechanical properties, and homogeneity. Silicon carbide (SiC) is one of the most widely incorporated reinforcing materials for MMC fabrication because of its superior properties and economical production in various forms, such as fibers [13,14], whiskers [15,16], and especially particles (SiCp), which are exceptionally affordable. The mechanical properties of SiCp and aluminum oxide (Al_2O_3)-reinforced Mg-based MMCs have been extensively studied by researchers [17–30]. Mg alloy sheets fabricated through rolling or extrusion and subsequent annealing can achieve excellent mechanical properties [31–33]. Due to global restrictions on CO_2 emissions, SiC particulate–reinforced Mg MMCs provide increased specific strength that contribute to weight reduction in the automobile industry. The particulate SiC–reinforced Mg alloy metal matrix composite (Mg/SiCp MMC) has rapidly replaced conventional materials in various industries, particularly in the automotive and aerospace industries, because of its ability to considerably reduce the weight of products. However, few researchers have studied AZ61 Mg MMCs processed through extrusion and subsequent annealing, or the effect of particle size or particle percentage of SiCp on the mechanical properties of such composites.

In this study, the effect of SiC particle size and particle percentage on the mechanical properties of Mg MMCs was investigated. The evolution of the microstructure and changes in material strength, ductility, and other mechanical properties caused by the subsequent annealing treatment of the extruded Mg MMC plates were also examined.

2. Materials and Methods

2.1. Materials

In this study, different sizes (10, 1, and 50 nm) and percentages (0.5, 1, and 2 wt %) of SiCp particles were added to AZ61 alloys to form AZ61/SiCp MMCs through the melt-stirring casting method for plate extrusion. The chemical composition of the AZ61 alloys is shown in Table 1.

Table 1. Chemical composition of the AZ61 alloys.

Al	Zn	Mn	Fe	Si	Cu	Ni	Mg
5.95	0.64	0.26	0.005	0.009	0.0008	0.0007	Bal.

2.2. Fabrication of the Ingot and Plate of the Mg MMCs and Their Heat Treatments

The AZ61/SiCp MMCs were fabricated using the melt-stirring technique, a schematic of which is shown in Figure 1. First, AZ61/SiCp MMCs ingots (diameter = 85 mm, height = 140 mm) were machined to be billets with a diameter of 75 mm and a height of 100 mm. After machining the ingots, but prior to the extrusion process, T4 treatment (homogenization) was performed on the billets for 10 h at 400 °C, which eliminated the β phase caused during casting and improved the strength of the extrusion process of the AZ61/SiCp MMC plates. The T4 treatment was performed using a 500-ton extruder (Gongyi, Taibei, Taiwan) with a single pushing cylinder at an extrusion temperature of 300 °C. An extrusion ratio of 31.56 was used during the extrusion process. Subsequently, T4 annealing treatment was performed on the extruded plates for 1 h at 250 °C, which improved their ductility and formability.

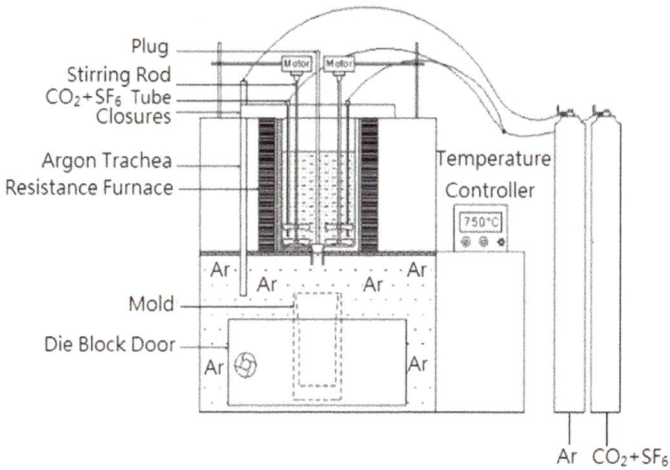

Figure 1. Schematic of the stir-casting melting furnace.

2.3. Morphological and Mechanical Properties

Microstructure characterization was performed with an optical microscope (OLYMPUS BX51, OLYMPUS, Tokyo, Japan), and the mean grain size was determined using the linear intercept method. For mechanical testing, an MTS Model 458 axial/torsional testing system was used according to standard ASTM B 557M-02a (Standard Test Methods of Tension Testing Wrought and Cast Aluminum- and Magnesium-Alloy Products). With a load of 500 gf, Vickers microhardness was measured using a Matsuzawa (Model MV-1) hardness tester (Akashi MVK-H1, Mitutoyo, Tokyo, Japan). The average hardness of each sample was obtained from nine tests and four specimens of each sample were tested under a strain rate of 1 mm/min. The morphologies were determined using scanning electron microscopy with energy-dispersive X-ray (SEM-EDX) and scanning transmission electron microscopy (SEM; FEG SEM, Hitachi S 3400, Tokyo, Japan).

3. Results and Discussion

3.1. Mechanical Properties

Figures 2–5 show the ultimate strength, yield strength, ductility, and hardness, respectively, of the AZ61/SiCp MMC extruded plates before annealing. The results show that the mechanical properties of the unannealed AZ61/SiCp MMC extruded plates increased when the percentage of particles increased; their mechanical properties (all except hardness) all increased when reinforcement particle size decreased, which caused significant grain size refinement. Among all the combinations, Mg MMCs with 1 wt % of SiCp/50 nm possessed the optimal mechanical properties. The ultimate tensile strength, yielding strength, ductility, hardness, and grain size of the AZ61/1 wt % SiCp/50 nm MMC plate were 331 MPa, 136.4 MPa, 43.1%, 62 HV, and 3.3 µm, respectively. Moreover, the ultimate tensile strength, yielding strength, ductility, hardness, and grain size of the AZ61/1 wt % SiCp/50 nm MMC plate increased by 6.4%, 3.4%, 83.4%, 2%, and 13.2%, respectively, compared with the SiCp-free Mg MMC plate (i.e., the AZ61 extruded plate).

Figures 6–9 show the ultimate strength, yield strength, ductility, and hardness, respectively, of the AZ61/SiCp MMC extruded plate after annealing. The results indicate that the ultimate strength and yield strength of the annealed AZ61/SiCp MMC extruded plates (all except the AZ61/SiCp/10 µm MMCs) increased when the percentage of particles increased; furthermore, their hardness decreased when particle sizes decreased, which is caused by significant grain size refinement.

Figures 5 and 9 reveal the microhardness trends of the plates. According to the Hall–Petch theory, the hardness of the unannealed AZ61/SiCp MMC extruded plate increased when the particle size increased because of the refinement of the grain size (Figures 10 and 11). Similarly, the hardness of the annealed AZ61/SiCp MMC extruded plate increased when the particle size decreased because of the refinement of the grain size (Figures 12 and 13).

The conventional Mg alloy was annealed to improve its ductility and formability. In the case of the SiCp-free AZ61 plates, annealing increased ductility by 73%, compared with that before annealing. However, in the case of the AZ61/SiCp MMC extruded plates, annealing both only increased the ductility of the AZ61/SiCp/10 μm MMC plates and decreased the ductility of the AZ61/SiCp/1 μm and AZ61/SiCp/50 nm MMC plates because adding submicron SiCp and nano SiCp reduced the number of effective nucleation sites.

After being subjected to annealing treatment at 250 °C for 1 h, the AZ61/SiCp/1 μm and AZ61/SiCp/50 nm MMC plates, irrespective of the quantity of nano or micro particulates, exhibited improvements in hardness and strength; however, ductility slightly decreased because of the precipitation of the β phase (see Section 3.2). Although the AZ61/0.5 wt % SiCp/10 μm MMC annealed plates improved in various properties, the amount of precipitate increased as the percentage of SiCp increased, which led to a drastic decrease in its strength and ductility.

The mechanical properties of the annealed SiCp-free AZ61 alloy plate were better than those of unannealed SiCp-free AZ61 alloy plate. Both annealed and unannealed AZ61/SiCp/1 μm and AZ61/SiCp/50 nm MMC plates exhibited higher strength, hardness, and ductility compared with the SiCp-free (i.e., AZ61) Mg alloy plate. However, the annealed AZ61/SiCp/10 μm MMC plate was not in this case because its yield strength did not improve, although its ultimate strength, hardness, and ductility did improve.

In short, adding submicron SiCp and nano SiCp particles to AZ61 MCC plates during processing results in excellent advantages, including strengthening the mechanical properties and increasing ductility of the plates. Annealing can further improve the ductility and formability of the Mg alloy and Mg MMCs with micron SiCp and reduce those of Mg MMCs with submicron SiCp and nano SiCp.

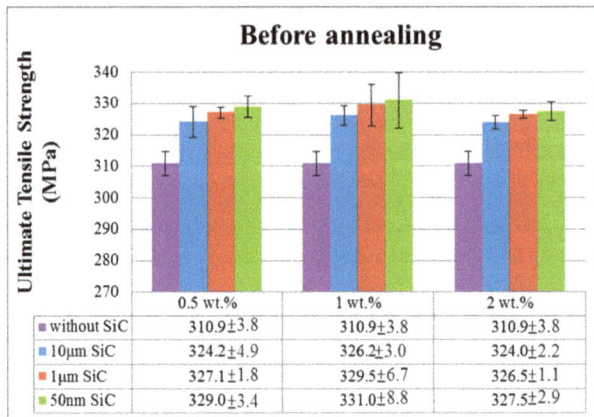

Before annealing			
Ultimate Tensile Strength (MPa)	0.5 wt.%	1 wt.%	2 wt.%
without SiC	310.9±3.8	310.9±3.8	310.9±3.8
10μm SiC	324.2±4.9	326.2±3.0	324.0±2.2
1μm SiC	327.1±1.8	329.5±6.7	326.5±1.1
50nm SiC	329.0±3.4	331.0±8.8	327.5±2.9

Figure 2. Ultimate strength of the AZ61/SiCp MMC extruded plates before annealing.

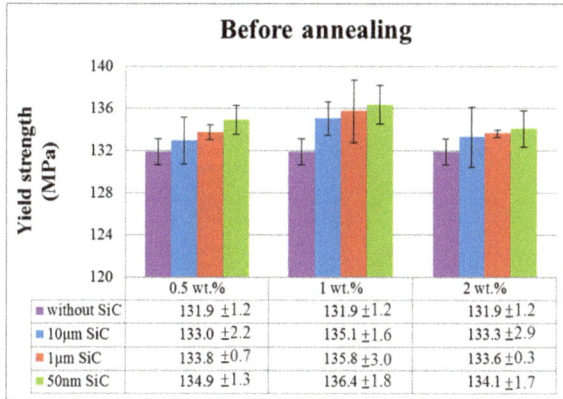

Figure 3. Yield strength of the AZ61/SiCp MMC extruded plates before annealing.

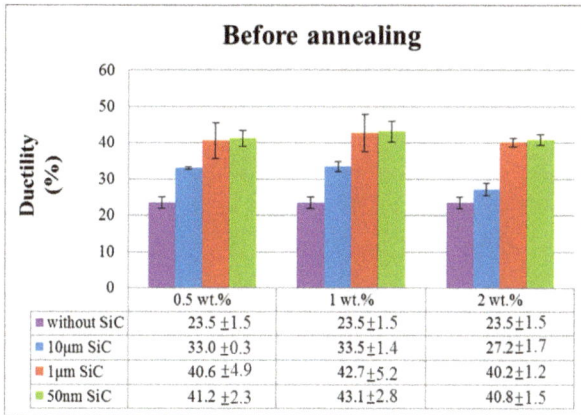

Figure 4. Ductility of the AZ61/SiCp MMC extruded plates before annealing.

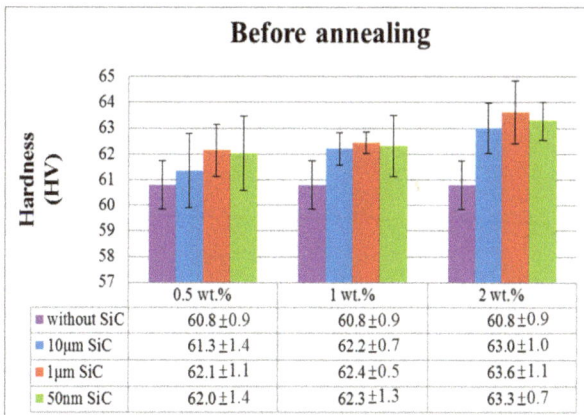

Figure 5. Hardness of the AZ61/SiCp MMC extruded plates before annealing.

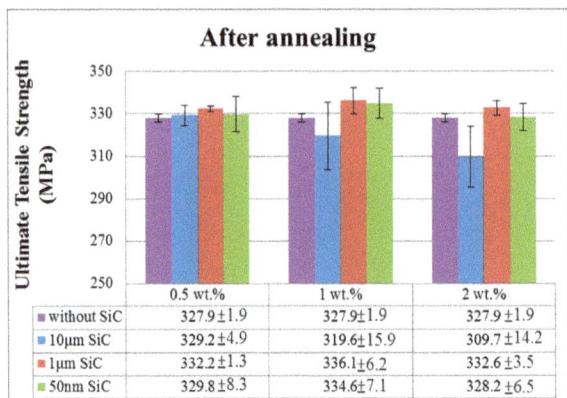

Figure 6. Ultimate strength of the AZ61/SiCp MMC extruded plates after annealing.

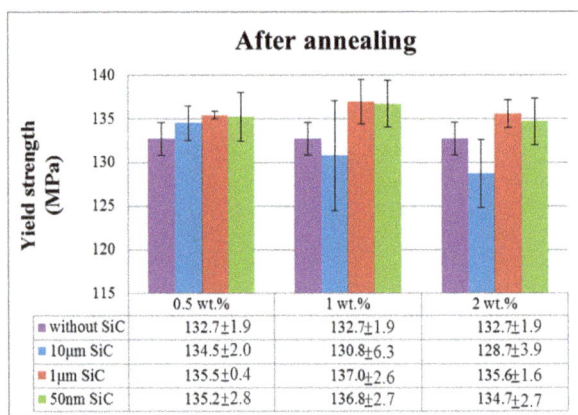

Figure 7. Yield strength of the AZ61/SiCp MMC extruded plates after annealing.

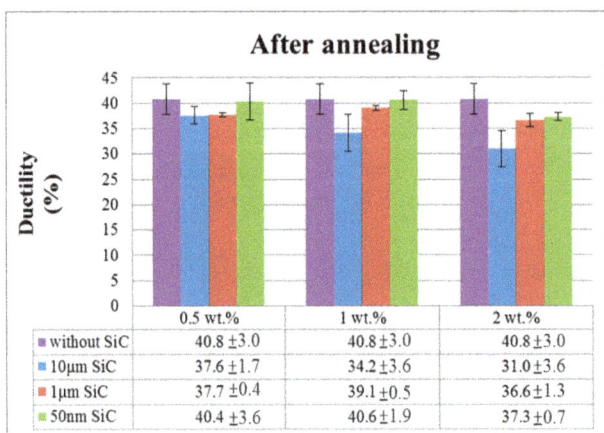

Figure 8. Ductility of the AZ61/SiCp MMC extruded plates after annealing.

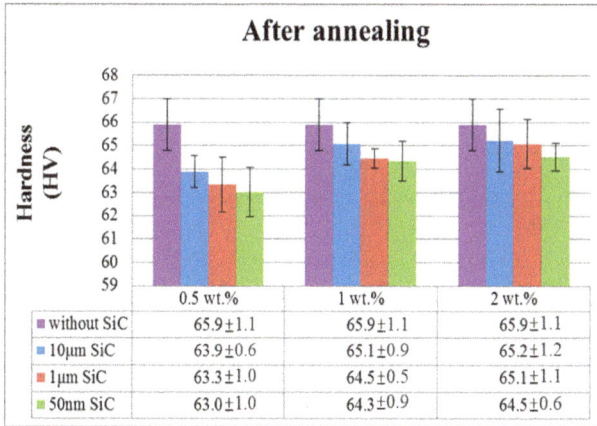

Figure 9. Hardness of the AZ61/SiCp MMC extruded plates after annealing.

3.2. Morphology and SEM Observation

Figure 10 presents the metallography of the AZ61/SiCp MMC extruded plates before annealing with the addition of different SiCp particle percentages and sizes. The average grain size was calculated through the linear intercept method (Figure 11). Compared with the average grain size of cast AZ61/SiCp MMCs, which is approximately 30–80 µm, the average grain size of the AZ61/SiCp MMC extruded plates was approximately 3–4 µm, which is substantially smaller. Overall, the average grain size decreased as the addition of SiCp particles increased and as the size of the SiCp particles decreased. The minimum average grain size obtained for the AZ61/SiCp MMC extruded plates was 2.9 µm, which diminished 24% compared with the raw AZ61 extruded plate. The results indicate that adding SiCp restricts grain growth during the extrusion process.

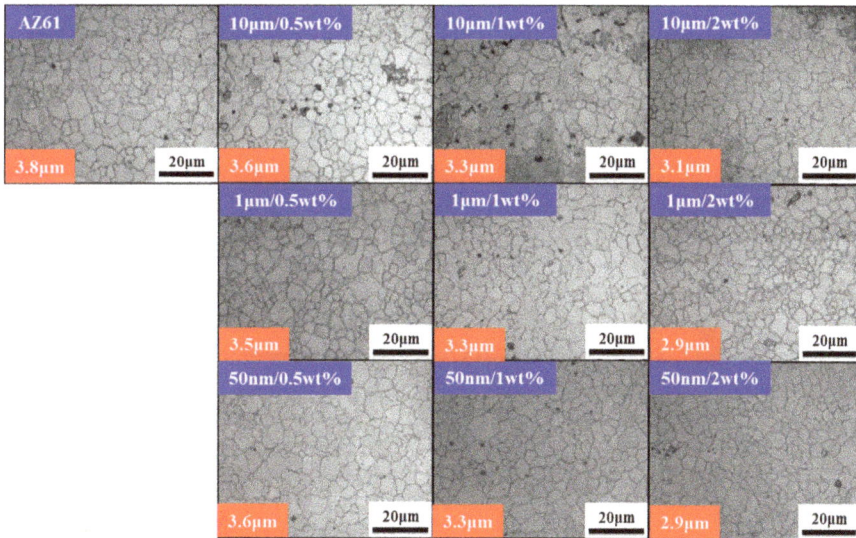

Figure 10. Metallography of the AZ61/SiCp MMCs extruded plates before annealing.

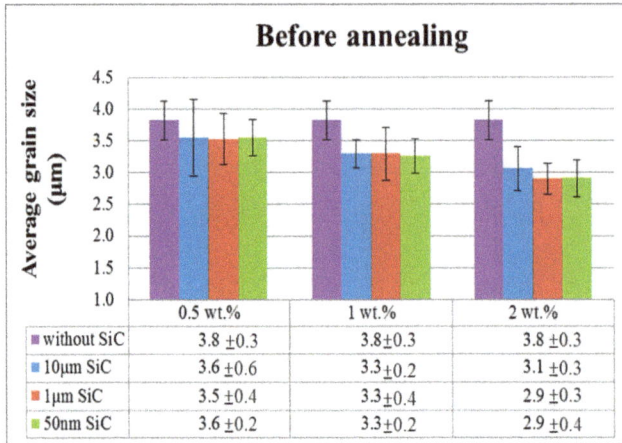

Figure 11. Average grain size of the AZ61/SiCp MMC extruded plates before annealing.

Figures 12 and 13 present the metallography and average grain size of the AZ61/SiCp MMC extruded plates after annealing with different percentages and sizes of SiCp particles, respectively. Prior to annealing, the average grain size of the AZ61/SiCp MMC extruded plates decreased as the percentage of added SiCp particles increased. Using reinforcement particles larger than 1 µm can initiate particle-stimulated dynamic recrystallization (particle-stimulated nucleation, PSN), and adding particles smaller than 1 µm mainly provides a grain boundary pinning mechanism that contributes to matrix grain refinement. The average grain size of the AZ61/SiCp MMC extruded plates after annealing was approximately 2.9–3.2 µm, which suggests that adding SiCp particles has no obvious effect on average grain size. Compared with the average grain size of the AZ61/SiCp MMCs before annealing, the average grain size of all samples after annealing decreased, except for the MMC with the addition of 2 wt % of SiCp. This occurred because the energy supplied by the annealing process could make the high strain zone grain results the static recrystallization. The situation is different for the average grain size of the AZ61/SiCp MMCs after annealing, where the average grain size increased as SiCp particles were added because the SiCp particles could not be the nucleation site in the annealing process. SiCp addition can restrict grain growth, but SiCp particles cannot be used directly as the nucleation site of an α-Mg alloy; thus, adding a strengthening phase reduces the effective nucleation site.

Figure 14 shows the images of the as-cast AZ61 Mg alloy; it mainly contains $Mg_{17}Al_{12}$ and α-Mg. The particles of SiCp were added to prepare the AZ61/SiCp MMC extruded plate before annealing (Figure 15). The existence of the SiCp particles was confirmed through EDX analysis, which revealed that the distribution was nonuniform. Compared with the as-cast AZ61 Mg alloy, SiC had a higher density and surface tension, which can cause the sinking and floating of SiC. Moreover, harder Mg_2Si produced through extrusion was observed in the matrix of the AZ61/SiCp MMCs prior to annealing, which can increase the hardness and strength of the MMCs compared with the as-cast AZ61 Mg alloy. Figure 16 presents the SEM images and EDX results of the AZ61/SiCp MMC extruded plates after annealing. These results reveal that the existence of the precipitation of Al-Mn and β phase $Mg_{17}Al_{12}$ is the principal reason why the mechanical properties changed (the yield strength and ultimate strength slightly decreased as SiCp addition increased after annealing, but their ductility declined).

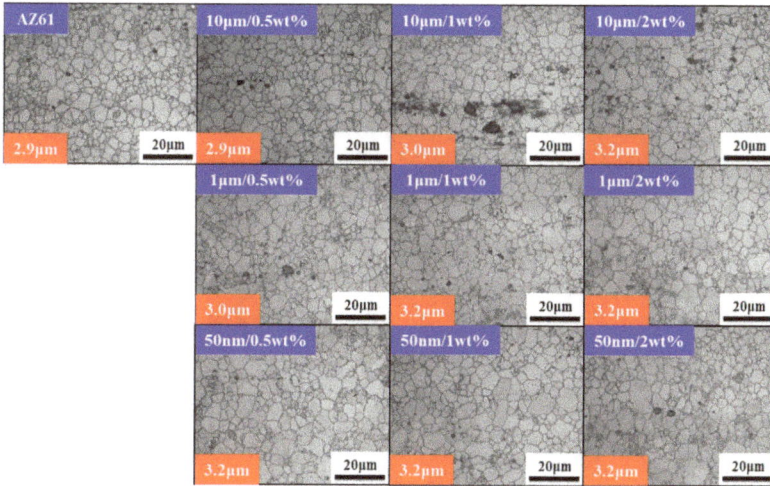

Figure 12. Metallography of the AZ61/SiCp MMC extruded plates after annealing.

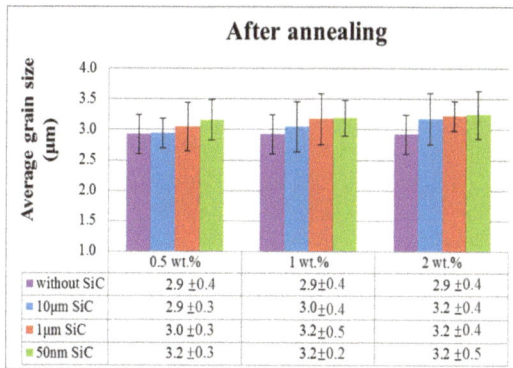

Figure 13. Average grain size of the AZ61/SiCp MMC extruded plates after annealing.

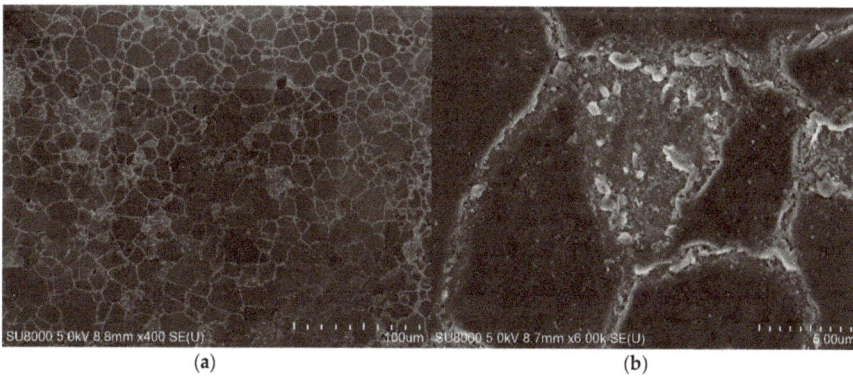

Figure 14. SEM images (**a**,**b**) of the dendrites on the edge of the as-cast AZ61 magnesium alloy with different magnifications, respectively.

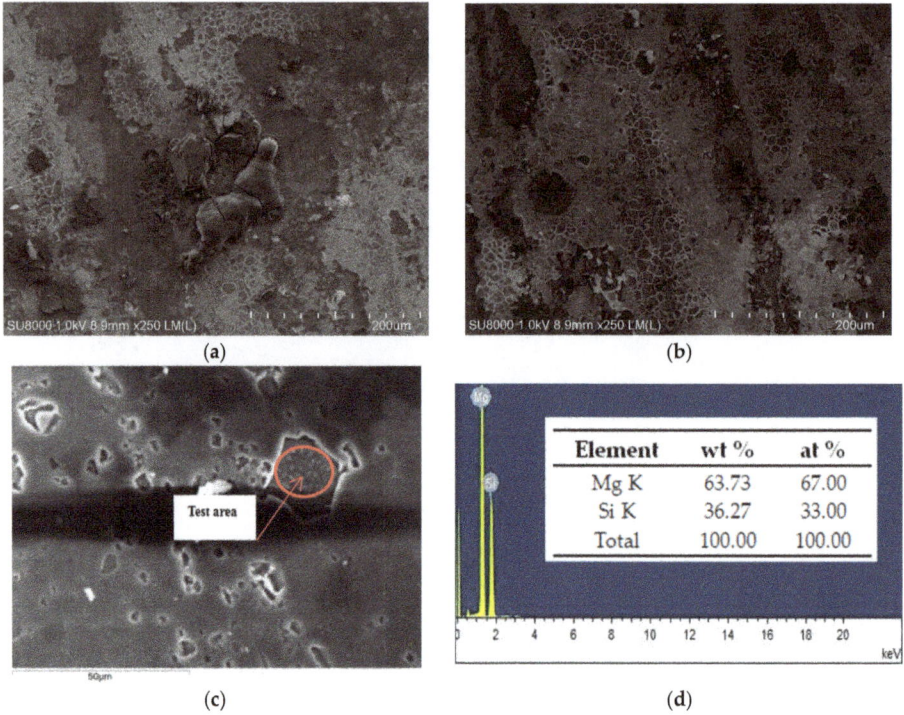

(a)

(b)

(c)

(d)

Element	wt %	at %
Mg K	63.73	67.00
Si K	36.27	33.00
Total	100.00	100.00

Figure 15. (**a**,**b**) SEM images of the AZ61/SiCp MMCs extruded plates before annealing (10 µm SiC with 5 wt %) with different magnifications; (**c**) Test area for EDX analysis of the AZ61/SiCp MMCs extruded plates before annealing (10 µm SiC with 5 wt %); (**d**) Element results from EDX analysis of the AZ61/SiCp MMCs extruded plates before annealing (10 µm SiC with 5 wt %).

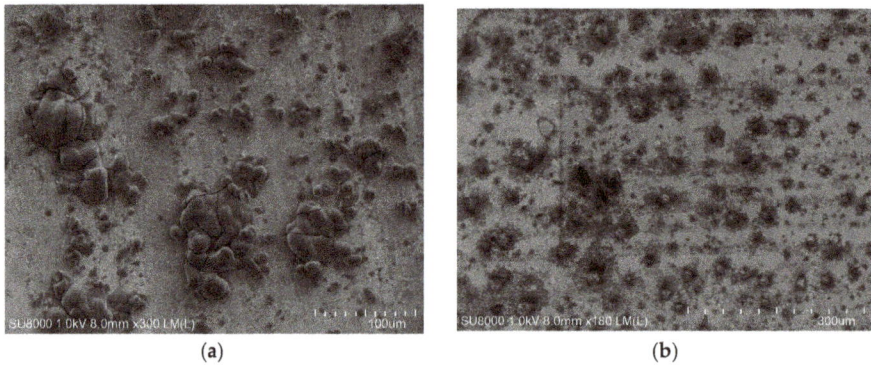

(a)

(b)

Figure 16. *Cont.*

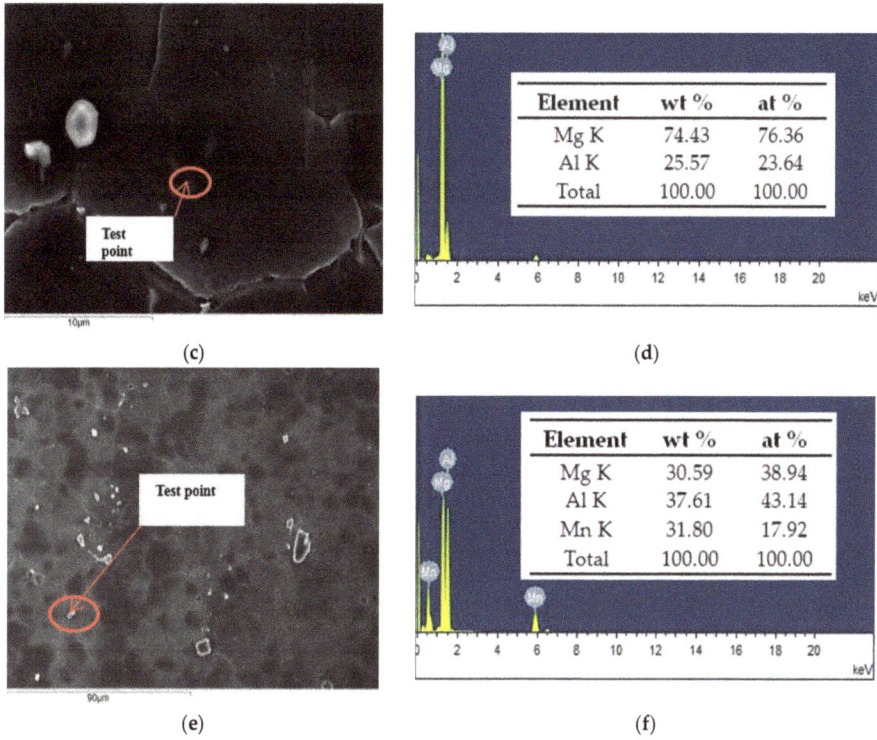

(c)

Element	wt %	at %
Mg K	74.43	76.36
Al K	25.57	23.64
Total	100.00	100.00

(d)

Element	wt %	at %
Mg K	30.59	38.94
Al K	37.61	43.14
Mn K	31.80	17.92
Total	100.00	100.00

(e)   (f)

Figure 16. (**a,b**) SEM images of the AZ61/SiCp MMC extruded plates after annealing (10 μm SiC with 5 wt %) with different magnifications; (**c,e**) Test point for EDX analysis of the AZ61/SiCp MMC extruded plates after annealing (10 μm SiC with 5 wt %); (**d,f**) Element results from EDX analysis of the AZ61/SiCp MMC extruded plates after annealing (10 μm SiC with 5 wt %).

3.3. Comparison of the Mechanical Measurements of Mg MMCs

The mechanical properties of the present composites and other composites are listed in Table 2. The present study's AZ61 MMCs with submicron SiCp and nano SiCp had higher ultimate tensile strength, but a reduced hardness compared with those of our previous work [34]. Additionally, the SiC/AZ61 MMCs in the present study had a similar hardness compared with that of Hong et al. [35] but exhibited lower hardness compared with that of Zulkoffli et al. [36] because of their sintering process. Casted AZ61 with 3, 6, and 18 vol % 10 μm SiCp AZ61 MMCs studied by Hu et al. [37] exhibited a strong yield strength of 168–186 MPa, but considerably smaller elongation of 1.7–4.5%, than did those of the present study. Casted AZ61 with 3, 6, and 9 vol % 10 μm SiCp AZ61 MMCs studied by Yan et al. [38] were very hard (79–115 HV) but had no other enhanced mechanical properties. Consequently, the SiC/AZ61 MMCs in the present study exhibited more complete mechanical property results than did those of other studies.

Table 2. Mechanical properties of the synthesized composites in the present study compared with those in other studies.

Notation	Specification	Yield Strength, [MPa] (% Change)	Ultimate Tensile Strength, [MPa] (% Change)	Hardness [HV] (% Change)	Elongation (% Change)	Reference
AZ61/BA	Extruded AZ61 Mg alloy before annealing	131.9 ± 1.2 (0)	310.9 ± 3.8 (0)	60.8 ± 0.9 (0)	23.5 ± 1.5% (0)	This study
AZ61/AA	Extruded AZ61 Mg alloy after annealing	132.7 ± 1.9 (+0.6)	327.9 ± 1.9 (+5.5)	65.9 ± 1.1 (+8.4)	40.8 ± 3.0% (+73.6)	This study
AZ61/10 μm SiC/BA	Extruded AZ61 with 0.5, 1, 2 wt % 10 μm SiCp AZ61 MMCs plate before annealing	133.0–135.1 (+1.6)	324.2–326.2 (+4.6)	61.3–63.0 (+2.2)	27.2–33.5% (+29.1)	This study
AZ61/10 μm SiC/AA	Extruded AZ61 with 0.5, 1, 2 wt % 10 μm SiCp AZ61 MMCs plate after annealing	128.7–134.5 (−0.2)	309.7–329.2 (+2.8)	63.9–65.2 (+6.2)	31.0–37.6% (+46.0)	This study
AZ61/1 μm SiC/BA	Extruded AZ61 with 0.5, 1, 2 wt % 1 μm SiCp AZ61 MMCs plate before annealing	133.6–135.8 (+2.1)	326.5–329.5 (+5.5)	62.1–63.6 (+3.1)	40.2–42.7% (+76.4)	This study
AZ61/1 μm SiC/AA	Extruded AZ61 with 0.5, 1, 2 wt % 1 μm SiCp AZ61 MMCs plate after annealing	135.5–137.0 (+3.3)	332.2–336.1 (+7.5)	63.3–65.1 (+5.6)	36.6–39.1% (+61.1)	This study
AZ61/50 nm SiC/BA	Extruded AZ61 with 0.5, 1, 2 wt % 50 nm SiCp AZ61 MMCs plate before annealing	134.1–136.4 (+2.5)	327.5–331.0 (+5.9)	62.0–63.3 (+3.0)	40.8–43.1% (+78.5)	This study
AZ61/50 nm SiC/AA	Extruded AZ61 with 0.5, 1, 2 wt % 50 nm SiCp AZ61 MMCs plate after annealing	134.7–136.8 (+2.9)	328.2–334.6 (+6.6)	63.0–64.5 (+4.9)	37.3–40.6% (+65.0)	This study
AZ61/4.5 μm SiC/T5	Extruded AZ61 with 1, 2, 5 wt % 4.5μm SiCp AZ61 MMCs tube after T5 treatment	136.0–145.0 (+4.9)	291.0–315.0 (+2.9)	67.2–75.6 (+2.7)	NA	[34]
AZ61/100 nmSiC/AA	AZ61 with 1 wt % 100 nm SiCp AZ61 MMCs after annealing	NA	NA	68–78 (+2.1)	NA	[35]
AZ61/20 μmSiC/sintered	Sintered AZ61 with 5 wt % 20μm SiCp AZ61 MMCs	NA	NA	71–96 (NA)	NA	[36]
AZ61/10μm SiC	Casted AZ61 with 3, 6, 18 vol % 10μm SiCp AZ61 MMCs	168–186 (+5.9)	NA	NA	17–45% (−48.7)	[37]
AZ61/10 μmSiC	Casted AZ61 with 3, 6, 9 vol % 10 μm SiCp AZ61 MMCs	NA	NA	79–115 (+48.7)	NA	[38]

( ) Brackets indicate % change with respect to the corresponding monolithic alloy and NA means no results reported.

4. Conclusions

The effect of reinforcement size and quantity (wt %) of SiCp on the AZ61 alloy synthesized through the stir-casting melting method followed by hot extrusion and annealing was investigated in this study. Our conclusions are as follows:

1. The mechanical properties of the annealed SiCp-free AZ61 alloy plate are better than those of the unannealed SiCp-free AZ61 alloy plate.
2. Both the annealed and unannealed AZ61/SiCp/1 μm and AZ61/SiCp/50 nm MMC plates exhibited greater strength, hardness, and ductility that did the SiCp-free (i.e., AZ61) MMC Mg alloy plate.
3. The annealed AZ61/SiCp/1 μm and AZ61/SiCp/50 nm MMC plates exhibited improvements in hardness and strength, but ductility slightly decreased because of the precipitation of the β phase.
4. Adding submicron SiCp and nano SiCp particles to AZ61 during processing resulted in the advantages of strengthening the mechanical properties and increasing ductility.
5. Annealing can improve the ductility and formability of Mg alloys and Mg MMCs with micron SiCp, but it weakens those properties in Mg MMCs with submicron SiCp and nano SiCp.
6. The SiC/AZ61 MMCs in the present study exhibited more complete mechanical property results compared with other studies.

Acknowledgments: The present study was supported by the National Natural Science Foundation of China (31300488) and Fujian Agriculture and Forestry University Fund for Distinguished Young Scholars (xjq201420). This manuscript was edited by Wallace Academic Editing.

Author Contributions: Song-Jeng Huang and Weigang Zhao conceived and designed the experiments; Yi-Jhang Wu and Cheng-Wei Kang performed the experiments; Song-Jeng Huang, Weigang Zhao, Yi-Jhang Wu, and Cheng-Wei Kang analyzed the data; Song-Jeng Huang and Weigang Zhao contributed reagents/materials/analysis tools; and Song-Jeng Huang and Weigang Zhao wrote the paper.

Conflicts of Interest: The authors declare no conflict of interest. The founding sponsors had no role in the design of the study; in the collection, analyses, or interpretation of data; in the writing of the manuscript, and in the decision to publish the results.

References

1. Rokhlin, L.L. The 7th International conference Magnesium alloys and their applications. *Metalloved. I Term. Obrab. Met.* **2008**, *3*, 22–23.
2. Schindler, S.; Mergheim, J.; Zimmermann, M.; Aurich, J.C.; Steinmann, P. Numerical homogenization of elastic and thermal material properties for metal matrix composites (mmc). *Contin. Mech. Thermodyn.* **2017**, *29*, 1–25. [CrossRef]
3. Abbott, T.B.; Rienass, G.; Zhen, Z.S. Property assessment of magnesium alloys for powertrain applications. *VDI Ber.* **2012**, *2158*, 321–332.
4. Wang, M.; Xiao, D.H.; Liu, W.S. Effect of Si addition on microstructure and properties of magnesium alloys with high Al and Zn contents. *Vacuum* **2017**, *141*, 144–151. [CrossRef]
5. Venkatachalam, G.; Kumaravel, A. Mechanical Behaviour of Aluminium Alloy Reinforced With SiC/Fly Ash/Basalt Composite for Brake Rotor. *Polym. Polym. Compos.* **2017**, *25*, 203–208.
6. Dai, J.H.; Jiang, B.; Peng, C.; Pan, F.S. Effect of mn additions on diffusion behavior of Fe in molten magnesium alloys by solid-liquid diffusion couples. *J. Alloys Compd.* **2017**, *710*, 260–266. [CrossRef]
7. Cui, S.; Wu, X.; Liu, R.; Teng, X.; Leng, J.; Geng, H. Effects of Te addition on microstructure and mechanical properties of AZ91 magnesium alloy. *Mater. Res. Express* **2017**, *4*, 016503. [CrossRef]
8. Volkova, E.F. Modern magnesium-base deformable alloys and composite materials (a review). *Met. Sci. Heat Treat.* **2006**, *48*, 473–478. [CrossRef]
9. Fridlyander, I.N. Modern aluminum and magnesium alloys and composite materials based on them. *Met. Sci. Heat. Treat.* **2002**, *44*, 292–296. [CrossRef]

10. Si, C.; Tang, X.; Zhang, X.; Wang, J.; Wu, W. Microstructure and mechanical properties of particle reinforced metal matrix composites prepared by gas-solid two-phase atomization and deposition technology. *Mater. Lett.* **2017**, *201*, 78–81. [CrossRef]

11. Tjong, S.C. Novel nanoparticle-reinforced metal matrix composites with enhanced mechanical properties. *Adv. Eng. Mater.* **2007**, *9*, 639–652. [CrossRef]

12. Grishina, O.I.; Serpova, V.M. Aspects of mechanical properties of metal matrix composites reinforced with continuous unidirectional fibers (review). *Trudy VIAM* **2017**. [CrossRef]

13. Shirvanimoghaddam, K.; Hamim, S.U.; Akbari, M.K.; Fakhrhoseini, S.M.; Khayyam, H.; Pakseresht, A.H. Carbon fiber reinforced metal matrix composites: Fabrication processes and properties. *Compos. Part A Appl. Sci. Manuf.* **2016**, *92*, 70–96. [CrossRef]

14. Yong, H.P.; Bang, J.; Oak, J.J.; Lee, Y.C.; Jung, K.H. The effect of the addition of Ti metal fiber and Al-Mg alloy binder on the mechanical properties of Al-Si/SiCp metal matrix composites fabricated by powder metallurgy. *J. Korean Inst. Met. Mater.* **2016**, *54*, 400–408. [CrossRef]

15. Iqbal, A.A.; Arai, Y.; Araki, W. Fatigue crack growth mechanism in cast hybrid metal matrix composite reinforced with SiC particles and Al_2O_3 whiskers. *Trans. Nonferrous Met. Soc. China* **2014**, *241*, S1–S13. [CrossRef]

16. Li, A.B.; Geng, L.; Meng, Q.Y.; Zhang, J. Simulation of the large compressive deformation of metal matrix composites with misaligned whiskers. *Mater. Sci. Eng. A* **2003**, *358*, 324–333. [CrossRef]

17. Huang, C.C. Effect of Reinforcement Size on the Mechanical Properties of as Cast and Extruded AZ61/SiCp Magnesium Matrix Composites. Master's Thesis, National Taiwan University of Science and Technology, Taipei, Taiwan, 2013.

18. Chen, T.J.; Jiang, X.D.; Ma, Y.; Li, Y.D.; Hao, Y. Grain refinement of AZ91D magnesium alloy by SiC. *J. Alloys Compd.* **2010**, *496*, 218–225. [CrossRef]

19. Chua, B.W.; Lu, L.; Lai, M.O. Influence of SiC particles on mechanical properties of Mg based composite. *Compos. Struct.* **1999**, *47*, 595–601. [CrossRef]

20. Zhang, B.H.; Zhang, Z.M. Influence of homogenizing on mechanical properties of as-cast AZ31 magnesium alloy. *Trans. Nonferrous Met. Soc. China* **2010**, *20*, 439–443. [CrossRef]

21. Li, J.Y.; Xie, J.X.; Jin, J.B.; Wang, Z.X. Microstructural evolution of AZ91 magnesium alloy during extrusion and heat treatment. *Trans. Nonferrous Met. Soc. China* **2012**, *22*, 1028–1034. [CrossRef]

22. Chen, Q.; Zhao, Z.; Shu, D.; Zhao, Z. Microstructure and mechanical properties of AZ91D magnesium alloy prepared by compound extrusion. *Mater. Sci. Eng. A* **2011**, *528*, 3930–3934. [CrossRef]

23. Hassan, S.F.; Gupta, M. Effect of length scale of Al_2O_3 particulates on microstructural and tensile properties of elemental Mg. *Mater. Sci. Eng. A* **2006**, *425*, 22–27. [CrossRef]

24. Agrawal, P.; Sun, C.T. Fracture in metal—Ceramic composites. *Compos. Sci. Technol.* **2004**, *64*, 1167–1178. [CrossRef]

25. Psakhie, S.; Ovcharenko, V.; Yu, B.; Shilko, E.; Astafurov, S.; Ivanov, Y.; Byeli, A.; Mokhovikov, A. Influence of Features of Interphase Boundaries on Mechanical Properties and Fracture Pattern in Metal-Ceramic Composites. *J. Mater. Sci. Technol.* **2013**, *29*, 1025–1034. [CrossRef]

26. Ma, G.; Han, G.; Liu, X. Grain refining efficiency of a new Al-1B-0.6C master alloy on AZ63 magnesium alloy. *J. Alloys Compd.* **2010**, *491*, 165–169.

27. Habibnejad-Korayem, M.; Mahmudia, R.; Pooleb, W.J. Enhanced properties of Mg-based nano-composites reinforced with Al_2O_3 nano-particles. *Mater. Sci. Eng. A* **2009**, *519*, 198–203. [CrossRef]

28. Huang, S.J.; Lin, P.C. Grain refinement of AM60/Al_2O_{3p} Megnesium Metal-matrix Composites Processed by ECAE. *Kov. Mater. Met. Mater.* **2013**, *51*, 357–366.

29. Hassan, S.F.; Gupta, M. Development of high performance magnesium nanocomposites using solidification processing route. *Mater. Sci. Technol.* **2004**, *20*, 1383–1388. [CrossRef]

30. Besterci, M.; Huang, S.-J.; Sülleiová, K.; Balloková, B.; Velgosová, O. Fracture analysis of AZ61-F magnesium composite materials at quasi-superplastic state. *Kov. Mater.* **2017**, *55*, 217–221.

31. Wu, X.; Yang, X.; Ma, J.; Huo, Q.; Wang, J.; Sun, H. Enhanced stretch formability and mechanical properties of a magnesium alloy processed by cold forging and subsequent annealing. *Mater. Des.* **2013**, *43*, 206–212. [CrossRef]

32. Miao, Q.; Hu, L.; Wang, G.; Wang, E. Fabrication of excellent mechanical properties AZ31 magnesium alloy sheets by conventional rolling and subsequent annealing. *Mater. Sci. Eng. A* **2011**, *528*, 6694–6701. [CrossRef]

33. Sun, H.F.; Li, C.J.; Yang, X.; Fang, W.B. Microstructures and mechanical properties of pure magnesium bars by high ratio extrusion and its subsequent annealing treatment. *Trans. Nonferrous Met. Soc. China* **2012**, *22*, 445–449. [CrossRef]
34. Huang, S.J.; Hwang, Y.M.; Huang, Y.S. Grain Refinement of AZ61/SiCp Magnesium Matrix Composites for Tubes Extruded by Hot Extrusion Processes. *Key Eng. Mater.* **2013**, *528*, 135–143. [CrossRef]
35. Hong, Y.; Xiaowu, H.; Qiao, N.; Lei, C. Aging behavior of nano-SiCp reinforced AZ61 magnesium matrix composites. *Res. Dev.* **2011**, *8*, 269–273.
36. Zulkoffli, Z.; Syarif, J.; Sajuri, Z. Fabrication of AZ61/SIC composites by powder metallurgy process. *Int. J. Mech. Mater. Eng.* **1970**, *4*, 156–159.
37. Hu, Q.; Jie, X.P.; Yan, H.; Zhang, F.Y.; Chen, G.X. Mechanical properties and damping capacity of AZ61 magnesium alloy matrix composites with SiC particulates. *Forg. Stamp. Technol.* **2008**, *33*, 106–109.
38. Yan, H.; Fu, M.F.; Zhang, F.Y.; Chen, G.X. Research on properties of SiCp/AZ61 magnesium matrix composites in fabrication processes. *Mater. Sci. Forum* **2007**, *561*, 945–948. [CrossRef]

MDPI

St. Alban-Anlage 66

4052 Basel

Switzerland

Tel. +41 61 683 77 34

Fax +41 61 302 89 18

www.mdpi.com

Metals Editorial Office

E-mail: metals@mdpi.com

www.mdpi.com/journal/metals

www.ingramcontent.com/pod-product-compliance
Lightning Source LLC
Chambersburg PA
CBHW051721210326
41597CB00032B/5560